Third International Congress of Histochemistry and Cytochemistry

New York, New York, August 18-22, 1968

SUMMARY REPORTS

Springer Science+Business Media, LLC

ISBN 978-0-387-90006-3   ISBN 978-1-4899-4579-2 (eBook)
DOI 10.1007/978-1-4899-4579-2

© 1968 by Springer Science+Business Media New York
Originally published by Springer-Verlag New York, Inc. in 1968
Softcover reprint of the hardcover 1st edition 1968

Library of Congress Catalog Card Number: 68-55397

Title No. 1536

THIRD INTERNATIONAL CONGRESS
OF HISTOCHEMISTRY AND CYTOCHEMISTRY

New York, New York—August 18-22, 1968

Congress Officers

President: Arnold M. Seligman (Baltimore)
Secretariat: Robert M. Rosenbaum (New York)
Program Chairman: David Glick (Palo Alto)
Organizing Committee Chairman: Russell J. Barrnett (New Haven)
Local Committee Chairman: Arline D. Deitch (New York)

Honorary Vice-Presidents

A. G. E. Pearse (England)   R. D. Lillie (U.S.A.)
O. Eränkö (Finland)   S. Seno (Japan)
J. Brachet (Belgium)   L. C. U. Junqueira (Brazil)
M. Vialli (Italy)   M. Wolman (Israel)
M. C. Bessis (France)   Z. Lojda (Czechoslovakia)
P. van Duijn (Netherlands)   M. Presnov (U.S.S.R.)
H. Holter (Denmark)   C. Velican (Romania)
T. H. Schiebler (West Germany)   I. Törö (Hungary)
B. Sylvén (Sweden)   A. A. Hadjiolov (Bulgaria)
                    M. J. Olszewska (Poland)

Program Committee

G. F. Bahr (Washington, D.C.)   R. L. Hunter (Davis, California)
T. Barka (New York)   M. L. Karnovsky (Boston)
R. J. Barrnett (New Haven)   A. Lazarow (Minneapolis)
E. P. Benditt (Seattle)   A. B. Novikoff (New York)
W. L. Doyle (Chicago)   G. E. Palade (New York)
E. Farber (Pittsburgh)   R. M. Rosenbaum (New York)
G. G. Glenner (Washington, D. C.)   A. M. Seligman (Baltimore)
D. Glick (Palo Alto)   L. M. Wattenberg (Minneapolis)

The Congress acknowledges support towards its scientific program by the following:

Abbott Laboratories
American Cancer Society
Burroughs Wellcome and Company (U.S.A.), Inc.
Ciba Pharmaceutical Company
Geigy Chemical Corporation
Hoffmann-LaRoche, Inc.
Merck Sharp and Dohme, Research Laboratories
Ortho Pharmaceutical Corporation
Charles Pfizer and Company, Inc.
Schering Corporation
G. D. Searle and Company
Smith Kline and French Laboratories
Ivan Sorvall, Inc.
The Wellcome Trust (Great Britain)

INTRODUCTION

The reports contained in this volume of abstracts comprise three categories of papers presented at the Third International Congress of Histochemistry and Cytochemistry. The format of the meeting called for a series of Plenary Sessions on each of the five mornings of the Congress as follows:

1. Recent Developments in Cytochemical Localization by Electron Microscopy.
2. Centrifugal and Electrophoretic Techniques in Histochemistry.
3. New Directions in Quantitative Cytochemistry by Physical Techniques.
4. Subcellular Structural-Functional Correlates.
5. Current Developments of Histochemistry in Pathology.

All the reports in these sessions were by invitation.

Four of the five afternoons were devoted to contributed papers from the platform, some of which represent invited reports. A third category represents papers submitted "by title".

In the present volume the three categories are not distinguished one from the other. All papers are listed alphabetically by the last name of the first author–the only division being papers received too late for inclusion in the main body of abstracts and published at the back of the volume. The position of specific papers in the overall program of the Congress and the names of session chairmen are listed in the separate program volume of the meeting.

All editorial work relative to the Congress, and specifically this volume, were undertaken by the staff of the Office of the Secretariat. In this respect I am grateful to the fine organizational and editorial work of Mrs. Dorothy K. Rosenbaum and the skilled secretarial efforts of Mrs. Pearl Wasser. Mrs. Jeanne Rich of Springer Verlag-New York, as representative of the publisher, was extremely cooperative in all respects.

Robert M. Rosenbaum, Ph.D.
Secretary General
Third International Congress of
Histochemistry and Cytochemistry

New York, July 15, 1968

MESSAGE OF WELCOME

The First International Congress of Histochemistry and Cytochemistry was held in Paris, France, in 1960 and the Second Congress was held in Frankfurt, Germany, in 1964. Both Congresses were unqualified successes and should be given credit for the support and enthusiasm which the Third Congress is receiving. The hard work of the Secretary General, Dr. Robert Rosenbaum, the effort of the chairman of the program committee, Dr. David Glick, as well as his section chairmen, and of the members of the executive committee and local committee, has brought us to what promises to be another successful and rewarding congress. I hope each one of you will find your week in New York enjoyable, stimulating, and enlightening.

Arnold M. Seligman, M.D.
President
Third International Congress for
Histochemistry and Cytochemistry
New York City—August 18-22, 1968.

THIRD INTERNATIONAL CONGRESS OF HISTOCHEMISTRY AND CYTOCHEMISTRY

August 18-22, 1968, New York, N.Y.

The Enzyme Histochemistry of Intestinal Metaplasia in Human Stomach

ABE, MUNEAKI, MASASUKE AKAMATSU, TAKAHARU MATSU-MOTO, NOBUO OHUCHI, and TOMICHI MASUYA (Dept. of Internal Medicine, Kyushu University School of Medicine, Fukuoka, Japan)

The tissue distribution of enzymatic activities in intestinal metaplasia stomachs exhibiting chronic gastritis was compared histochemically with that of the small intestine in man.

Examined enzymes were: Alkaline phosphatase (AlP), Acid phosphatase (AcP), Leucine aminopeptidase (LAP), Succinate dehydrogenase (SDH), Lactate dehydrogenase (LDH), NADH-diaphorase (NADHD), Glucose-6-phosphate dehydrogenase (G6PDH).

The reaction pattern of surface and foveolar cells in intestinal metaplasia was the same as that of villus cells of the small intestine, exhibiting high activities of AlP, LAP, SDH, LDH, NADHD and moderate activities of G6PDH and AcP.

Activity of AlP was localized to the striated border and juxtaapical portion of cytoplasm.

In crypts of intestinal metaplasia, moderate activities of SDH, LDH, and NADHD were seen, while no or little activities of AlP, G6PDH and LAP were observed.

This reaction pattern was similar to that in crypts of the small intestine. In addition, intense activities of SDH, LDH and NADHD were detected at the base of crypts of metaplasia. However, these patterns of activities of the enzymes were altered in irregularly proliferated areas.

Activities of these enzymes varied occasionally from place to place, suggesting multifocal irregular proliferation of tissue.

In the striated border, reactive figures of AlP was variable, showing thinning of reactive zone and dot-like positive granules in some cells.

From these observations, it will be stressed that enzymatic reaction patterns in intestinal metaplasia are influenced by the state of cell proliferation.

1

Localization of Fibrinolytic Activity in Vessel Wall and its Correlation with the Blood Coagulation System

ABE, T., K. NAKASHIO, M. KAZAMA, and M. MATSUDA (Dept. of Medicine and Surgery, University of Tokyo School of Medicine, Hongo, Bunkyo-ku, Tokyo, Japan)

In order to study the correlation between blood (intrinsic) and tissue (extrinsic) fibrinolytic activities (FAs), the FA, particularly its localization in blood vessel walls, was scrutinized.

The fibrin film method, of Todd or Kwaan, was applied together with some modifications having more demonstrable features of lysis in the fibrin membrane. As the original method had some disadvantages, such as that tissue sample and fibrin film had a certain discrepancy in their staining conditions, including pH, concentration of hemalum agent, etc., fluorescein isothiocyanate (FITC) labeled human fibrinogen was utilized as the substrate. This could give a special color in itself without staining. The staining condition was exclusively arranged for tissue material. Besides specimens from various kinds of tissue, the canaliculating portion of occlusive thromboangitis in human and organizing Teflon implants in the aorta in dogs were pursued by means of our new method. At the same time, the FA of blood samples which contacted with certain limited areas of the patched vessel was also tested to see the correlation between the FAs of grafted aorta tissue and stagnating blood.

In heart tissue, FA was developed at sites of endocardium, epicardium, or blood vessel in connective tissue of myocardium with no lysis in muscle tissue. In the aorta, remarkable lysis was seen at the vasa vasorum in adventitia and slightly in the endothelium. As a general rule, liver tissue showed infrequent lysis at the capsule or Glisson's sling but in a specimen of cirrhosis, large areas of lysis were observed in enlarged mesenchymal tissues, particularly at newly formed vessels. The FA of kidney proceeded mainly from the endothelium of the pelvis and medulla as well as in the archform vessel at the border between medulla and cortex. Lung, pleura, trachea and bronchus also showed lysis at the sites of vessels. The specimens of patched dog aorta developed rash proliferation of epithelium and connective tissue around the Teflon patch, and FA followed this organization. In the canaliculation process of obliterating thrombus, the FA of the endothelium seemed to play an important role in dissolving fibrin coagula.

The fibrinolytic system in tissues was shown to be at sites of a blood vessel-connective tissue unit. The FA in the vessel wall played a role of liaison between intrinsic and extrinsic systems giving an important basis for understanding the patho-physiological features in the living bodies. The origin of this activity seemed to be located in endothelium cells.

Quantitative Histochemistry of ³H-Cholesterol and ¹²⁵I-Labelled Plasma-Protein Influx in Rabbit Aortic Wall

ADAMS, C. W. M., S. VIRÁG, R. S. MORGAN, and C. C. ORTON (Dept. of Pathology, Guy's Hospital Medical School, London University, London, S.E. 1. England)

It has long been considered that lipids enter the arterial wall in a lipoprotein transport vehicle. In support of this view, immunofluorescence, immunoelectrophoretic and histochemical studies indicate that plasma protein and β-lipoprotein are present in human atherosclerotic plaques. Recently, however, some doubt has been cast on this notion because a number of investigators have reported that the entry rate of cholesterol into the arterial wall does not match that of other lipoprotein components.

In order to test the hypothesis that cholesterol does not enter the arterial wall in a lipoprotein vehicle, we injected normal and atheromatous rabbits with ³H-cholesterol, ¹²⁵I-globulin, ¹²⁵I-albumin and ³²PO₄. At intervals of 1-4 days, the animals were killed and their aortas were sectioned from inside-outwards on a thermoelectric microtome (modified Linderstrøm-Lang technique). Radioactivity in the resulting multiple layers of aorta was determined both by scintillation-counting and by autoradiography.

The results show an inside-outward gradient for cholesterol entry, and a reversed (outward-inward) entry for ¹²⁵I-globulin and ¹²⁵I-albumin in both normal and moderately atheromatous aortas. In some severely atheromatous aortas, however, the labelled plasma proteins showed an inside-outward gradient (i.e. they matched the cholesterol gradient).

These results suggest that plasma lipoproteins are not the vehicles that carry cholesterol across the normal and slightly diseased arterial wall. Nevertheless, after severe atheroma has become established, plasma proteins sometimes appear to leak directly into the lesion from the lumen. We tentatively speculate that phospholipid or protein, synthesized within the arterial wall, may be the vehicle for cholesterol transport across the normal vascular wall.

Cytochemical Studies with Several Insect Viruses

ADAMS, JEAN R. and THEODORE A. WILCOX (Insect Pathology Laboratory, Agricultural Research Center, Beltsville, Md.)

Cytochemical investigations were conducted on tissues from five insect species infected with three types of viruses. The viruses used were nuclear and cytoplasmic polyhedra and granulosis of the following host

4

species: fall armyworm, *Spodoptera frugiperda* (J. E. Smith); cabbage looper, *Trichoplusia ni* (Hübner); corn earworm, *Heliothis zea* (Boddie); almond moth, *Cadra cautella* (Walker); and Indian-meal moth, *Plodia interpunctella* (Hübner). Healthy and diseased tissues were treated with DNase, RNase, pepsin, trypsin, lipase, and pronase.

Observations will be discussed and illustrated on cytochemical differences in healthy and diseased tissues and in replication cycles of the viruses in the tissues.

Histochemical Investigation of Synaptic Vesicles at Cholinergic Sites: An EM Study

AKERT, KONRAD (Brain Research Institute, University of Zürich, Zürich, Switzerland)

The effect of zinciodide-osmium staining of nerve terminals was examined in peripheral and central regions of the nervous system. The so-called clear cynaptic vesicles turn black with this stain. The synaptic membranes, the basement membranes of junctional folds, and the mitochondria remain indifferent. No reaction is seen in the synaptic gap. Dark cored vesicles of the 800 to 1500 Å range remain equally indifferent. On the other hand, positive reactions were found at typical cholinergic sites: the motor endplate terminals and nerve terminals in the subfornical organ of mammals. The selectivity of this new vesicular staining method is demonstrated in various other regions where either cholinergic or adrenergic transmission is known to occur.

Étude Morphocytochimique de la Siderose dans la Maladie de Cooley

ALBERTI, RACHELE (Istituto di Anatomia e Istologia Patologica dell'Università di Ferrara, Italy)

L'étude histochimique systématique des organs des sujets atteints par la maladie de Cooley agès entre 16 mois et 12 années a montré une siderose constante, d'intensité variable entre les différents cas, mais surtout evidente dans le foie, pancreas, rate, moelle osseuse, thyroïde, surrénale et hypophyse. Dans les cas non transfusés la siderose hepatique est faible et surtout hépathocytaire, tandis que dans les cas politransfusés elle est très evidente dans les cellules de Kupffer. Des résultats pareils nous a montré la recherche conduite sur les autrea organs et nous verrions par conséquence à la conclusion que: la siderose, qui dépénde de les damages du métabolisme du fer, qui sont charactéristiques de la maladie de Cooley, est faible essentiellement parenchimateuse; 2) la massive siderose qui s'accompagne à une intense partecipation du SRI et à la sclerose, dépende de la quantité de sang transfusé et de l'intensité de l'hémolyse.

Quantitation of Plasma and Tissue Esterase Zymograms: Application to The Study of Physiological and Pathological Changes *

ALLEN, ROBERT C., DOROTHY J. MOORE, and RICHARD L. TYNDALL (Biology Division, Oak Ridge National Laboratory, Oak Ridge, Tenn.)

Study of plasma and tissue esterases, separated by discontinuous electrophoresis in acrylamide gel and quantified with an automated microdensitometer (1), has provided an important additional approach to the investigation of physiological and pathological changes. During erythropoiesis in mice made hypoxic in silicone membrane chambers, two of the esterase isoenzyme bands (six and seven) decrease progressively in specific activity to roughly one-half of normal within seven to fourteen days respectively; on the other hand, band 12 increases concomitantly along with the increase in hematocrit. Upon removal of mice from the chamber to normal atmospheric oxygen levels, all three esterases return to normal specific activity. Marked alterations in specific activity of certain esterases are apparent also in both Rauscher virus-infected tissue cultures and in spleens of similarly infected, leukemic mice. Quantitative techniques have also shown the proliferative capacity of donor spleen cells containing a specific esterase marker in irradiated recipients lacking the esterase marker. Decreases in specific activity of albumin esterases have been observed also, with little or no associated change in the albumin concentration in protein pherograms from human patients with severe uremia and nephritis. A major problem encountered in all of these studies has been accurate densitometric measurement of either protein pherograms or esterase zymograms in the haptoglobin and macroglobulin region, owing to lack of adequate separation and band resolution with a resultant baseline drift. This has required that each band be isolated individually with the optics of the microdensitometer for reasonably accurate densitometry. To overcome this problem and to increase quantitative accuracy, an improved source of power for electrophoretic separation has been employed. With this, peak widths at one-half peak height have been reduced up to 50% with resulting improvement in resolution, especially of minor components, along with virtual elimination of baseline drift in esterase zymograms. As a result, three additional esterases in mouse plasma which are only partially resolved using conventional power supplies, have been readily resolved and quantified. In addition, separation times have been reduced by approx-

6

imately one-third. These procedures and their significance for the quantification of isoenzymes separated in acrylamide gel will be discussed.

References:

(1) Allen, R. C. and G. R. Jamieson: Anal. Biochem. *16*, 450 (1966).

* Research Sponsored by the U.S. Atomic Energy Commission under contract with the Union Carbide Corporation.

Variations in Substructural Preservation of Skeletal Muscle Treated with Glutaraldehyde Fixatives *

ANDERSON, PAUL J. and SUN K. SONG (Mount Sinai School of Medicine, New York, N. Y.)

Purification of commercial glutaraldehyde by distillation or repeated washing with charcoals of high surface area eliminates material with maximal absorbance at 235 mμ (probably of polymeric or oligomeric origin) while preserving the dialdehyde responsible for absorbance at 280 mμ (1). Samples of skeletal muscle exposed to purified glutaraldehyde yielded higher recoveries of several oxidoreductases, transferases and hydrolases than samples exposed to untreated glutaraldehyde (2) but the quality of fixation obtained by immersion of tissue samples in different glutaraldehyde mixtures has not been assessed.

Since glutaraldehyde fixation seems to involve a cross-linking reaction with reactive groups of tissue proteins, the quality of fixation would be expected to vary with the sample size and duration of fixation, and the purity, concentration, penetration rate, and osmolality of the glutaraldehyde mixture. These factors were therefore tested by immersing rat skeletal muscle in glutaraldehyde mixtures of various composition and comparing the preservation of contractile elements and membranous components by electron microscopy.

Best preservation was observed in tissues exposed for 2 hours to 2.5% purified glutaraldehyde in 0.135M Sorensen buffer (A_{280} = 0.03; A_{235} = 0.001; mOs = 631). Good preservation was also obtained with untreated glutaraldehyde of approximately the same composition but with high absorbance at 235 mμ (A_{280} = 0.04; A_{235} = 0.085; mOs = 631).

Lowering or raising the glutaraldehyde concentration (1% and 5%) and osmolality (395 mOs and 1286 mOs) resulted in poor definition of sarcoplasmic reticulum, mitochondria, and synaptic membranes. Definition was improved by prolonged exposure to glutaraldehyde (12-24 hours) or by post treatment of briefly fixed tissues with osmium tetroxide. Post treatment, however, was always associated with narrowing of interfibrillar spaces and shortening of sarcomeres, sometimes by as much as 30%.

Prolonged fixation with purified glutaraldehyde (6-18 hours at 4°C) gave excellent preservation of membranous components and did not prevent the histochemical demonstration of enzymatic activity at the motor end-plate.

References:

(1) Fahimi, H. D. and P. Drochmans: J. Micr. *4*, 725 (1965).··
(2) Anderson, P. J.: J. Histochem. Cytochem. *15*, 652 (1967).

---

* Supported by USPHS Grant No. NB 05041.

Ethylenediamine (EDA) Condensation Reaction for the Histochemical Demonstration of Catecholamines *

ANGELAKOS, E. T. and M. P. KING (Boston University School of Medicine, Boston, Mass.)

A new method was developed for the demonstration of tissue catecholamines based on the EDA reaction (Weil-Malherbe, 1952-54). This reaction is more specific than the formaldehyde method (Falck and Hillarp, 1961-62) but less specific than the trihydroxyindole method (Angelakos and King, 1965-67). Catecholamines are transformed into fluorescent products following exposure of tissues to EDA dissolved in isopropanol. Ascorbic acid stabilizes the reaction. Alternatively, EDA fumes can be used. Previous exposure to iodine potentiates the fluorescence and improves specificity of the reaction. Fluorescence spectra (obtained in a microspectrofluorometer) from known noradrenaline (NA), adrenaline (A) and dopamine (DA) containing structures exposed to EDA were the same as those from "models" of pure solutions of the corresponding amines with fluorescence maximal at 470 mμ (NA), 500 mμ (A), 490 mμ (DA). This new method may be useful as an adjunct to those previously available for the differentiation of catecholamines from other monoamines and especially for the identification of DA for which it seems to be particularly sensitive.

---

* Supported by grants from the National Science Foundation, GB 4386; the Life Insurance Medical Research Fund, and USPHS Career Development Award K3-HE 15,457.

Autoradiography with the Retention of Soluble Labelled Compounds

APPLETON, TIMOTHY C. (State University of New York, Downstate Medical Center, Department of Pathology, Brooklyn, N. Y.)

Techniques of autoradiography for the localization of soluble labelled compounds are becoming well established at the light microscope level.

8

In order to localize soluble compounds to specific cell structures it is necessary to extend the technique for use in the electron microscope. Preliminary results have indicated that it is possible to cut ultrathin (less than 0.1.μ) sections of unfixed biological material onto dry glass or diamond knives using a refrigerated ultramicrotome operating with very low cutting speeds and at ultra-low temperatures (-60 to $-90°$ C.). Such sections were picked up onto cold ($-15°$ C.) formvar coated grids using the temperature differential method employed in the LM technique (1,2,3), freeze-dried and fixed in osmium vapour before viewing in the electron microscope. No extraction of soluble compounds occurred (3). Experiments are now in progress to apply ultrathin frozen sections to photographic emulsions so that soluble compounds can be located at the LM level. The results and prospects for EM soluble compound auto-radiography will be discussed.

References:

(1) Appleton, T. C.: J. Roy. Micr. Soc. *83*, 277 (1964).
(2) Appleton, T. C.: J. Histochem. Cytochem. *14*, 414 (1966).
(3) Appleton, T. C.: 12th Symposium Gesselschaft fur Histochemie, Ghent, 1967, Acta Histochemica Suppl. VIII, *in press.*

Ultrastructural and Cytochemical Studies on Cellular Autophagy

ARSTILA, ANTTI U. and BENJAMIN F. TRUMP (Duke University, Durham, N. C.)

The formation of autophagic vacuoles (AV) is described in (1) glucagon-treated rat liver cells *in vivo* and in glucagon-treated rat liver slices *in vitro;* (2) isolated flounder kidney tubules *in vitro;* (3) testosterone-treated ductus deferens *in vivo;* and (4) in HeLa cells *in vitro.* In some experiments the effects of metabolic inhibitors such as actinomycin, puromycin, cycloheximide, dinitrophenol, and azide on autophagy were determined. The relation of AV to other cytoplasmic organelles (in sections or density gradient fractions) was studied by electron microscopic histochemical methods (acid phosphatase, aryl sulfatase, non-specific esterase, glucose-6-phosphatase, inosine diphosphatase, adenosine-triphosphatase and thiamine pyrophosphatase. In some studies, acid phosphatase or thiamine pyrophosphatase were demonstrated on the same section with aryl sulfatase. The relation of autophagic vacuoles to pre-existing heterolysosomes was studied in rat liver *in vivo* by pre-loading the heterolysosomes with ferritin or *in vitro* loading with I^{125}—labelled polyvinylpyrrolidone. These studies indicate that AV originate as acid hydrolase-free, double-membrane limited autophagosomes, by enclosure of a small portion of cytoplasm by cytoplasmic sacs devoid of ribosomes.

In later stages, these were transformed to single membrane-limited acid hydrolase-containing lysosomes by fusion with primary and/or secondary lysosomes and disappearance of the inner limiting membrane presumably by acid hydrolase digestion. Protein synthesis is apparently not necessary either for the formation of the limiting membranes or the hydrolytic enzymes while the formation of AV seems to be inhibited by inhibitors of energy metabolism. Autolysosomes contained both aryl sulfatase and acid phosphatase whereas Golgi vesicles, presumed to represent primary lysosomes, seemed to be carriers for one enzyme only.

Steroidogenic Activity in Testis Interstitial Cells Grown "in Vitro"

AURELI, G., A. LAURIA, and M. RIZZOTTI (Istituto di Anatomia degli Animali Domestici-Istologia ed Embriologia, Università di Milano, Italia)

Remarkable differences are present in testis interstitial cells from various species, both during ontogenesis and in their behaviour in culture (grown in liquid medium on coverslip following testis dissociation by means of trypsin).

These elements dedifferentiate towards the end of foetal life, remain so unchanged in the newborn and differentiate during prepuberal life (e.g., cattle). $\Delta 5 - 3\beta$ – Hydroxysteroid dehydrogenase ($\Delta 5 - 3\beta$ – HSD) activity can be elicited histochemically only when interstitial cells are differentiated. In cultures of testis from bovine eight to nine months foetus and from newborn calf, cells are fibroblast-like and no $\Delta 5 - 3\beta$ – HSD activity is demonstrable; when interstitial cells are already differentiated, on the other hand, testis cultures show numerous strongly positive elements, even after a few hours.

Interstitial cells can keep their features from foetal to adult life (e.g., swine). $\Delta 5 - 3\beta$ – HSD activity can be always observed, also in testis cultures.

Equidae foetal interstitial cells are epithelioid in shape and very numerous, and reach their maximum development towards the middle of pregnancy. Towards the end of foetal life, many elements show chromolipoid pigments which, at birth, stuff all cells ("xanthochrome" cells). "Xanthochrome" cells characterize Equidae testis in the prepuberal period until the appearance of definitive Leydig cells. $\Delta 5 - 3\beta$ – HSD activity appears transitorily towards the end of foetal life, is thoroughly absent in "xanthochrome" cells, and is obviously observable in Leydig cells. Horse foetus (4-10 months) testis cultures show no $\Delta 5 - 3\beta$ – HSD activity and many elements assume some morphological and histochemical features typical of "xanthochrome" cells excepting the pigment colour which is present only in 10 months foetuses. When grown *in vitro,* "xanthochrome" cells show: a) deep modification of their morphology and of the histophysical characteristics of the pigment, which, however, keeps its histochemical characteristics: PAS-positivity, sudanophilia and fluorescence; b) the appearance of $\Delta 5 - 3\beta -$ HSD activity.

These observations show that the cells studied maintain, *in vitro,* the peculiarity they showed in the organism at the moment of their explanting excepting horse testis interstitial cells where the *in vitro* modifications seem to reflect the unique ontogenetic processes taking place in the organism over a long period of time.

Dry Mass Determinations with the Electron Microscope as an Adjunct to Cell Biological Studies

BAHR, G. F. (Armed Forces Institute of Pathology, Washington, D. C.)

1) Rat liver mitochondria have been isolated and purified by conventional differential centrifugation (1) and subsequently subjected to differential centrifugation on a zonal density gradient. Fractions containing large and small mitochrondria were harvested and studied for dry mass distribution by quantitative electron microscopy using a recently designed integrating densitometer IPMII (Carl Zeiss, Oberkochen/Wuert., West Germany). The median dry mass ratio of small to large mitochondria was found to be close to 1:2. DNA determination in these fractions render a ratio of only 1:1.6 for the two fractions studied. Possible implications will be discussed.

2) In collaboration with Dr. Charles Kiddy of the Department of Agriculture, Beltsville, Maryland, bull spermatozoa were separated into slow and fast fractions by sedimentation on storage media for semen. Isopycnic centrifugation of the fractions on sucrose gradients revealed no differences in buoyant density. Dry mass determinations by quantitative electron microscopy rendered consistently smaller values for the slow fraction than for the fast fraction by 3 to 6 per cent. The weight differences are suggested to reflect a partition in XX and XY bearing sperm cells.

References:

(1) Glas, U. and G. F. Bahr: J. Cell Biol. *29,* 507 (1966).

Evaluation of Electron Micrographs by a Taxonomic IntraCellular Analysis System (TICAS)

BAHR, G. F., G. L. WIED, and P. BARTELS (Armed Forces Institute of Pathology, Wash., D. C., University of Chicago, Chicago, Ill. and University of Arizona, Tucson, Ariz.)

The application of pattern recognition techniques to light microscopic images has in recent years led to encouraging results and fundamental insights into the capabilities and requirements of computer assisted analysis of biologic images. Groups of collaborators in the laboratories of Mendelsohn, Lipkin, Ledley, and Wied have demonstrated to us that the methodologies and machines at hand already permit automatic identification of certain cell types and of the chromosomes of an idiogram. Beyond the fast automatized recognition of morphologically distinct objects, it has been possible to tell morphologically indistinct objects apart with very low classification error (1).

The principles of pattern recognition in light microscopy can, with little difficulty, be applied also to electron micrographs. A scanning microdensitometer with features for analog to digital conversion of measurements and incremental recording of data on computer compatible magnetic tape has been used in the assessment of electron micrographs of whole mounted unstained tissue culture cells. The system provides sensitive indication of threshold effects of cellular radiation injury as well as means for numerical definition of chromatin rearrangement in cell division.

Conditions of electron microscopic photography were maintained in such a manner that both pattern analysis as well as quantitative determination of dry mass could be carried out in one operation.

References:

(1) Wied, G. L., P. H. Bartels and G. F. Bahr: Acta Cytol. *12* (1968).

Comparative Enzyme Histogenesis of Spontaneous and Experimentally Induced Myocardial Lesions

BAJUSZ, E. (Bio-Research Institute, Cambridge, Mass.)

Studies on the morphogenesis and enzyme histogenesis of various necrotizing cardiomyopathies have shown that, depending on the nature of the eliciting agent, myocardial lesions differ from each other not only in rate of development and in certain characteristics (histologic), but also in early enzymatic changes, including alterations occurring in glycogen metabolism of the injured muscle cells (1, 2). While use of conventional stains establishes no clear distinction between primary (or 'metabolic') cardiac necroses and those resulting secondarily from circulatory anoxia, some histochemical techniques appeared to be applicable to differential diagnosis of these 2 types of myocardial lesions (3, 4). The existence in an inbred strain of Syrian hamsters of a hereditary, degenerative cardiomyopathy (5) provided an opportunity to extend investigations along these lines, especially with respect to similarities and differences in the genesis of spontaneous and experimentally induced heart-muscle injuries.

Anoxic-type myocardial lesions were induced in LSH (London School of Hygiene) hamsters by surgical occlusion of the left coronary artery and injection of a single toxic dose (1.2 mg/100 gm b.w., s.c.) of epinephrine, while metabolic necrosis was provoked by feeding animals of this healthy strain a K-deficient diet. The histochemistry of all these induced lesions was analyzed comparatively with that of the spontaneous heart-muscle degeneration regularly seen in the cardiomyopathic strain (BIO 14.6). In these animals the myocardial pathology originates in a genetically determined, metabolic defect; the focal lesions develop with no vascular involvement and are, on routine histologic sections, indistinguishable from the spotty myolysis normally elicited by epinephrine overdosage. Nevertheless, the spontaneous myocardial lesions differ histochemically in many ways from all experimentally evoked lesions so far studied. Among prenecrotic changes developing after occlusion of the coronary artery and injection of epinephrine, rapid loss of phosphorylase activity, an equally speedy depletion of glycogen reserve, and early decline in reactions for monoamine oxidase, cytochrome oxidase and succinic dehydrogenase were outstanding and all were restricted to myocardial areas where necrotic lesions later developed. In the hearts of animals kept on K-deficient diet there was no relation between activity of phosphorylase and amount of stainable, labile fraction of glycogen; progressive reduction of oxidative enzyme activity was diffuse throughout the myocardium and correlation between histochemical abnormalities and appearance of focal lesions could not be ascertained. In the cardiomyopathic hamsters, reduced enzymatic staining was seen only in myocardial fibers that had already shown microscopic evidence of damage with routine stains. Furthermore, significant increase or decrease in glycogen content of muscle cells was not caused by the hereditary metabolic defect itself when it spared the structural integrity of myocardial fibers. However, early changes in activity and distribution of lysosomal hydrolases (6) were detected during development of spontaneous myocardial lesions only. Enlargement and aggregation in the juxtanuclear region of granules showing intense acid phosphatase, β-glucuronidase and aryl sulfatase activities were seen in fiber segments during the initial stages of degeneration. In fibers with advanced stages of myolysis, lysosomes could no longer be identified by staining for these hydrolases: diffuse and weak acid phosphatase and esterase activities were occasionally apparent in and around the disintegrating muscle cells, while reaction for β-glucuronidase and aryl sulfatase remained negative. The question whether the lysosomes represent the primary site of damage by the genetically transmitted metabolic abnormality responsible for myocardial degeneration cannot be answered with confidence on the basis of present observations. Nevertheless, there is little doubt that only an understanding of the metabolic aberrations involved in the genesis of each myocardial lesion will enable us to draw etiological conclusions from histochemical observations.

References:

(1) Bajusz, E. and G. Jasmin: Acta Histochem. *18*, 222 and 238 (1964).
(2) Bajusz, E.: *In:* Methods and Achievements in Experimental Pathology, vol. 2, pp. 172-233 (Bajusz and Jasmin, edit.) Chicago: Year Book Med. Publ. (1967).
(3) Bajusz, E. and G. Jasmin: Experientia *20*, 373 (1964).
(4) Bajusz, E. and G. Jasmin: Am. Heart J. *69*, 83 (1965).
(5) Bajusz, E. *et al.:* Ann. N.Y. Acad. Sci. *138*, 213 (1966).
(6) Abraham, R. *et al.:* J. Histochem. Cytochem. *15*, 596 (1967).

Electron Microscopic Localization of Glycogen in the Organ of Corti

BALOGH, KÁROLY (Harvard Medical School, Boston, Mass.)

Cochleas of various mammalian species were dissected and fixed in cold (4° C) absolute ethanol saturated with picric acid or in cold 2.5% buffered glutaraldehyde. Dehydrated blocks of tissue were embedded in Epon, cut and stained with a modified periodic acid—Schiff (PAS) reaction. Under the light microscope, glycogen presented as PAS-positive granules that were digestible with alpha-amylase. No appreciable loss of glycogen was observed with either method of fixation, as compared with freeeze-dried cochleas. The distribution pattern of glycogen showed striking differences: the hair cells of guinea pigs were very rich in glycogen, but cats had none. In mice, Deiters' cells were also full of glycogen, whereas the hair cells contained only moderate amounts.

For electron microscopy, glutaralydehyde-fixed specimens were stained with the periodic acid—thiosemicarbazide method of Seligman *et al* (1). The glycogen particles were visualized in the hair cells of guinea pigs; these particles were completely digestible with alpha-amylase.

Acoustic stimulation (white noise, 100 decibels for 30 min.) produced a moderate depletion of glycogen in the hair cells of guinea pigs; these changes were reversible, however. The significance of these observations will be discussed.

References:

(1) Seligman *et al.:* J. Histochem. Cytochem. *13*, 629 (1965).
(2) Rosa and Johnson: J. Histochem. Cytochem. *15*, 14 (1967).

Some New Instruments for Quantitative Cytochemistry

BARER, R. (Dept. of Human Biology and Anatomy, The University, Sheffield)

Instruments currently developed and tested in this laboratory include:
(1) A sensitive electronic method of measuring optical path differences

using a Pockels Effect light modulator in conjunction with Fresnel rhombs as achromatic quarter wave elements. This opens up new possibilities for using wavelength variations for mass measurements in interference microscopy.

(2) A new type of image scanning and integrating microdensitometer (developed with Dr. D. J. Goldstein) with special compensating system.

(3) A new prototype scanning and integrating microdensitometer developed by Vickers Instruments Ltd. This can be used for either absorptiometry or integrated mass measurement by interference microscopy.

The Effect of Isoproterenol on Salivary Glands: A Model of Induced Cell Proliferation

BARKA, TIBOR (Mount Sinai School of Medicine, New York, N. Y.)

Isoproterenol, a drug acting primarily on β-receptors, greatly stimulates DNA synthesis and mitotic activity in parotid and submaxillary glands in rats and mice. Its daily administration leads to a hyperplastic and hypertrophic enlargement of these organs (1-5). This stimulation offers a rather unique model for cell proliferation for the following reasons: The 'inducer' is a low molecular weight, synthetic compound with known metabolism, and it is effective in a single dose; the stimulation is dose dependent, predictable, and can be inhibited by an organ and species—specific suppressor. Furthermore, the low rate of cell proliferation in the salivary glands of adult rats or mice favors comparison between control and stimulated cell populations.

A single dose (16 mg per 100 g body weight) of isoproterenol stimulated the incorporation of tritiated thymidine 4-13 fold with a maximum of 27 hours after administration of the drug. With 3 doses of the drug up to 30-fold stimulation of thymidine incorporation was achieved within 36 hours in female animals. As shown autoradiographically, more than 10% of the acinar cells were stimulated to enter synchronously into the S-phase. The stimulatory effect of the drug could be prevented by the administration of a 'suppressor' isolated from the gland of intact animals. The 'suppressor' was partially purified from 120,000 x g supernatants of sonicated submaxillary gland homogenates using Sephadex G75 and DEAE cellulose chromatography. The 'suppressor' was either absent from the glands of rats chronically treated with isoproterenol or it was greatly reduced in amount in such animals. The purified 'suppressor' fraction displayed high proteolytic activity measured by using synthetic trypsin substrates. Circumstantial evidence suggests that the 'suppressor' is enzymatically active. Direct proof of this awaits further purification of the 'suppressor.' Since the 'suppressor' was most active

18-20 hours after the administration of isoproterenol, at the time when stimulation of RNA synthesis was maximal, it may act at the level of transcription. Available evidence is compatible with the hypothesis that isoproterenol induces cell proliferation by acting on a genetic level.

References:

(1) Barka, T.: Exptl. Cell Res. *37*, 662 (1965).
(2) Barka, T.: Exptl. Cell Res. *39*, 355 (1965).
(3) Barka, T.: Exptl. Cell Res. *48*, 53 (1967).
(4) Baserga, R.: Life Sci. *5*, 2033 (1966).
(5) Selye, H., R. Veilleux, and M. Cantin: Science *133*, 44 (1961).

Quantitative Enzyme Cytochemistry by the Labeled Inhibitor Method [*]

BARNARD, ERIC A. and JANUSZ KOMENDER (Molecular Enzymology Unit, State University of New York, Buffalo, N. Y.)

For certain enzymes for which there is available a suitable inhibitor that binds tenaciously or reacts stoichiometrically, this inhibitor can be applied in isotopic form to cells or tissue (living or fixed) and, after suitable treatment, autoradiographs will reveal the sites of the enzyme in question. Grain-counting (after ^{3}H labeling) gives the relative concentrations of the enzyme at various sites. After ^{32}P or similar labeling, counting of beta tracks in nuclear track emulsion gives the absolute number of these enzyme molecules per cell or structural element, as illustrated by the track measurements on the acetylcholinesterase (AChase) content of differentiating megakaryoblasts (1).

The investigations required to establish the basis of the method for measurement of a given enzyme will be outlined. The important factors include specificity: this is obtained by using protection by substrate, or applying other specific affinities of the enzyme active center. After covalent reaction, e.g. of di-isopropylfluorophosphate (DFP), all non-reacted inhibitor has been shown to be removed by exchange with excess unlabeled inhibitor. Checks on the linearity of quantitative response in the autoradiography are also important. In favorable material, quantitation can also be tested by bulk counting of isotope in known numbers of the processed cells, and has given excellent agreement.

Rates of reaction, ligand affinity constants, and other characteristics of the enzyme *in situ* can also often be measured thus. Illustration will be given from recent work on a protease of mast cells. Reaction with ^{32}P-DFP and track-counting showed that $6(\pm 0.3 \text{ S.E.}) \times 10^8$ enzyme sites per rat mast cell react rapidly (2). The reaction was strongly retarded by chymotrypsin substrates and not by substrates for non-specific esterase.

16

Serotonin, but not histamine, also binds strongly. EM autoradiography (3) showed that this enzyme is only in the mast cell granules. Mouse peritoneal mast cells contain about 9 x 10⁸ molecules per cell, but masto-cytoma (Furth) cells growing in similar mice have only 1.2 x 10⁸. The mastocytoma cells in culture show a similar amount. The reaction rate of DFP at these enzyme sites is the same in living and formalin-fixed cells. Reaction with specific irreversible inhibitors showed the active center is similar to that of chymotrypsin, and that no zymogen is apparent *in situ*. Isolation of the granules from the mastocytoma cells showed that they contain the enzyme. Reaction with ³²P-DFP in living cells was also followed by biochemical isolation of the labeled enzyme as a pure product. The mouse mast cell also contains a trypsin-like enzyme. The main DFP-reactive component has been isolated similarly without the prior inhibition: it is a chymotrypsin-like esterase and protease, whose properties accord with those found by the *in situ* technique.

This application to mast cells can be used to follow them *in vivo* (after reinjection of labeled living cells) and to follow the fate of the enzyme. It also provides quantitative information on this protease in different types of mast cell *in vivo*.

The most useful reagent applied so far in this general methodology is DFP, which, with suitable techniques, can localize each of a number of esterase molecules, in certain sites. However, a number of other suitable inhibitors are, in principle, available. One such, ³H-amethopterin, has been shown to be of value in measuring relative amounts of folate re-ductase in various cell types (4), using a similar technique.

References:

(1) Darzynkiewicz, Z., A. W. Rogers, and E. A. Barnard: J. Histochem. Cyto-chem. *14*, 915 (1967).
(2) Darzynkiewicz, Z. and E. A. Barnard: Nature *213*, 1198 (1967).
(3) Budd, G. C., Z. Darzynkiewicz, and E. A. Barnard: Nature *213*, 1202 (1967).
(4) Darzynkiewicz, Z., A. W. Rogers, E. A. Barnard, D. H. Wang, and W. C. Werkheiser: Science *151*, 1528 (1966).

---

* Supported by a grant (DRG-867) from the Damon Runyon Memorial Fund, and (GM-11754) from the National Institutes of Health, U.S.P.H.S.

Measurement of Acetylcholinesterase Molecules at Motor Endplates, and Their Correlation with Impulse Transmission *

BARNARD, ERIC A. and JAN WIECKOWSKI (Molecular Enzymology Unit, State University of New York, Buffalo, N. Y. and Medical Academy, Warsaw, Poland)

The strategy outlined in the associated abstract (1) has been applied to AChase at motor endplates. Using labeled DFP, the numbers of reactive sites at the endplates of four types of mouse muscle have been deter-mined; it has been shown in several cases, using protection by specific reversible inhibitors, and also using the specific reactivation by pyridine-2-aldoxime methiodide (2-PAM), that one-third of this total of sites are AChase (2, 3). By EM autoradiography, the AChase molecules were also measured and shown to be primarily along the post-synaptic membranes of the junction 3, 4). With rat muscles, the numbers have now been found to be fairly similar: 10. 8 ($\pm$ 0.2 S.E.) x 10^7 (sternomastoid) and 0.6 ($\pm$ 0.3 S.E.) x 10^7 (diaphragm) total sites per endplate, with one-third of these molecules again AChase. A unimodal narrow distribution is found in the population of each endplate type. The method is of value for comparing AChase contents of endplates in different muscle types and different animal species.

Properties that have been examined for AChase *in situ* in the endplate include reaction rates with DFP, active center ligand affinities, and the "ageing" phenomenon, i.e., loss of 2-PAM-reactivation capacity. The two-thirds of the DFP-reactive sites that do not react as AChase are not merely protected by membranes, since the lipid-soluble benzoyl analogue of 2-PAM gives the same final degree of reactivation.

Rat nerve-muscle preparations, in Tyrode solution at 37°, have been treated with labeled DFP (0.5 — 1 μg/ml). 70-80% of the DFP-reactive sites, and of the AChase sites, have become blocked at the point where single twitch potentiation becomes significant. Still lower residual levels of AChase permit tetanic contraction, which disappears before full DFP reaction of AChase. Hence only a fraction of the AChase seen at the junction is essential for normal impulse transmission. The non-AChase DFP-reactive sites, on the other hand, can be fully labeled without effect on impulse transmission.

References:

(1) Barnard, E. A. and J. Komender: these Abstracts.
(2) Rogers, A. W., Z. Darzynkiewicz, E. A. Barnard, and M. M. Saltpeter: Nature *210*, 1003 (1966).
(3) Barnard, E. A. and A. W. Rogers: Ann. N.Y. Acad. Sci. *144*, 584 (1967).
(4) Salpeter, M. M.: J. Cell Biol. *32*, 379 (1967).

---

* Supported by a grant (GM-11754) from the National Institutes of Health, U.S.P.H.S.

Fine Structural Localization of Enzymes Concerned with Lipid Metabolism

BARRNETT, R. J. (Dept. of Anatomy, Yale University Medical School, New Haven, Conn.)

(no abstract submitted)

Non-Specific Esterase and Acetylcholinesterase in Brain Subcellular Fractions *

BARRON, K. D., A. H. KOEPPEN, and J. BERNSOHN (Neuropsychiatric Research Laboratory and Neuropathology Research Section, Veterans Administration Hospital, Hines, Ill., and Department of Neurology and Psychiatry, Northwestern University Medical School, Chicago, Ill.)

Homogenates of rat cerebrum in 0.32M sucrose were subjected to differential centrifugation. The crude mitochondrial pellet was treated with ice-cold distilled water to release trapped cytoplasm and contained organelles by "osmotic shock" (Laatsch, 1961; DeRobertis, 1962). After osmotic shock, further fractionation was carried out by density equilibrium centrifugation in a discontinuous sucrose gradient (0.8, 0.9, 1.0, 1.2M). The fractions were assayed for non-specific esterase (NsE) with alpha-naphthyl acetate (NA) as substrate and acetylcholinesterase (AChE) with acetylthiocholine as substrate. From 15-20% of the NA hydrolysis was inhibited by 5×10^{-6}M eserine sulfate. The fractions were examined by electron microscopy which revealed a myelin fraction of high purity, two fractions rich in synaptic membranes, a mixed fraction containing membranes and light mitochondria and a pure mitochondrial pellet.

Debris, cell sap, and the microsomal fraction contained more than 60% of the total activity of both enzymes. About 13% of the total was present in the isodensity bands of which approximately one-half was contained in the myelin fraction. The myelin fraction was washed repeatedly in distilled water or 0.32M scurose and as much as 75% of the activities of both enzymes could be removed thereby. This finding suggests non-specific adsorption of the enzymes assumedly due to protein-protein or protein-lipid interactions. Evidence for a degree of non-specific adsorption of enzymes to other membrane fractions was obtained also. The possibility of artificial redistribution should be considered in evaluation of enzyme localizations obtained by cell fractionation procedures. However, the membrane fraction retained by 0.9M sucrose appeared to possess enzymes which were more tightly bound and may reflect the *in situ* condition. The relative quantitative distributions of both AChE and NsE were similar, which may suggest that these enzymes have some interrelated biochemical role. The 0.9M sucrose fraction had the highest relative specific activity for both enzymes.

---

* The work was supported in part by Public Health Service Grants NB-04191 and NB-04722.

Recent Developments in Cell Discrimination Techniques

BARTELS, PETER H., G. F. BAHR, and G. L. WIED (University of Arizona, Department of Microbiology, Tucson, Ariz., Armed Forces Institute of Pathology, Washington, D. C., University of Chicago, Departments of Obstetrics and Gynecology and Pathology, Chicago, Ill.)

Attempts to discriminate between cells with only insignificant morphologic but functional differences have led to the development of a number of descriptive techniques as sub-routines of the basic TICAS cell discrimination program.

These are the application of equiprobable extinction range contours which in both their static and dynamic employment lead to highly specific cell type and cell state patterns; and the use of multi-dimensional decision spaces, with Euclidean distances used as metric in the decision rules.

A method is described to determine the optimum number of dimensions for the decision space of a cell discrimination problem at hand.

Finally, an information theoretical approach to cell identification is introduced in which the sequence of line scan data is analyzed for their characteristic distribution of transitional probabilities. The applying amounts of information carried by each source symbol are computed and can serve as a guide in systematic descriptor development. Computations of this type also constitute the basis for the design of two dimensional image pattern of encoded symbols which can be population-averaged and used as cell type standards in the computer recognition of cell states or types.

Semi-Quantitative, Regional Estimation of Protease Activity in Bone and its Relation to Factors Influencing Osteolysis *

BÉLANGER, LEONARD F. (Department of Histology and Embryology, Faculty of Medicine, University of Ottawa, Ottawa, Canada)

The Adams and Tugan (1) procedure making use of exposed and processed photographic plates, has shown that proteolytic activity is mainly located to mature osteocytes (2) in fresh bone (3) or in bone demineralized in isotonic EDTA (4). This process, however, was not reliable towards providing a variation index related to physiological or pathological protease-stimulating or depressing factors. Along with the histochemical test, a comparable incubation was carried out on thick cross-sections of bone taken at various levels. The area of protease activity in the photographic plate was projected at x13 on filter paper and

20

delineated. Relative weights of paper cut-outs have revealed greater activity in proximal metaphysis of tibia as compared to other regions. For a given area, the activity was enhanced by PTE (2) and by PTH (5). It was considerably decreased by hypophysectomy (5), by a diet deficient in phosphate (4) and by actinomycin D (5). When calcitonin was given concurrently to PTH, the stimulating effect of the latter was apparently inhibited (5).

References:

(1) Adams, C. W. A. and N. A. Tugan: J. Histochem. Cytochem. *9*, 469 (1961).
(2) Bélanger, L. F. and B. B. Migicovsky: J. Histochem. Cytochem. *11*, 734 (1963).
(3) Bélanger, L. F.: Second Internat. Congress Histo-Cytochemistry, Springer-Verlag, Berlin, Abstr. p. 156 (1964).
(4) Bélanger, L. F., I. Clark, L. Krook, and C. Gries: J. Histochem. Cytochem. *13*, 404 (1965).
(5) Bélanger, L. F. and H. Rasmussen: Proc. 3rd Conf. Parathyroid and Thyrocalcitonin (Calcitonin), Excerpta Medica, Amsterdam (*in press*).

* With financial support from the Medical Research Council of Canada and the Canada Department of Agriculture.

Morphogenetic and Histochemical Study on Cooleyan Osteopathology

BELTRAMI, CARLO ALBERTO (Istituto di Anatomia ed Istologia Patologica, Università di Ferrara, Italia)

In Cooleyan subjects, the marrow first appears hyperplastic, there is variable accumulation of glycoproteic material and a regressive nucleo-cytoplasmic alteration of erythroblasts. Afterwards, a progressive reticulo-hystiocytary hyperplasia appears which manifests itself in a strengthening of the reticular-hystiocytary network and often in aspects of the saurosic hystiocytosis.

The hyperplasia of the bone marrow causes lytic lesions of bone tissue; to these are constantly associated phenomena of osteogenesis, resulting in deposition of an osteoid rich in interlacing fibers, with the features of the embrional osteoid. The periosteum shows very often the characteristics of an immature mesenchymal tissue.

The marked participation of the mesenchymal structures with activation of osteogenetic processes which show patterns of embryonic osteogenesis distinguish the Cooleyan osteopathy from others, which are secondary to hyperplasia of the bone marrow.

Within the marrow tissue, there are often areas of intense concentration of the reticular network in close connection with nests of epithelioid cells, so that foci of myelofibrosis are formed, sometimes occupying a whole space and comprising also neoformed bone tissue.

With intensification of processes of osteogenesis, a progressive thickening of neoformed bone tissue occurs, with gradual shrinking of the marrow spaces and sometimes phenomena of ivorisation.

Microspectrofluorometry of Dopamine and Noradrenaline in Tissue Sections

BJÖRKLUND, A., B. EHINGER, and B. FALCK (Institute of Anatomy and Histology, Department of Histology, University of Lund, Lund, Sweden)

In the histochemical method of Falck and Hillarp for the demonstration of certain catechol and indol derivatives in tissue sections, there has hitherto not been any possibility of differentiating directly between the two important amines, dopamine and noradrenaline, since they both give rise to fluorophores with similar ffuorescence characteristics. However, Corrodi and Jonsson have recently shown in model experiments that the fluorophore derived from noradrenaline upon rather vigorous treatment with hydrochloric acid, will transform into a fully aromatic isoquinoline, while the fluorophore formed from dopamine will remain as a 6,7-dihydroxy-3,4-dihydroisoquinoline. These two isoquinolines have widely different excitation characteristics: the excitation peak of the dopamine fluorophore is at 360-370 $m\mu$ and that of noradrenaline is at 330 $m\mu$. It has now been shown that this reaction can be performed also in tissue sections with milder conditions and without introducing any non-specific fluorescence. The sites of the two fluorophores can then readily be analyzed in a microspectrofluorometer. Dopamine has been found to be characterized by the appearance of a peak at 370 $m\mu$ upon HCl treatment, and this peak is not affected by prolonging the treatment. The presence of noradrenaline is suggested by the gradual, relative decrease of the excitation curve at 370-410 $m\mu$ and a concomitant, marked, relative increase of the excitation peak at approximately 330 $m\mu$ upon prolonged HCl treatment. The method has been successfully applied to a number of structures. As examples, dopamine was studied in the bovine liver mast cells and in the varicose nerve terminals in the pedal ganglion of the fresh-water bivalve, *Anodonta piscinalis*, while noradrenaline was studied in the adrenergic nerve terminals of the guinea-pig vas deferens and in the vascular nerves of the bovine liver.

References:

(1) Corrodi, H. and G. Jonsson: J. Histochem. Cytochem. *13*, 484 (1965).

A Method for the Fine Structural Demonstration of Monoamine Oxidase Activity

BOADLE, MARGARET C. and FLOYD E. BLOOM (Department of Pharmacology, Yale University School of Medicine, New Haven, Conn.)

Guinea pig kidneys fixed briefly with 2% paraformaldehyde and sectioned on a cryostat at 15 μ and 80 μ, were reacted at 37° C in a medium adapted from Glenner *et al.* (1) containing 5 mg tryptamine hydrochloride, 1 mg of thiocarbamyl nitro-blue tetrazolium (2), 8% sucrose, and 0.8 mg Na_2SO_4 in 4 cc of 0.04 M Na_2HPO_4 buffer pH 8, for 60 minutes or more. In control experiments, tryptamine was omitted from the media. After being rinsed in buffered sucrose for 30 minutes, sections were then exposed to 1% buffered osmium tetroxide at room temperature or at 40° C for 1 hour. Light microscopic examination of 15 μ sections showed reactivity mainly within proximal tubule cells. Fine structural observations on sections incubated in the complete media revealed a uniform increase in electron opacity on the outer limiting membrane of the mitochondria of proximal tubule cells. There was no increase in electron opacity on mitochondrial cristae or on any other plasma or intracellular membrane. The electron opacity of the reaction product could be increased by exposure to osmium tetroxide at higher temperatures, but above 40° C the tissue disintegrated. Treatment with osmium tetroxide vapor at 60° C did not increase the electron opacity above that produced by osmium tetroxide at 40° C. In control sections, there was no increase in electron opacity of the outer mitochondrial membrane when compared to plasma membranes. Araldite 502 was the embedment of choice as Vestopal and Maraglas dissolved the formazan.

References:

(1) Glenner, G. G., H. J. Burtner, and G. W. Brown, Jr.: J. Histochem. Cytochem. *5*, 591 (1957)
(2) Seligman, A. M., H. Ueno, Y. Morizono, H. L. Wasserkrug, L. Katzoff, and J. S. Hanker: J. Histochem. Cytochem. *15*, 1 (1967).

---

* Supported by U.S.P.H.S. Grant MH 12380.

Labelling of Bacterial Toxins as a Tool for the Study of Uptake of Soluble Antigens by Leucocytes

BONA, C. (Instit. "Dr. I. Cantacuzino," Bucharest, Romania)

Interactions between leucocytes and bacterial toxins are obviously involved in the process of immunogenesis. The formation of specific antitoxins is an attribute of the transmission to immunological competent cells, of the antigenic information carried by toxins. The fact that bacterial toxins are soluble (i.e. may be found as macromolecules in solution) has led us to assume that they might be incorporated into leucocytes by means of pinocytosis (described for the first time by Lewis) (1). In order to ascertain this assumption, we investigated leucocytes submitted to the action of thermostable endotoxin obtained from S. *typhimurium*, of thermolabile endotoxin (neurotoxin) from *Serratia marcescens*, crystallized diphtheria toxin (3100 Lf/mg N, 0.625 mg protein/ml), purified Dick erythrotoxin 10^6 rabbit STO/mg N, 0.9 protein/ml) in optimum concentrations of toxins which might not kill leucocytes within a time limit of three hrs.

The process of uptake was followed in three ways: (*1*) labelling the toxins with fluorescein isothiocyanate, in presence of 0.5 M bicarbonate-carbonate buffer pH 9, acetone and Cellite 423. The ratio protein of toxin/isothiocyanate was of 1 mg/0.05 mg (2). Observations were performed in fluorescence microscopy. (2) labelling the toxins with uranyl-acetate of PH 6.5. Observations were performed in electron microscopy. (*3*) fluorescence microscopy examination of the cells put in contact with non-labelled toxins and subsequently treated with fluorescein-labelled specific antitoxic sera.

Observations in phase contrast microscopy showed that leucocytes exposed to bacterial toxins exhibit an intense membrane activity, the formation of invaginations and even of short-lived channels. Under fluorescence and electron microscopy we observed the presence of labelled toxins within the pinocytic vacuoles. After 15 minutes of contact between leucocytes and the labelled toxin, a very few small fluorescent vacuoles appear in the cytoplasm, close to the cell membrane; fine channels become apparent, consisting of 2-3 fluorescent droplets. The number of vacuoles increases significantly after 30 minutes and after 45 minutes the vacuoles appear accumulated around the nucleus. The pinocytosis rate differs from one toxin to another (2,3).

The utilization of cytochemical methods and biological tests for the study of toxin activity (Schwartzmann phenomenon for the endotoxin, Fraser test for the diphtheria toxin) also supports our opinion that bacterial toxins are taken up in leucocytes, by pinocytosis.

Although the pinocytosis of toxins by leucocytes does not appear as a specific immunological process, it seems highly probable that, during pinocytosis, leucocytes possess the ability to discriminate foreign matter, or even altered autologous macromolecules (4).

References:

(1) Lewis, W. H.: Bull. Johns Hopkins Hosp. *49*, 17 (1931).
(2) Mesrobeanu, I., C. Bona, L. Mesrobeanu: Exptl. Cell Res. *36*, 434 (1964).
(3) Mesrobeanu, I., C. Bona, L. Ioanid, L. Mesrobeanu: Exptl. Cell Res. *42*, 490 (1966).
(4) Bona, C., A. Sulica, M. Dumitrescu, D. Vranialici: Nature *213*, 824 (1967).

24

A Study of Species Differences in Mammalian Kidney Relative to the Distribution of Enzymatic Activities

BRADSHAW, M., L. MONUS, and S. STROMAN (H. R. Laboratories, Greenvale, N. Y.)

Activities of many enzymes show a reproducible regional distribution along the course of the nephron and collecting ducts. Within the same species there is a corresponding distinctive zonal distribution of activity in cortex and medulla with a precise pattern of localization for each enzyme. The interspecies variation of this pattern is evident in newborn kidneys as well as in adult kidneys (1).

One of the most striking examples of species differences is the definite amount of formalin-resistant adenosine triphosphatase activity localized within the glomerulus of the rat kidney whereas none is histochemically active within the glomerulus of the rabbit kidney. Enzymes, such as 5'-nucleotidase, are active in the cells of the collecting ducts of the inner medulla in one species and, in other species, the activity of these enzymes cannot be localized, histochemically, in the same cells.

Functionally, species differences are observed in the transport of urea by the kidney (2) and in osmotic diuresis (2,3). After manitol, urea and sodium chloride loading in dogs and rabbits, fractional urinary sodium excretion is of such magnitude that the investigators state that the reabsorptive capacity at sites proximal to the distal tubules are damaged. In the rat, however, fractional reabsorption remains more or less constant over a considerable range of load increments (3).

Variations in the patterns of enzymatic activities in, as well as functional differences between, the species demonstrate the hazards inherent in applying data from one species to another.

References:

(1) Wachstein, M. and M. Bradshaw: J. Histochem. Cytochem. *13*, 44 (1965).
(2) Gottschalk, C. W.: Am. J. Med. *36*, 670 (1964).
(3) Giebisch, G. and E. E. Windhager: Am. J. Med. *36*, 643 (1964).

The Localisation of Norepinephrine Binding Sites in Peripheral Nerve Using Electron Microscope Autoradiography

BUDD, G. C. and M. M. SALPETER (Cornell University, Ithaca, N.Y.)

Following an intravenous injection of tritiated norepinephrine (^{3}HNE) the highest proportion of bound radioactivity in homogenates from peripheral tissues with an extensive adrenergic innervation is in the microsomal fractions (1-3). The labeled fractions are presumed to be derived in part from agranular and granular vesicles of adrenergic nerves. Autoradiography has also shown that [3]HNE can be firmly bound in peripheral and central nerve terminals. It has therefore been assumed, but not demonstrated, that the norepinephrine is bound in granular vesicles (4,5).

Electron microscope autoradiograms were prepared of pineal gland and adrenal capsule from mice which received [3]HNE intravenously. NTE emulsion (Kodak) was used under conditions which gave an expected resolution of 1000 Å (6). The observed distribution of developed grains relative to whole nerve terminals was analysed. The results resembled theoretical curves for the expected distribution assuming a uniform distribution of radioactivity within the nerve terminals. Further analysis revealed that the grain density over small agranular and granular vesicles (470 Å diameter, occupying 2/3 the volume of nerve terminal) was significantly greater than over ground axoplasm. No distinction could be made between the agranular and granular vesicles. The grain density relative to a population of larger granular vesicles (1000 Å diameter) was considered separately. Directly over these larger vesicles the grain density was not significantly above that for the ground axoplasm, but there was a peak in grain density 500-1500 Å outside this type of vesicle corresponding to an encapsulating zone of radioactivity. Control animals injected with reserpine up to 8½ hours before [3]HNE did not incorporate the label and thus, the exogenous [3]HNE which is incorporated into peripheral sympathetic nerve terminals does not accumulate within the large granular vesicles. Most of the firmly bound radioactivity is distributed throughout the nerve, much of it in association with small granular and agranular vesicles.

References:

(1) Potter, L. T. and J. Axelrod: J. Pharmacol. *142*, 291 (1963).
(2) Taylor, P. W., Jr., C. A. Chidsey, and K. C. Richardson: Biochem. Pharmacol. *15*, 681 (1966).
(3) Austin, L., I. W. Chubb, and B. G. Livett: J. Neurochem. *14*, 473 (1967).
(4) Wolfe, D. E., L. T. Potter, K. C. Richardson, and J. Axelrod: Science *138*, 140 (1962).
(5) Aghajanian, G. K. and F. E. Bloom: Science *153*, 308 (1966).
(6) Salpeter, M. M., L. Bachmann, and E. E. Salpeter: (*in preparation*).

Histochemistry of Hormonally Activated Elements of the Regenerating Skin

BUKHONOVA, A. I. (Laboratory of Neuroendocrinology, Institute of Medical Radiology of AMS USSR, Obninsk, USSR)

Results of a study on reparative processes in standard skin wounds of white rats subjected to treatment with pituitary, adrenal, sex and thyroid hormones in various combinations with total gamma-irradiation (500, 600, 700 r) are summarized. The presence and activity of polysaccarides, RNA, SH-groups, succindehydrogenase and alkaline and acid phosphatases were studied during subsequent phases of the reparative reactions in cells of wound exudation, in granulation and connective tissues and in young epithelium. Microscopic pictures and the histochemical indexes of three different zones were compared, namely, the unchanged part of intact skin, the middle zone bordering on the wound and the newly formed regeneration area.

Proliferation and differentiation of new tissues under conditions of posttraumatic regeneration depend on a hormonal balance that is reflected in number, appearance and histochemical properties of various cells. Microscopic structure of biopsies of the wounds shows that at the normal ratio of hormonal factors in an organism, the cells of regenerating and unlesioned skin are not similar in their properties and reactions to hormonal treatment.

Glucocorticoid hormones of the adrenal cortex speed fibroblast differentiation which does not facilitate development of granulation tissue and results eventually in reduction of new connective tissue and change of rate and character of epithelization of the wound surface. Glucocorticoids decrease content of RNA and glycogen in all cells of these zones. However, content of SH-groups, activity of succindehydrogenase and acid phosphatase increase, especially in parts of the frontier zone.

ACTH gives a similar effect on the regenerating structures. Another hormone of the pituitary gland, STH, stimulates proliferation and growth of new connective tissue elements and the epithelial layer. Epidermal cells of skin from the hypertrophied frontier zone and regenerating parts apart from the most basal layer contain much glycogen and RNA. STH causes low activity of succindehydrogenase, phosphatases and SH-groups. DOCA and, in part, thyroid hormone exert influence analogous to STH.

STH and DOCA appeared to be more effective at the period of proliferation of new tissues. Influences of these hormonal substances have similar character in the irradiated animals also. They activate proliferation of the new elements followed by increase of DNA, RNA and glycogen and decrease of the acid phosphatase and SH-groups. Stimulation of exudate cells and granulation tissue by these hormones is revealed by the first period of radiation disease.

Hormones of the periphery glands, especially glucocorticoids, stimulate mainly activity of the tissue enzymes catalyzing differentiation of regenerating structures. Elevation of the glucocorticoid level disturbs the gradient of interrelations between main elements of the new connective tissue and results in inhibition of regeneration of the epithelial structures. It is observed especially in animals subject to total gamma-irradiation.

The Ultrastructure and Localization of ATPase Activity of Tetrahymena pyriformis Cilia

BURNASHEVA, S. A. and G. A. JURZINA (Bach Institute of Biochemistry of Academy of Sciences, Moscow, USSR)

High resolution electron micrographs have revealed that fibers of cilia of *Tetrahymena pyriformis* are composed of near-spherical subunits of about 35-40 Å in diameter. These subunits are arranged in straight longitudinal arrays to compose the walls of the fibers. The latter appears to have 13 subunits per subfiber when viewed in cross section. The lead salt method introduced by Wachstein and Meisel (Am. J. Clin. Pathol. 27, *13*, 1967) for the cytochemical demonstration of ATPase activity was used to determine sites of activity in cilia from *Tetrahymena pyriformis*. Reactions were run with unfixed cilia or cilia prefixed in glutaraldehyde. Sites of ATPase activity were visualized with the electron miscroscope. The results have shown that the lead phosphate precipitate is localized in the areas between the central and the double outer fibers of cilia and on the membranes of cilia. Stronger distribution of the precipitate was found along the length of the part of outer fibers, running into the basal body and within the kinetosome. The distribution of the precipitate was not found within or along the central fibers of cilia.

Embryonal Metabolism *

BURT, ALVIN M. (Vanderbilt University, Nashville, Tenn.)

Glucose metabolism is a major source of energy during embryonic development. Enzymes associated with both glycolytic and pentose cycle metabolism are present in the early stages. The levels of activity of these enzymes undergo marked changes during development. However, the regulatory mechanisms remain obscure (1). Microchemical studies on small tissue samples and discrete cellular populations provide an opportunity to correlate changes in enzymatic activity with other biochemical, physiological, and morphological parameters of development. In the embryonic spinal cord, G6DH activity (an "index" of pentose cycle activity) is at a maximum during early development, and the ontogenetic pattern for this enzyme in both the proliferative neural epithelial and the differentiating mantle layers parallels that of mitotic activity. However, the enzymatic activity of the amitotic mantle layer is greater than that of the neural epithelial layer. These relationships and the results of other studies have led to the suggestion that increased pentose cycle activity is associated with both rapid proliferation and early neural differentiation. The increased activity may provide a source of pentose for nucleic acid synthesis and a source of TNPH for the reductive synthesis necessary for both processes (2).

A study of 6-phosphofructokinase and aldolase activity suggests an overall increase in glycolytic metabolism during neurogenesis. An initial increase precedes the rapid decline in pentose cycle activity and a second increase is coincident with the rapid increase in spontaneous motor activity and immediately precedes the rapid increase in choline acetyltransferase activity (an "index" of molecular differentiation). The increased glycolytic metabolism during the period of functional differentiation may reflect an increase in the energy requirements associated with functional activity (3). Increased catabolic metabolism is a characteristic of many developing tissues and is not restricted to the central nervous system.

References:

(1) Papaconstantinou, J.: *In:* The Biochemistry of Animal Development, pp 58-113 (Weber, edit.) New York: Academic (1967).
(2) Burt, A. M.: Develop. Biol. *12*, 213 (1965).
(3) Burt, A. M.: J. Exptl. Zool. *165*, 317 (1967).

---

* Supported by Grants GM-11733 and GM-14709, U.S.P.H.S.

Cytochemical Study of Bile Canaliculi in Experimental Livers *

CHANG, JEFFREY P., PETER F. SCHATZKI, and TAKUMA SAITO (The University of Texas M. D. Anderson Hospital and Tumor Institute at Houston, Houston, Texas)

Reports from this laboratory indicated that a rapid loss of adenosine triphosphatase (ATPase) in bile canaliculi occurred within 48 hours after rats were fed the carcinogen, 3'-methyl-4-dimethylaminoazobenzine. However, ATPase had increased activity in these organelles of fasted animals and no change in control rats. Studies were therefore initiated to examine other enzyme systems under various experimental conditions Sprague-Dawley rats were fed the carcinogen or poisoned with carbon tetrachloride. Untreated animals had their bile ducts partially or fully ligated. The liver tissues were then processed for localization of ATPase, 5'-nucleotidase, alkaline phosphatase, or thiamine pyrophosphatase at the light and electron microscopic levels. In addition, lanthanum nitrate, ruthenium red, or horse radish peroxidase were used to demonstrate the relationship between bile canaliculi and sinusoids.

ATPase was reduced drastically in the center vein area of liver of rats treated with the carcinogen or carbon tetrachloride. The other enzymes studied either had an insignificant loss or were not affected. By the use of a modified *section freeze-substitution* technique (2), the bile canaliculi were distinctly demonstrated continuing from center vein to portal vein of the lobule.

Lanthanum technique has been the most successful one for demonstration of relationships between bile canaliculi and spaces of Disse. In normal liver, lanthanum was clearly deposited on the cell surface as well as in bile canaliculus and its adjacent intercellular space. The tight junction near the canaliculus was the limiting factor for movement of lanthanum. In livers with ligated bile ducts, the tight junction appeared shorter than normal and its adjacent intercellular space was dilated. These observations were substantiated by ruthenium and peroxidase techniques.

These data, plus that being accumulated, would elucidate the structure of these cell organelles and the response of certain canalicular enzymes to experimental conditions.

References:

(1) Chang, J. P., D. N. Ward, and K. Ichinoe: Acta Unio Intern. Contra Cancrum *19*, 560 (1963).
(2) Chang, J. P. and S. H. Hori: J. Histochem. *9*, 292 (1961).

---

* Supported by grants CA05312 from USPHS and IN-43-13 from ACS.

Histochemical Aspects of the Pre-tetrapod Adrenal

CHAVIN, WALTER (Wayne State University, Detroit, Mich.)

Historically, the adrenal histochemistry of fish has been a controversial field. Application of modern methodology dealing with both cortical and medullary tissue has revealed that evolutionary patterns of the adrenal components were established early in the course of vertebrate evolution. The results from the available species in the evolutionary sequence (lamprey, shark, skate, reed fish, bichir, sturgeon, bowfin, garpike, goldfish, Australian lungfish, African lungfish) demonstrating the presence of lipids, cholesterol, phospholipids, ascorbic acid, Δ^5 3-β-hydroxysteroid dehydrogenase, catecholamines, etc., will be reviewed. The surprising feature in this evolutionary study of pre-tetrapod vertebrates is that the adrenal generally has remained fairly constant, both in histochemical composition and intimate relation to major venous drainage of the trunk.

Attempts to visualize autoradiographically and quantitate extremely small amounts of various hydroxysteroid dehydrogenases with the use of a tritiated tetrazolium salt (Nitro BT) will be described.

Cytochemical Determination of Enzymatic Activities of PK Cells Infected with RNA and DNA Viruses

CHYLE, M., B. KORYCH, Z. LOJDA, and F. PATOCKA (Department of Medical Microbiology and Immunology and 1st Department of Pathology, Charles University Medical Faculty, Prague, Czechoslovakia)

Enzymatic activities of an established line of pig kidney cells (PK) were studied after infection with representatives of different groups of RNA and DNA viruses. Enzymatic activities were assayed by cytochemical methods at the close of the first cycle of virus replication as determined by immunofluorescent techniques, and after 72 hours incubation under agar overlay. The latter cultures are useful for the assessment of activities in cells during the early stages of viral infection before gross morphological changes are apparent.

The dependence of cell response to input of virus and cytochemical reactions of infected cultures previously treated by the fluorescent antibody technique were studied.

The possible specificity of enzymatic activities of PK cells interacting with different viruses is discussed.

Synthesis, Transport, and Storage of Proteins in Phagocytes

COHN, Z. A. (The Rockefeller University, New York, N. Y.)

The two major phagocytic cells of mammals are the neutrophilic granulocyte and the macrophages of the reticuloendothelial system. Both cell types are characterized by the presence of large numbers of membrane bounded cytoplasmic organelles with properties of lysosomes. These structures are intimately involved in the degradation of both soluble and particulate molecules taken up by heterophagy. By means of morphological, biochemical, and cytochemical methods, the formation and properties of granulocyte and macrophage lysosomes will be characterized and compared. Granulocyte lysosomes are formed in the maturing myelocyte and are derived from the packaging of enzymes within Golgi-derived membranes. These represent storage granules, not subsequently synthesized in the mature cell, and activated upon fusion with a phagosome. Macrophage granules are in most cases secondary lysosomes or digestive bodies. They are derived from pinocytic vacuoles which subsequently acquire acid hydrolases in the vicinity of the Golgi complex. In this cell, the vesicular elements of the Golgi or primary lysosomes, appear to transport newly synthesized hydrolases to a vacuole of endocytic origin.

The Interrelationships of Cells in the Alveolar Membrane of the Rat Lung

CONNING, D. M. (Industrial Hygiene Research Labs., Imperial Chemical Industries Ltd., Cheshire, England)

Three distinct cell types may be identified in the alveolar membrane, using enzyme histochemistry and electron microscopy.
They are:
1. The basic alveolar cell (Type I).
2. The so-called "giant alveolar cell" (or Type II) with the characteristic inclusions.
3. The alveolar macrophage.
The proportions of these cells are relatively constant in the lungs of untreated specific pathogen free rats, but vary when the animals are subjected to pulmonary irritants. Analysis of the lungs during inhalation studies involving a variety of gaseous and particulate materials, suggests that there is an inverse relationship between Type II cells and alveolar macrophages during early exposure, but that they both increase at the expense of the Type I cell during later stages. It may be suggested that both cells are derived from Type I cells and further that the type of differentiation depends upon the external stimulus presented.

The Effects of Triton-X-100 on the Mobility of Desmoesterases of the Agouti (Dasyprocta aguiti) in Disc and Starch Gel Electrophoresis *

COUTINHO, HÉLIO B., JÁCIA T. ROCHA, and BENJAMIN F. JALES (University of Pernambuco, Brazil)

It is well established in histochemistry that only soluble esterases (lyoesterases) are fractioned in gel electrophoresis while the unsoluble esterases (desmoesterases) remain at the origin (Eränkö, Harkonem, Kokko, and Raisanen: J. Histochem. Cytochem. 12, 570, 1964). The use of the non-ionic detergent Triton-X-100 as a tool for improvement of the mobility of desmoesterases of Rhynchosciara angelae in larval development was proposed by Coutinho, Katchburian, and Pearse (J. Clin. Path.: 19, 617, 1966).
Triton-X-100 was added to homogenates of stomach, duodenum, jejunum, large intestine, liver, subanal gland, kidney, and lung of the agouti (D. aguti). Disc and starch gel electrophoresis were performed

according to the methods proposed by Davis (Ann. N. Y. Acad Sci.: *121*, 44, 1964 and Smithies (Biochem. J.: *61*, 629, 1955), respectively. The alpha naphthyl butyrate was used as substrate and the diazonium salt Blue RR (C. I. 37085) as coupler.

The comparison of the zymograms of the Triton-X-100 treated samples with the ones in which no detergent was added, revealed that more bands were observed when the non-ionic detergent was used. Since after the Triton-X-100 treatment, no esterases remained at the origin, we consider the detergent to be effective to improve the electrophoretic mobility of desmoesterases of the agouti tissues.

---

* Supported in part by grants TC 7485, Conselho Nacional de Pesquisas, Brazil, and LD1)B, W. K. Kellogg Foundation, U.S.A.

Gel Electrophoresis of Pepsinogen and Pepsin

CUNNINGHAM, LEW and RICHARD HEITSCH (Department of Anatomy, Marquette Medical School, Milwaukee, Wisc.)

The standard disc electrophoresis method has been modified for use in studies of the heterogeneity of crystalline porcine pepsinogen. The usual buffers (1), pH 6.7, 8.3, 8.9, are used since this range of pH values is favorable for stability of pepsinogen. The concentration of acrylamide in the small-pore gel is increased to 15% (2). To minimize diffusive spreading during the fixing-staining period, the "discontinuous" electrophoresis is performed in a sheet of gel only 0.1 mm thick. Proteins are then precipiated (3) and stained with 0.3% crystal violet adjusted to pH 8.5 by addition of 0.001 N "tris" base. (The acid stain, amido black, is unsatisfactory.)

Pepsinogen is apparently a more compact molecule than pepsin in the mildly alkaline separation gel, and runs faster than pepsin. But crystalline pepsinogen contains a trace of a protein identified by its electrophoretic behaviour as pepsin. Another impurity has been observed but not identified. We are attempting to reverse (4) the alkali-inactivation of pepsin after the electrophoresis in order to demonstrate proteolytic activity in the putative pepsin band by a substrate film method (5).

References:

(1) Davis, B. J.: Ann. N.Y. Acad. Sci. *121*, 404 (1964).
(2) Reisfeld, R. A., U. J. Lewis, and D. E. Williams: Nature *195*, 281 (1962).
(3) Marston, H. R.: Biochem. J. *17*, 851 (1923).
(4) Northrop, J. H.: J. Gen. Physiol. *14*, 713 (1931).
(5) Cunningham, L.: J. Histo. Cytochem. *15*, 292 (1967).

Characterization of RNA from Isolated Polytene Chromosomes

DANEHOLT, B. and J.-E. EDSTRÖM (Dept. of Histology, Karolinska Institute, Stockholm, Sweden)

Nuclear components from the dipteran, *Chironomus tentans* have been microisolated from salivary glands labelled *in vitro* and the extracted RNA analyzed by sedimentation analysis and gel electrophoresis. It was found possible to use fixed material for investigation of high molecular weight RNA. Nucleoli, chromosomes, Balbiani rings, and nuclear sap have been used by the investigations. The nucleoli are characterized by a series of ribosomal RNA precursors which can be demonstrated by use of different labelling times in combination with chase experiments. The chromosomes contain rapidly sedimenting RNA with slow migration rate in gel electrophoresis. There is also RNA in the 4-5 S range in all of the chromosomes. The low molecular weight RNA is missing in the Balbiani rings investigated but the other RNA is similar in the Balbiani rings and in other regions of the chromosomes. Data indicate that this RNA is reduced in size in the nucleus. The nuclear sap contains RNA with properties similar to those of the chromosomes. Ribosomal RNA components have not been observed in the sap after labelling times up to 5 hrs.

Saccharides in Chick Embryo Allantoic Epithelium with Ruthenium Red Stain

DANILOVA, L. V. and K. D. ROKHLENKO (Laboratory of Electron Microscopy of Academy of Sciences of the USSR, Moscow, USSR)

Recently, a new method, ruthenium red staining of extracellular acid mucopolysaccharides, was introduced in electron microscopy (Luft, 1956). We employed this method for staining of chick embryo allantoic epithelium.

It is known from light microscopic investigations that there are many granules stained as acid mucopolysaccharides in allantoic epithelium (Myslivećkova, 1958). We observed on electron microscopic preparations, an issue of granules from allantoic cells on the luminal surface into the allantoic cavity (Danilova, 1964). An amorphous or thin granular dense layer was found on the luminal surface of allantoic cells which have microvilli. The layer seems to be a product of allantoic cell secretion and may be a result of issue of the granules.

The layer is stained very intensively with ruthenium red and is composed of acid mucosubstances. A small amount of ruthenium red penetrates cells and intercellular spaces although we did not observe the penetration of the stain through the tight junction or zonula occludens on ultrathin sections. The more intensive staining was observed in the intercellular spaces of desmosomes and also on lateral surfaces of cells. A small amount of ruthenium red was lain on the other membranes of granules and on the membranes of endoplasmic reticulum.

References:

(1) Danilova, L. V.: *In:* Electron and fluorescent microscopy of the cell. pp. 26-30 (Acad. Sci. USSR) Moskow-Leningrad (1964).
(2) Luft, J. H.: J. Biophysic. and Biochem. Cytol. *2,* 799 (1956).
(3) Mysliveckova, A.: Ceskoslovenska Morphol. *6,* 3 (1958).

Phosphatase Localization in a Development System

DAUWALDER, M., W. G. WHALEY, and J. E. KEPHART (The Cell Research Institute, The University of Texas of Austin, Austin Texas)

Cytochemical techniques for the localization of inosine diphosphatase, thiamine pyrophosphatase, and acid phosphatase (enzymatic activities often found to be associated with the Golgi apparatus) (1) have been applied to the developing root tip of *Zea mays.* It has been shown that the progressive morphological specialization of the Golgi apparatus for participation in the production of a largely carbohydrate secretory product in the development of the root cap, epidermis, and phloem cells is paralleled by an increase in the amount or the activity of a particular inosine diphosphatase in the cisternae of the Golgi apparatus, whereas the apparatus in other cell types is unreactive. Thiamine pyrophosphatase was found in the Golgi apparatus of most cell types of the root but was notably absent in regions of the root cap although the apparatus shows marked functional activity (vesicle production, inosine diphosphatase activity, and accumulation of H^3-glucose). Acid phosphatase, generally accepted as a lysosomal marker, was found in only a few specific cell types. The localization was usually in a single cisternae at the face of the Golgi apparatus concerned with the production of secretory veiscles and associated regions of what may be smooth endoplasmic reticulum (see Novikoff *et al.,* 1964). No structures clearly identifiable as "lysosomes" were found.

References:

(1) Novikoff, A. B. and S. Goldfischer: Proc. Nat. Acad. Sci. *47,* 802 (1961).
(2) Novikoff, A. B., E. Essner, and N. Quintana: Fed. Proc. *23,* 1010 (1964).

Actinomycin Induced Changes in Cell Culture *

DEITCH, ARLINE D., STANLEY G. SAWICKI, and GABRIEL C. GODMAN (Dept. of Microbiology, Columbia University, New York, N. Y.)

Doses of actinomycin D (AMD) between 0.1 and 1.0 mg/10^5 cells applied for 1 hour are lethal to a large proportion of HeLa cells. The rate of killing is maximal between 4 and 8 hours. Moribund cells are in interphase, and cells in DNA synthesis are the most susceptible. The role of lysosomes in the rapid lethal effect is illustrated by altered localization of acid phosphatase. Despite sustained reduction in RNA synthesis of 70-90% reached within an hour after treatment, HeLa cells surviving AMD treatment undergo marked increases in area, volume and dry mass of 2-4 times initial or control values by the 3rd day. This chondrial proliferation leads to increased concentration of mitochondria per cell (1). Primary cultures of human amnion, which are much more resistant to the cytocidal effects of AMD, also have profoundly reduced RNA synthesis and exhibit marked mitochondrial proliferation without, however, significant increases in total cell volume or mass. The proliferated mitochondria form remarkable networks; they have histochomically demonstrable dehydrogenase and ATPase activity like those of untreated cells.

After a latent interval proportional in duration to the dose of AMD, reproductive recovery of a surviving fraction of the original HeLa population ensues. Continued AMD injury is manifested in late survivors by the occurrence of abnormal mitoses and aberrant mono and polynuclear giant cells. Eventually, normal appearing cells dominate the recovering population.

References:

(1) Deitch, A. D. and G. C. Godman: Proc. Nat. Acad. Sci., 1607 (1967).

---

* Supported by grants AI-05708 and GM-14864 from the USPHS.

Tentatives de Localisations des Steroides sur Coupes de Testicule de Lézards avec l'Histospectrographie de Fluorescence

DELLA CORTE, FRANCESCO (Dept. of Endocrinology, Istituto di Istologia ed Embriologia, Università di Napoli, Italia)

Nous avons déjà ècrit en 1962 (Della Corte-Liaci) que les spectrogrammes de la fluorescence émise par coupes de testicules de rats et de cobayes traitér avec acide sulfurique à chaud, sont superposables à ceux émis par la testostérone pure également traitée.

Maintenant, nous avons exécuté les spectrogrammes de la lumiére de fluorescence émise par coupes cryostatiques des testicules de trois groupes de lézards: un groupe en repos sexuel, un autre en activité et un troisiéme apres injection de LH. Les coupes ont été recouvertes avec acide sulfurique à différentes concentrations et chauffées à differentes températures (pour les détailes techniques voir Della Corte Liaci 1962).

Ensuite on a exécuté les spectrogrammes de la lumière de fluorescence émise par les cristaux de testostérone, de propionate de testostérone, d'oestrone, d'oestradiol 17-β, de progestrône, mélangée à l'acide sulfurique et chauffées aux mêmes températures que les coupes.

Nous montrerons en projection les spectres photografiés sur les tissus et ceux photografiés sur les hormones pures.

References:

(1) Della Corte, F. and L. Liaci: Ann. Histochem. Suppl. 2, 151 (1962).

A New Diazo Reagent for the Histochemical Demonstration of Conjugated Bilirubin

DESMET, V. J., A.-M. BULLENS, J. DE GROOTE, and K. P. M. HEIRWEGH (Laboratorium voor Histochemie en Cytochemie and Laboratorium voor Leverfysiopathologie Academisch Ziekenhuis St. Rafaël, Leuven, Belgium)

The selective histochemical demonstration of conjugated and total bilirubin in cholestatic tissues represents a useful tool for studying the mechanisms of bile pigment accumulation in different forms of cholestasis.

A new technique is presented for the selective histochemical demonstration of conjugated bilirubin in tissue sections, using the diazonium salt of ethyl anthranilate. The specific coupling of this reagent with conjugated bile pigments was demonstrated by Van Roy and Heirwegh (1). The authors studied the applicability of this reagent in a histochemical reaction on tissue sections of cholestatic liver and kidney, and several control test materials: films and sections of 3% agar blocks containing either 5 gm/l bilirubin or 5 gm/l conjugated bilirubin, and sections of Gunn rat renal papillae, which are known to contain unconjugated bile pigment.

Stability of the final azo-bilirubin pigment could be obtained by treatment of the stained sections with a zinc (II)—ammonia solution.

The results obtained with the ethylanthranilate-reagent were compared to those obtained with a recently described method, for conjugated and total bilirubin using the diazoniumsalt of 2,4-dichloraniline (2,3), and to those obtained with one of the classical oxidative methods for demonstrating bilirubin in tissue sections (4).

The specific staining of conjugated bile pigments with ethylanthranilate is not dependent on the reaction time, Moreover, there is no diffusion artefact under the prescribed working conditions. The method is applicable on fresh cryostat sections and frozen sections of cold formol-calcium fixed tissue. Up to now, no satisfactory staining of total bilirubin could be obtained with the discussed reagent.

While the method using 2,4-dichloraniline seems to be the best available technique for the histochemical demonstration of *total* bilirubin, the ethylanthranilate procedure proves to be the most selective method available for the demonstration of *conjugated* bile pigments.

References:

(1) Van Roy, F. and K. P. M. Heirwegh: Biochem. J., *in press.*
(2) Raia, S.: Nature *205,* 304 (1965).
(3) Raia, S.: *In:* Bilirubin Metabolism, pp. 285-289 (Bouchier *et al.,* edit.) Oxford: Blackwell Scientific Publications (1967).
(4) Hall, M. J.: Amer. J. Clin. Path. *34,* 313 (1960).

Tetrazolium Salts as Reagents in the Histochemical Localization of Epimerases

DICULESCU, I., DOINA ONICESCU, and G. SZEGLY (Catedra de Histologie, Facultatea de Medicina, Bucharest, Romania)

Nitro-BT and Tetra-nitro BT were used in the histochemical localization of epimerases which activate hexoses and their derivatives and also some amino acids. The histochemical and biochemical data which render possible the differentiation of epimerases from the respective dehydrogenases are reported. The mechanism of the epimerization process and the fundamental principles of the utilization of tetrazolium salts in the histochemical localization of this new group of enzymes are discussed.

Isozymes as Tools for the Geneticist

DOANE, WINIFRED W. (Department of Biology, Yale University, New Haven, Conn.)

Research in the last decade has established that multiple molecular forms of most, if not all, enzymes may be found at various levels of biological organization from the interspecific, and higher, to the intracellular level of the individual. Geneticists, above all, have profited from this realization and have made extensive use of isozymes as research tools for studying the heredity of diploid organisms, including man (e.g., 1, 2). They are indebted to histochemists for providing the background of staining techniques needed to demonstrate enzyme variants separated in a variety of electrophoretic support media. The awareness that multiple forms of enzymes performing similar biochemical functions may exist within the cytoplasm of a single cell represents a significant advance in histochemistry. This field should profit even more from the fruits of genetical research into problems of synthesis and control of these gene products.

Investigations utilizing multiple forms of α-amylase in *Drosophila* illustrate the advantages of employing isozymes to study the genetics of a higher organism. They were used in linkage experiments to locate a structural gene for amylase (*Amy*) in *D. melanogaster* both genetically and cytogenetically (3,4,5). Banding patterns produced by homozygous *Amy* strains and their hybrids suggest the enzyme is a monomer or oligomer. The *Amy*[1] pattern, with one major band, is most common and considered ancestral. Recombinational data (4,5) suggest homozygous strains producing two major amylases have a duplication of the *Amy* locus. (Minor bands derivative of major ones may appear.) Physicochemical properties of the isoamylases favor this interpretation. Nine out of eleven known patterns are of this kind. Thus, five electrophoretically distinct amylases may occur in various combinations of two together. Not all enzyme variants have an altered net charge. Heat inactivation studies exposed two 'silent' variants (4,7), bringing the total for the species to seven.

A method (6) to quantitate amylases separated by disc electrophoresis from minute amounts of crude tissue extracts revealed tissue specificities as well as ontogenetic changes in activity associated with individual bands. These specificities have important evolutionary implications, especially in consideration of double-banded strains where the duplicated genes are estimated to lie only 0.008 map units apart (5). These findings also bear on problems of cellular differentiation and genetic control mechanisms.

Amylase patterns of 42 species of *Drosophila*, from 18 species groups, indicate phylogenetic relationships associate with enzyme mobility (7). In no other species was evidence found for duplicated amylase loci, although widespread polymorphism was observed. Amylase variants appeared in six out of 11 species sampled from natural populations; five cases of genetic polymorphism in a given population were revealed. This is not surprising in view of the high estimates of genetic diversity and polymorphism in *D. pseudoobscura* (8).

Other examples of the application of isozymes to genetic analyses will be discussed. These include studies of somatic cell genetics, phylogenetic relationships based on dissociation-reassociation of subunits of isozymes, and evidence of genetic control mechanisms among the vertebrates, especially mammals.

References:

(1) Shaw, C. R.: Science *149*, 936 (1965).
(2) Harris, H.: Proc. Roy. Soc. Lond., B, *164*, 298 (1966).

(3) Kikkawa, H.: Jap. J. Genet. *39*, 401 (1964).
(4) Doane, W. W.: Amer. Zool. *7*, 780 (1967).
(5) Bahn, E.: Hereditas *58*, 1 (1968).
(6) Doane, W. W.: J. Exp. Zool. *164*, 363 (1967).
(7) ———: unpubl.
(8) Lewontin, R. C. and J. L. Hubby: Genetics *54*, 595 (1966).

Dynamics of Nucleic Acids, Certain Enzymes and Polysaccharides of Mammal Chorion and Amnion

DONSKIKH, N. V., V. D. NOVIKOV, M. YA. SUBBOTIN and N. I. TSIRELNIKOV (Medical Institute, Novosibirsk, USSR)

DNA content has been determined cytophotometrically and RNA content —histochemically in different stages of pregnancy in human and albino rat chorion and amnion epithelium. DNA content in cytotrophoblast nuclei corresponds to the diploid and tetraploid set of chromosomes. In chorion syncitiotrophoblast, diploid nuclei reveal themselves in all stages. In early periods in human and albino rat, amnion epithelium trimodal distribution of DNA (2n, 4n, 8n) is recorded; in the last term of pregnancy, mainly diploid nuclei are registered. An insignificant per cent of polyploid nuclei show predominately amitosis. The diagram of DNA synthesis dynamics obtained as a result of treatment of data of cytotrophoblast nuclei cytophotometrically has stepped characteristics. A considerable increase in RNA in syncitiotrophoblast is recorded in the first trimester of pregnancy and a regular decrease in later periods. In staining with toluidine blue, alcian blue and the Hale method, it has been shown that the amount of mucopolysaccaride in human and pig chorion and amnion stroma sharply decreases with increasing pregnancy. This decrease is especially marked within the first fifteen weeks of pregnancy. In pig placenta and in human embryo, the decrease in the amount of mucopolysaccaride goes on regularly during the whole pregnancy. Hyaluronidase activity of placental extracts and amniotic fluid increases in parallel with the decrease in the amount of mucopoly-saccaride. In all cases, a gradual increase in the amount of the revealed PAS—positive material in the stroma and decrease in the epithelium of the organ is observed. The question of application of ultrasound in histochemistry of mucopolysaccharides is discussed.

Histochemical and Fine Structure Study of Choroid Plexus Epithelium of Adult Fowl

DOOLIN, PAUL F. and WESLEY J. BIRGE (Neuropathology Research Section, Veterans Administration Hospital, Hines, Ill., Department of

40

Pathology, Loyola University Stritch School of Medicine, Hines, Ill., and the Division of Science and Mathematics, University of Minnesota, Morris, Minn.

In the localization and intensity of toluidine blue and PAS staining, adult plexus epithelium reveals a picture similar to embryonic epithelium (1, 2) but shows greater individual cellular variation.

Cytoplasmic RNA in adult plexus epithelium varies from moderate amounts distributed rather uniformly throughout cuboidal cells to a heavy subnuclear condensation in columnar cells. The latter displays less-dense RNA basophilia in the supra-nuclear region with a restricted narrow zone immediately adjacent to the apical plasmalemma not stained by toluidine blue.

A diastase-resistant PAS positive reaction in adult plexus epithelium varies from light to moderate or sometimes heavy in a narrow band at the apical margin. Marked staining at the junctional-complex region is indicated.

A complex cell surface of micro-villi and cilia is evident at the ventricular interface. Aberrant cilia (8 + 2) are occasionally noted. Interdigitated plasmalemmas are present on the lower lateral cell peripheries. Pinocytotic vesicles with a hirsute surface bordering the cytoplasm are noted in the apical and basal plasmalemmas in addition to the lateral plasma membrane at the level of the nucleus. Binucleate epithelial cells occasionally appear.

Granular E.R. in the sub-nuclear region varies from a dense whorled profusion in some cells to a relatively sparse population of cistern profiles. In the apical area, granular E.R. extends to but is restricted from a narrow marginal band of delicate fibrillar material adjacent to the apical surface. Membrane-free ribosomes are found throughout the cell including the marginal band at the ventricular surface. Inclusions, presumably lipid, are noted in the apical zone.

Frequently, the cisterns of granular E.R., showing an enlargement ranging from moderate to enormous, contain an electron dense substance. At the extreme enlargement, E.R. profiles, oval in contour, occupy a major portion of the cell volume with the largest profiles in the basal area. Mitochondria are greatly reduced in number in cells where the E.R. is expanded in contrast to a dense mitochondrial population in cells exhibiting a heavy concentration of unexpanded E.R. Dilated E.R. profiles occupy the area from the basement membrane to the apical margin. No connection of the cell surface with these dilated cisterns is evident.

The Golgi complex is usually located in the supra-nuclear region. It is frequently found, however, in the subnuclear zone in cells with normal profiles of E.R. and cells with enormously expanded E.R. Paired centrioles with striated rootlets are also detected in the subnuclear region of epithelial cells with narrow or expanded E.R. cisterns.

Most plexus epithelial cells reveal light to moderately dense hyaloplasm after glutal and OsO_4 fixation. A small population of epithelial cells, however, has a much darker or denser hyaloplasm and a higher concentration of mitochondria than the lighter counterpart. Multi-vesicular bodies and single-membrane-limited dense bodies exist in both types of cells but dark cells appear to contain more dense bodies than the light cells. Micro-tubules and filaments, present in light cells in both apical and basal areas, were not detected in dark cells.

References:

(1) Doolin, P. F. and W. J. Birge: Intern. Neurochem. Conf. Oxford, Absu p. 23, Permagon (1965).
(2) Birge, W. J. and P. F. Doolin: 8th Intern. Congr. Neurol., Vienna, pp. 2-6, (1965).

Protein Synthesis and Transport in Neurons

DROZ, BERNARD (Département de Biologie, Commissariat à l'Energie Atomique, Saclay, France)

Light and electron microscope radioautography following the administration of tritium labeled amino acids indicates that the nerve cell is a site of intense protein metabolism. Neuronal proteins are continuously broken down and renewed. They fall within two classes at least: "sedentary" proteins sojourn about 2 weeks in the perikaryon and "migratory" proteins pass, in less than one day, from the nerve cell body into its cytoplasmic expansions (1).

The earliest radioautographic reaction pinpoints sites of protein *synthesis,* that is the perikaryon and the base of dendrites. Most of the newly-formed proteins are elaborated by the ribosomes of the Nissl substance.

Later radioautographic reactions, occurring in previously unlabeled structures, signal the *transport* of radioactive proteins synthesized elsewhere. Thus, a fraction of newly-synthesized proteins migrates to and accumulates in the Golgi complex. It is probable that the Golgi complex gives rise to protein and glycoprotein components of lysosomes and "coated" vesicles (2). Another fraction, which bypasses the Golgi complex, might participate to the edification of neurofilaments and neurotubules present in the neuroplasm (3).

Then, newly-formed proteins are engulfed into the axon hillock, and later into the axon proper along which they are transported toward nerve endings at a rate of 0.6 - 0.9 mm per day in adult and 2.0 - 2.5 mm per day in young growing rats. Therefore, "migratory" proteins, which appear to originate from and migrate into a series of various cell organelles (neurofilaments, neurotubules, mitochondria, etc.), share finally a common pathway of transport: they are conveyed along the axon. The problem of whether an exceedingly small amount of protein is synthesized locally in axonal processes and nerve endings (4,5) will be discussed. Nevertheless, nerve endings are ultimately invaded by large amounts of labeled protein mostly related to mitochondria and regions rich in synaptic vesicles.

In conclusion, the axons are continuously supplied with new proteins and cell organelles manufactured in the nerve cell body. Thus, the incessant traffic of components from the perikaryon toward the nerve endings might account for both the maintenance and the plasticity of neural circuits.

References:

(1) Droz, B. and C. P. Leblond: J. Compar. Neurol. *121,* 325 (1963).
(2) Droz, B.: J. Microscopie *6,* 419 (1967).
(3) Droz, B.: J. Microscopie *6,* 201 (1967).
(4) Koenig, E.: J. Neurochem. *14,* 437 (1967).
(5) Barondes, S. H.: J. Neurochem. *13,* 721 (1966).

Measurement of Protein Turnover in Mitochondria by Means of Quantitative Electron Microscope Radioautography

DROZ, B. and M. BERGERON (Département de Biologie, Commissariat à l'Energie Atomique, Saclay, France)

When working on a whole organ with biochemical techniques, the turnover rate of protein ascribed to a given cell organelle is not necessarily specific for a definite cell type since a great variety of cells are included. Electron microscope radioautography may help to determine the kinetics of disappearance of labeled protein in organelles of a definite cell type. This possibility was tested in a rather homogeneous organ, the liver, to facilitate a comparison with biochemical experiments. Rats were given repetitive injections of lysine-H^3 and phenylalanine-H^3, twice a day for four days, and sacrificed after various time intervals. Epon embedded sections of liver were cut at equal thickness with an automatic Ultrotome and radioautographed. At each time interval, the concentration of the radioactivity in mitochondria was determined and expressed as the number of silver grains per 100 μ^2.

The distribution of the silver grains counted over mitochondrial profiles fitted a theoretical Poisson distribution. Consequently the radioactive proteins are randomly distributed in the mitochondrial population. The half-life of mitochondrial protein was then calculated by plotting the log of the radioactivity concentration against time after the injections. A half-life of 9.4 days was derived from the radioautographic data whereas biochemical results obtained from a similar material provided half-lives ranging from 8.5 (1) to 10.3 (2) days.

In conclusion, quantitative electron microscope radioautography is a suitable method for estimating the order of magnitude of protein turnover in cell organelles.

References:

(1) Beattie, D. S., R. E. Basford, and S. B. Koritz: J. Biol. Chem. *242* 4584 (1967).
(2) Fletcher, M. J. and D. R. Sanadi: Biochim. Biophys. Acta *51*, 356 (1961).

Histochemistry of Organogenesis

DRUKKER, J. (Anatomical-Embryological Institute of the University of Amsterdam, The Netherlands)

A main topic in chemical embryology has been the question of whether it is possible to find correlations between morphological changes and alterations in chemical properties in developing systems.

On the occasion of the session on "Chemodifferentiation" discussions, as related to the different contributions, will be centered around three main themes. In illustrating these themes, ample use has been made of the contributions and discussions during the symposium "Histochemistry of Morphogenesis," January 8th and 9th, 1968 in Amsterdam.

1) Chemomorphology during the phase preceding the morphodifferentiation of an organ anlage:

Myosin is already present in early myoblasts, these cells still have the capacity to undergo mitosis.

β-granules are already found in some cells of the fore-gut prior to the formation of the pancreatic bud.

In the embryonal pharynx, the site of the presumptive laryngeal bud is indicated by a high concentration of glycogen.

The localization of the anlage of the ear capsule can be recognized by a low content of glycogen as compared with the concentration of glycogen in the surrounding mesenchyma.

2) Chemomorphology after an organ anlage has appeared:

Acetylcholinesterase occurs in the neural tube and in the somites.

Catecholamines are to be found in the primary sympathetic chain and in the developing C-cells of the thyroid gland, even before these cells have reached their final localization.

Alkaline phosphatase is present in the metanephrogenic mesoderm after contact with the ureteric bud. It reaches a high level also in the developing nervous system. This phase of morphogenesis is not only characterized by the generation of new cells with specific properties but also many degenerative changes may be found. In many instances (e.g., Wolffian duct) these degenerating cells show a high content of acid phosphatase; in other organs where degenerative changes play a major role in morphogenesis (e.g., central nervous system) no such changes have been reported. In many cases a high content of acid phosphatase has been seen simultaneously with progressive changes.

3) Chemomorphology in the stages after the completion of primary morphogenetic changes:

The adult gross morphology is reached, but further histogenetic changes may take place. Generally, "onset of function" is described to occur in this phase. It is very difficult to determine the exact stage at which an organ starts functioning. Glands may be taken as an example. The accumulation of specific glandular products could be taken as a sign that the gland is able to manufacture its typical secretions; it does not prove that the organ is physiologically active Some glands act under the influence of a releasing factor. More-over, the responsiveness of the target organs must be taken into account. In many organs (e.g., kidney, gut) the appearance of certain enzymes has been related to the onset of function. It will be extremely difficult to prove this correlation. The central nervous system presents its own specific problems. The neural tube, after having gone through the proliferation phase, is not able to function properly. The dendrites, necessary to accomplish interneuronal connectivity, develop only in later stages. The outgrowth of dendrites occurs simultaneously with changes in localization and activity of certain perikaryal enzymes.

Generally it may be stated that:

a) It is difficult to correlate morphological with chemical ontogenetic changes.

b) It is dangerous to make a distinction between chemodifferentiation as an ontogenetic event and chemical changes related to the physiology of a given organ.

A New Application of a Silver and Cholinesterase Procedures to Demonstrate the Structural Pattern and Enzymatic Activity of Central and Peripheral Nervous Tissue in the Same Histological Preparation [*]

DUARTE-ESCALANTE, OVIDIO (Duke University Medical Center, Durham, N. C.)

Using a combined procedure to demonstrate nervous structure and cholinesterase activity in the same histological preparation, the material from mammals (cat, dog) in normal condition and under pharmacological treatment (MAO-1 drugs, monoamine precursors, amphetamine) and surgical preparations (denervation, transplants) were studied. The fresh tissue is processed in small pieces prefixed in a cooled solution of formaldehyde in sucrose, or by fixing in the same solution as with cryostat-sections from frozen tissue. The sections are incubated with the substrates acetylthiocholine and iso-OMPA. As controls, alternate sections are prepared suppressing either one or two substrates of the incubating medium and by adding physostigmine to the medium. After the development of true cholinesterase activity, the sections are selected and treated with a buffered solution of silver nitrate at a pH 6.0.

Consistently the results show, at the sites of enzymatic activity, a fine reddish-brown precipitate which can be seen directly with the light or phase-contrast microscope. At the same time, these reveal a remarkable delicate fiber arborization along the nerve trunks around and in the neurons, and between and at the smooth muscles.

With the experimental preparations and correlating the catecholamine fluoresence developed under the paraformaldehyde gas treatment, a thesis on the interrelationship between the cholinergic and adrenergic mechanisms in peripheral autonomic nervous system is developed.

---

* Supported by National Institute of Mental Health, 5 TI MH-8394-04.

Histochemical Demonstration of Aldolase in Skeletal Muscle

DUBOWITZ, VICTOR (Department of Child Health, University of Sheffield, Sheffield, England)

Early methods for aldolase lacked specificity and recent attempts have been made to provide a more reliable method (Nepvaux & Wegmann, 1962; Abe & Shimizu, 1964; Lake, 1965). The technique depends on the conversion by aldolase of fructose-1, 6-diphosphate to glyceraldehyde-3-phosphate and dihydroxacetone-phosphate; oxidation of the glyceraldehyde phosphate by NAD-dependent glyceraldehyde phosphate dehydrogenase, and the use of a tetrazolium salt as electron acceptor in the reduction of the NAD.

The reaction is thus dependent on 3 enzyme systems: aldolase, glyceraldehyde-3-phosphate dehydrogenase and $NADH_2$ tetrazolium oxidoreductase (diaphorase).

While the method can be tested readily for specificity by omitting substrate or NAD from the incubating medium, it is more difficult to assess the accuracy of localization and intensity of the reaction. Mammalian skeletal muscle is a useful tissue for this purpose.

Animal and human skeletal muscle contains at least two types of fibres (type 1 and type 2), as well as intermediate fibres. Type 1 fibres have a high content of oxidative enzymes, such as succinate dehydrogenase, and also react strongly for $NADH_2$ diaphorase, but are weak in glycolytic enzymes such as phosphorylase. Type 2 fibres are rich in phosphorylase but weak in the oxidative enzymes (Dubowitz & Pearse, 1960).

In a study of aldolase by a modification of the above methods in various animal and human skeletal muscles, there was a consistently strong reaction and a normal checkerboard pattern of strongly and weakly reacting fibres.

In serial studies with phosphorylase and $NADH_2$ diaphorase, the type 1 fibres gave a consistently stronger reaction for aldolase than the type 2 fibres.

In contrast to these results, Prewitt & Salafsky (1967) have shown by biochemical methods a higher content of aldolase in fast skeletal muscle, which has predominantly type 2 fibres, than in slow muscle, which has predominantly type 1 fibres.

Thus, it would appear that the histochemical method does not give an accurate reflection of the aldolase activity in a particular muscle fibre. This is directly comparable to the NAD-linked lactate dehydrogenase reaction, which also shows a falsely high concentration in type 1 fibres and low in type 2 (van Wijhe et al., 1963). The reaction probably is limited by the level of $NADH_2$ diaphorase present. As in the case of lactate dehydrogenase, this can be reversed also in the aldolase reaction by the use of an intermediate electron acceptor such as phenazine methosulphate, after which type 2 fibres give a stronger reaction.

References:

(1) Abe, T. and N. Shimizu: Histochemie *4*, 209 (1964).
(2) Dubowitz, V. and A. G. E. Pearse: Nature (Lond.) *185*, 701 (1960).
(3) Nepvaux, P. and R. Wegmann: Ann. Histochim. 7, 21 (1962).
(4) Prewitt, M. A. and B. Salafsky: Amer. J. Physiol. *213*, 295 (1967).
(5) van Wijhe, M., M. C. Blanchaer, and W. R. Jacyk: J. Histochem. Cytochem. *11*, 505 (1963).

Substructure of Chromosomes and Chromosomal Fibers, as Determined by Quantitative Electron Microscopy

DUPRAW, E. J. and G. F. BAHR (University of Maryland Medical School, Baltimore, Md., and Armed Forces Institute of Pathology, Washington, D.C.)

Human and honeybee chromosomes prepared as whole mounts by surface spreading-critical point drying (DuPraw, 1966) are ideal objects for determining relative or absolute dry masses by electron scattering measurements (Bahr and Zeitler, 1965). The morphological units in these chromosomes are bumpy, 200-300 Å chromatin fibers, each of which is dissected by weak trypsin solution into a single 30-60 Å subfibril and ultimately to a single 20 Å DNase-digestible filament (DuPraw, 1965). Electron scattering measurements have now revealed that the 30-60 Å subfibril has a more symmetrical mass distribution and a higher density than the 200-300 Å fiber. Preliminary evidence is consistent with a model in which the smaller (type A) subfibril corresponds to a single DNA helix wrapped with an equal weight of protein in a β-configuration (e.g. very lysine-rich histone), whereas the larger (type B) fiber is built up by supercoiling of the type A subfibril together with addition of more protein (e.g., arginine-rich histones). The lower density of the type B fibers is probably due to a higher percentage of protein, lowering the relative proportion of DNA.

For a type B fiber the mass per unit length is approximately proportional to its radius squared, giving a line on which the value for a 230 Å fiber is about 6×10^{-20} gm/Å. Since a DNA helix has a mass of 3.26×10^{-22} gm/Å, one to four extended helices would account for only 0.54 to 2.16 percent of such a fiber. On the other hand, it is well established that isolated metaphase chromosomes contain 15 to 20 percent DNA, requiring a DNA packing ratio of at least 28:1 in a 230 Å fiber (extended helix length: fiber length). Such a high packing ratio probably requires a supercoiling mechanism.

The cross-sectional masses of entire chromatids and their constituent fibers have been compared. A fully condensed honeybee chromatid (Fig. 1 in DuPraw, 1965) has a cross-sectional mass equivalent to 223 to 282 type B fibers. By contrast, the number of fiber equivalents in stretched chromatids or in extended centromeres is much lower. In one human chromatid pair (Fig. 4 in DuPraw, 1966) the number of type B fibers passing through each centromere cannot exceed 5 to 7. In another human chromatid (Fig. 5 in DuPraw, 1966) a stretched arm region has a cross-sectional mass equivalent to 9 to 14 type B fibers. Finally, in one 6-$\times$-12 chromatid (Fig. 1 in DuPraw, 1966) a stretched long-arm region contains about 27 to 34 type B fiber equivalents. These data provide strong confirmation for a folded fiber model of chromosome structure (DuPraw, 1965) and demonstrate the effectiveness of dry mass determinations with the electron microscope (Bahr and Zeitler, 1965).

References:

(1) DuPraw, E. J.: Nature *209*, 577 (1966).
(2) Bahr, G. F. and E. Zeitler: *In:* Symposium on Quantitative Electron Microscopy, pp. 217-239 (Bahr and Zeitler, eds.). Williams & Wilkins, Baltimore, Md. (1965).
(3) DuPraw, E. J.: Nature *206*, 338 (1965).

Tissue Reaction to Cardiac Arrest and Perfusion *

ECKNER, FRIEDRICH A. O., EUGENE H. BLACKSTONE, and PETER V. MOULDER (The Congenital Heart Disease Research and Training Center, Hektoen Inst. for Medical Research, and the Departments of Surgery and Pathology of the University of Chicago School of Medicine, Chicago, Ill.)

Anoxic cardiac arrest, hypoxic coronary perfusion, and total body hypoxic perfusion were studied in normal dogs by serial full thickness myocardial biopsies and serial liver biopsies. The histochemical examination of the frozen and dried tissues included glycogen, phosphorylase and transglycosidase, cytochrome oxidase, succinic dehydrogenase, alkaline phosphatase as well as SH groups (DDD) and protein groups reactive with DNFB. The findings were correlated with the clinical state (ability to resuscitate the heart) and the metabolic state (cross coronary bed lactate, pyruvate, glucose, and oxygen extraction). Thirty minutes of anoxic cardiac arrest resulted in focal loss of glycogen in heart and liver with corresponding inability to demonstrate phosphorylase and transglycosidase. The patchy areas of glycogen loss became larger in the six hour recovery period while the easily resuscitated heart reverted from its anaerobic phase post-arrest to normal aerobic metabolism. Liver glycogen was diffusely lost during the recovery and the alkaline phosphatase reaction was more pronounced. Longer anoxic arrest led to larger areas of myocardial glycogen loss after arrest and complete loss in the recovery phase with difficulty in resuscitation, and unpredictable switches between aerobic and anaerobic metabolism occurred. Again the enzymes were unchanged. The liver was not different from the 30 minute arrest group. Hypoxic perfusion of coronaries led to rapid dynamic failure, loss of myocardial glycogen and demonstrable phosphorylase activity. However, the transcoronary metabolism remained aerobic with decrease in number of contractions and their force. The same metabolic state, but with increased blood sugar levels was found in total body hypoxic perfusion during which the myocardial glycogen remained unchanged for two and one-half hours. The other histochemical parameters were unaltered. Liver glycogen was progressively lost, first periportally and later centrilobularly, with a midzone still containing some glycogen. The failure to resuscitate an arrested heart was thus related to a lack of substrate rather than to a structural alteration of the myocardial fibers.

* This work was supported by grant number HE-07605-06 from the National Heart Institute, National Institutes of Health, Bethesda, Md.

Cytoscreener and its Use in Sputum Cytology

EHRLICH, M. P. (Ca Detection Laboratories, New Hyde Pk., N. Y.)

An electronic cytoscreener has been designed for the cytologic screening of cellular material. The instrument utilizes a scanning ultraviolet microspectrophotometer which examines cells flowing through a capillary approximately 70 microns in diameter. Special techniques are utilized which afford sufficient signal-to-noise ratio to allow the accumulation of information regarding many cell parameters. The information generated is fed into a special-purpose computer which utilizes pattern recognition techniques to identify each cell. If the required cell identification criteria are met, then additional comparisons are used to determine whether the cell is classified as a "normal" or "abnormal" cell. The cell count output is fed into a binary code digital readout which registers the total number of "normal" and "abnormal" cells observed. An "ambiguity" register is also included in the output system to count certain cells which do not fall within the other two categories. Some experimental results will be presented.

Peptidases of the Anterior Pituitary Gland

ELLIS, STANLEY, J. KEN McDONALD, and P. X. CALLAHAN (NASA, Ames Research Center, Moffett Field, Calif.)

Clearly identifiable peptidases in the pituitary gland consist of an endopeptidase, monoamino acid aminopeptidases, and dipeptidyl aminopeptidases.

The endopeptidase, usually referred to as proteinase I (1) acts optimally at pH 4, and hydrolyzes the B chain of insulin with a specificity closely similar to that of pepsin and cathepsin D. Growth hormone and prolactin are inactivated by proteinase I.

The *monoamino acid* aminopeptidases (2) which have been identified in the pituitary gland consist of a dipeptidase, an aminotripeptidase, and an aminopolypeptidase, all of which have optimal activity between pH 7 and 8. Dipeptidase is inactivated by thiol compounds and EDTA, but the activity can be restored by Mn^{++}, Co^{++}, and Zn^{++}. Purified pituitary dipeptidase hydrolyzes most neutral and basic dipeptides but shows little action on those containing glycine. On β-naphthylamides (βNA) and p-nitroanilides, the specificity is much narrower since the Arg and Lys derivatives are rapidly hydrolyzed with little action occurring on the neutral and acidic derivatives. Aminotripeptidase hydrolyzes tripeptides such as Ala—Gly-Gly, Lys—Gly-Gly, and Ala—Ala-Ala but does not hydrolyze either monoamino acid or dipeptidyl arylamides. Whereas dipeptidase and aminotripeptidase are restricted to the hydrolysis of di- and tri-peptides, aminopolypeptidase appears to cleave only larger peptides by releasing NH_2-terminal amino acids from hexa-, penta- and tetra-alanine to the tripeptide stage. Through the consecutive action of these peptidases, the larger peptides can be completely degraded by crude pituitary extracts to amino acids. In contrast to the restricted specificity of dipeptidase, aminopolypeptidase hydrolyzes not only the βNA of Lys and Arg but also those of Phe, Met and Leu; moreover, the aminopolypeptidase is strongly activated by thiol compounds and inhibited by puromycin.

Three *dipeptidyl* aminopeptidases can be differentiated in the pituitary gland by their specificity in the hydrolysis of dipeptidyl-βNA and through their response to modifiers. Dipeptidyl arylamidase I (3) has a broad specificity, requires thiol groups and chloride ions for full expression of activity, and cleaves, at pH 4, βNA derivatives at the following rates: Gly-Arg > Ser-Met > Ala-Ala > Ser-Tyr > His-Ser > Leu-Ala; but Lys-Ala and Arg-Arg are not hydrolyzed. The enzyme shows amidase activity on the corresponding amides and, in the presence of hydroxylamine, transferase activity. N-terminal dipeptides are cleaved from the N-terminal decapeptide of adrenocorticotropin and tetra-alanine. Dipeptidyl arylamidase II (4) cleaves at pH 5.5 the dipeptide from Lys-Ala-βNA and, at a slower rate, from Arg-Ala-βNA; but the substrates of arylamidase I are not hydrolyzed nor are thiol groups or chloride ions required for activity. Cations such as Na^+, $Tris^+$, and puromycin are highly inhibitory. The narrower specificity of arylamidase II on β-naphthylamides is paralleled by a restricted specificity on dipeptide esters; only Lys-Ala and Ser-Met methyl esters have been found to be susceptible substrates. N-terminal dipeptides are cleaved, optimally at pH 4, from the tripeptides Ala-Ala-Ala, Ser-Met-Glu and Met-Met-Ala; di- and tetrapeptides are not attacked. Dipeptidyl arylamidase III (5) is thiol activated and can be specifically identified by its cleavage of the dipeptide from Arg-Arg-βNA at pH 9. NH_2-terminal dipeptides are cleaved from neutral and basic peptides containing a minimum of four amino acids, i.e., tetra-, penta- and hexa-alanine. By differential centrifugation of homogenates prepared in 0.25 M sucrose, arylamidases I and II were located in the lysosomes and arylamidase III in the cell sap. The concerted action of proteinase I and the aminopeptidases would appear to account for a major portion of the intracellular degradation of proteins and peptides within the pituitary cell.

References:

(1) Ellis, S.: J. Biol. Chem. *235*, 1694 (1960).
(2) Ellis, S. and M. Perry: J. Biol. Chem. *241*, 3679 (1966).

(3) McDonald, J. K., S. Ellis, and T. J. Reilly: J. Biol. Chem. *241*, 1494 (1966).
(4) McDonald, J. K., T. J. Reilly, B. B. Zeitman, and S. Ellis: J. Biol. Chem., *In Press*.
(5) Ellis, S., and J. M. Nuenke: J. Biol. Chem. *242*, 4623 (1967).

A Cytophotometric Analysis of Histones in Aging

ENESCO, HILDEGARD E. (McGill University, Montreal, Canada)

This study was undertaken to determine whether a change occurs in the amount of histones present in the cell nucleus during aging. Previous cytophotometric measurements of Feulgen stained nuclei of tissues from 3 months and 27 months old male Sherman rats had indicated that there is no difference in the DNA content of nuclei from young adult rats as compared to nuclei of aged rats (1). Thus, cell aging could not be attributed to DNA loss. No cytophotometric measurements of histone concentration in young as compared to aged animals had ever been made. On the one hand, both DNA and histones are usually present in constant and equalized amounts in the cell nucleus. On the other hand, histones are known to modify DNA activity, and functional changes in histones had been reported in aged animals which have led to theoretical interpretations of the importance of histones in the aging process (2). The purpose of the present measurements is therefore to determine whether quantitative changes in histone concentration as measured by Fast Green dye binding accompany aging at the cellular level. For this study, tissues of 3 month and 27 month old male Sherman rats and of 3 month and 24 month old female C57BL/6J mice were used. Tissues were fixed in formalin and stained specifically for histones using the alkaline Fast Green method (3). Cytophotometric measurements were made at the absorption peak of the dye, 640 mμ, using the one-wave-length plug method (4). Preliminary measurements on rat brain and liver indicate that there is no statistically significant difference between the histone concentration of nuclei of the young adult and the aged rats. Additional data on several tissues, including renewal, post-mitotic and slowly growing tissues, of young and old mice will be available at the time of presentation.

References:

(1) Enesco, H. E.: J. Geront. 22, 445 (1967).
(2) Von Hahn, H. P.: J. Geront. 21, 291 (1966).
(3) Alfert, M. and I. Geschwind: Proc. Nat. Acad. Sci. 39, 991 (1953).
(4) Swift, H. and E. Rasch: *In:* Physical Techniques in Biological Research, Vol. III pp. 353-400 (Oster *et al.*, edit.) New York: Academic Press (1956).

Histochemistry of Human Myopathies

ENGEL, W. KING (Medical Neurology Branch, National Institute of Neurological Diseases and Blindness, National Institutes of Health, Bethesda, Md.)

Myopathy is difficult to define precisely, but by use has come to designate muscle abnormality not consequent to impaired innervation. (It is debatable whether cachetic and disuse atrophies are myopathic or neurogenic.) In a myopathy, muscle may appear to be the only abnormal tissue or may be involved secondary to dysfunction of a distant tissue, e.g., hyperthyroidism. Our previous clinical pathologic correlation studies of muscle biopsy histochemistry are detailed in a recent report (Engel, Pediat. Clin. N. Amer., *14*, 963-995, 1967).

In this talk, our most recent as well as some previous histochemical studies will be presented. Discussed will be a few of the more common but less tangible myopathies, such as Duchenne dystrophy (including carriers), myotonic dystrophy, and collagen-vascular diseases (e.g., dermatomyositis). Less common but with more tangible abnormalities are phosphorylase deficiency and rod (nemaline) myopathy, as well as the abnormality called "tubular aggregates" (formerly termed "mitochondrial aggregates"). We have proposed that some disorders often considered to be myopathies may actually be due to abnormal motor innervation. These include central core disease, type I fiber hypotrophy, and type II fiber atrophy. The histochemistry of pertinent experimental animal "models" will be mentioned. Human and animal disorders that preferentially affect one histochemical fiber type will be emphasized.

Effect of Inhibitors of DNA Synthesis on the Mitotic Cycle

EPIFANOVA, O. I., L. YA. LOMAKINA, and V. V. TERSKIKH (Institute of Molecular Biology, USSR Academy of Sciences, Moscow V-312, USSR)

The present study was undertaken to investigate the effect of several inhibitors of DNA synthesis on the mitotic cycle in connection with the specificity of their action upon the cells in S-period. The experiments were performed on human amnion and Chinese hamster lines of cells grown as monolayers. A radioautographic technique using ^{3}H-thymidine and ^{3}H-deoxycytidine was employed. Two kinds of inhibitors were chosen: 5-fluoro-2'-deoxyuridine (FUdR), amino- and amethopterin (10^{-7}–10^{-6}M) which effect the metabolic pathways of DNA biosynthesis; ultraviolet light (100 erg/mm²) and mitomycin C (3×10^{-6}M) which damage the DNA molecules.

It was found that all inhibitors produced a similar effect on the progression of cells through the mitotic cycle. They brought about an accumulation of cells in the S-period and had no effect on the duration of other periods. The depression of DNA synthesis resulted in subsequent cessation of mitoses. When cells were removed from the inhibitors, DNA synthesis resumed and the recovery of the mitotic activity began. In the experiments with FUdR, amino- and amethopterin, a partially synchronized cell growth has been achieved. The results obtained provide strong evidence that these inhibitors in concentrations generally used for synchronization do not collect cells at the G_I—S boundary. This must be kept in mind when using these agents in experiments on the mechanisms of the onset of DNA synthesis.

Fine Structural Localization of Glucose-6-phosphatase in Renal Proximal Tubules *

ERICSSON, JAN L. E. and STEN JAKOBSSON (Dept. of Pathology at Sabbatsberg Hospital, Karolinska Institutet Medical School, Stockholm, Sweden)

Biochemical evidence is consistent with the view that the renal tubules contain glucose-6-phosphatase (G-6-Pase). Light microscopic histochemical studies have indicated that this enzyme is present in the proximal convoluted tubules of mammalian kidneys. Due to the sensitivity of G-6-Pase to aldehyde-containing and other fixatives, the localization of this enzyme has been studied mostly in fresh frozen sections in which structural preservation of the cells is suboptimal.

A method for the fine structural demonstration of G-6-Pase allowing acceptable preservation through brief perfusion fixation with glutaraldehyde was recently worked out in the laboratory for studies of liver cells. This method has been applied for the demonstration of G-6-Pase in the proximal convoluted tubules of rat kidney.

Rats were anesthetized with Nembutal and their left kidney was perfused with 1.5% glutaraldehyde in 0.1M cacodylate buffer for one to ten minutes. The fixative was subsequently washed out of the tissue by perfusion with buffer. ~ 50 μ sections (frozen or non-frozen) were incubated at pH 6.7 in a medium containing G-6-P and lead ion for 15 to 30 minutes. The sections were subsequently washed in buffer and were post-fixed in OsO_4. The following controls were included: (1) incubation in a medium lacking the substrate; (2) incubation in media containing 0.01, 0.05 or 0.1 M NaF; (3) incubation of heat-inactivated sections; and (4) incubation in a medium containing fructose-1, 6-diphosphate.

In sections incubated in the G-6-P medium, final product was specifically precipitated on profiles of the endoplasmic reticulum (ER), including its local specializations: "the paramembranous cisternal system" along lateral plasma membranes; profiles closely surrounding the microbodies; and "smooth-surfaced conglomerates of tubules and vesicles." In addition, lead phosphate precipitate was present on the apical plasma membrane (external surface of the microvilli), and on apical tubular invaginations, apical dense tubules, apical vesicles, and some apical vacuoles. With NaF in the incubation medium, reaction product was absent from the ER. However, it was still present on the apical plasma membrane and, occasionally, on apical vesicles, vacuoles and tubules. A similar localization of precipitate occurred when G-6-P was substituted with fructose-1, 6-diphosphate in the incubation medium (NaF not present). Final product was absent from heat-inactivated sections and sections incubated in a medium lacking the substrate.

In homogenates of rat kidney cortex, biochemical measurements showed an approximately 70% inhibition of G-6-Pase with 0.01 M NaF. This inhibition appeared to be similar in different membrane subfractions of the homogenate, i.e., those containing mainly plasma membrane and those composed, in large part, of ER membranes (and identified by biochemical criteria). G-6 Pase activity in liver homogenates showed the same degree of inhibition with NaF as the kidney fractions. There was negligible hydrolysis of fructose-1, 6-diphosphate when incubation was performed under the same conditions as for G-6-Pase.

The findings are compatible with the hypothesis that specific activity of G-6-Pase is present in all portions of the ER in renal proximal tubule cells. Apparent histochemical hydrolysis of G-6-P in the brush border and associated structures in the apical cytoplasm may—at least partially—be explained by the presence of a different and aldehyde-resistant hydrolytic enzyme, perhaps a more nonspecific "polyphosphatase."

---

* Supported by a grant from the Swedish Cancer Society (project nr. 75-K68-02XA).

The Study of Peculiarities of Proliferation and Maturation in a Culture of Malignant Fibroblasts of Mice during Enzymic Removal of Different Intracellular Ingredients

ERISTAWI, K. D., L. K. SHARASHIDZE, and N. S. STURUA (Institute of Experimental and Clinical Surgery, Tbilisi, USSR)

We have studied the induction of proliferation and maturation and the cytochemical peculiarities of cellular elements in monolayer of three day cultures of malignant fibroblasts of mice after trypsin, ribonuclease, hyaluronidase, amylase and lipase treatment. The enzymes were used in solutions with concentrations from 0.1 mg/ml to 3.0 mg/ml. The incubation times were from 1 to 3 hours. For estimation of the proliferation process, the mitotic index in all its phases was calculated in promille. For estimation of the maturation process, we calculated average indices of relative quantity of young and mature forms of cellular elements in six day cultures. To reveal the cytochemical peculiarities of cells treated with enzymes, we used reactions defining RNA and DNA, proteins (including the amino acids—tryptophane, tyrosine and histidine), glycogen, neutral and acid mucopolysaccharides, lipids, oxido-reductive, glycolytic and hydrolytic enzymes.

The results obtained prove that partial cytochemically registrable removal of various chemical components from the cells due to the effect of the corresponding enzymes leads to disturbance of proliferation and maturation processes in malignant fibroblasts of mice.

It is shown that trypsin delays transition of the cells from the postsynthetic period into mitosis but does not prevent the further course of mitosis. The ribonuclease treatment promotes transition of the cells from the postsynthetic period into mitosis, but blocks the mitotic process in prophase. The treatment with hyaluronidase, amylase and lipase sharply delays the transition of cells from the postsynthetic period to mitosis and all the phases of mitosis. In all these cases of treatment there has been observed a considerable retardation of the maturation processes.

Ultrastructural Cytochemistry of the Renal Juxtaglomerular Cells

FABRIS, G., G. M. MARIUZZI and I. NENCI (Istituto di Anatomia e Istologia Patologica and Cattedra di Istochimica Normale e Pathologica, Universita' di Ferrara, Ferrara, Italy)

Cytochemical study of renal juxtaglomerular cells by modification of the silver-methenamine reactions applied to electron microscopy, the specificity of which was controlled by blocking and digesting methods, has shown positive reactions in the secretory granules. According to the histochemical data, the secretory granules are neutral glucidic groups.

Fine Structural Localization of Exogenous Peroxidase in Rat Liver *

FAHIMI, H. DARIUSH and MORRIS J. KARNOVSKY (Departments of Pathology, Harvard Medical School and Boston City Hospital, Boston, Mass.)

The presence of large amounts of glycogen in normal rat liver interferes seriously with the fine structural localization of peroxidase in this tissue. Preliminary attempts of applying Karnovsky's 3,3-diaminobenzidine technic to sections of glutaraldehyde-formaldehyde fixed rat liver followed by embedding in Epon revealed that large vacuoles and bubbles formed in the vicinity of glycogen deposits inside the liver cells. Such vacuoles interrupted the fine structural integrity of cells and interfered seriously with the proper penetration of Epon and subsequent ultramicrotomy. Further analysis revealed that hydrogen peroxide was primarily responsible for the formation of large bubbles around the glycogen deposits. Therefore, fasted (24-36 hours), glycogen depleted Wistar/Furth rats were studied at various intervals after the intravenous injection of horseradish peroxidase (Sigma type II) and adequate fine structural preservation of tissue and peroxidase activity was obtained. The methods of fixation and staining for peroxidase were essentially similar to that described previously by Graham and Karnovsky (J. Histochem. Cytochem. *14*: 291, 1966).

Sixty to ninety seconds after injection, peroxidase was present in hepatic sinusoids and in the spaces of Disse covering the hepatic microvilli and extending into the microvillar pits. Peroxidase was also present in occasional large and small vesicles just below the hepatic microvilli. In Kupffer cells peroxidase was noted in large and small vacuoles, some of which resembled lyosomes. The electron dense tracer also penetrated into the narrow intercellular space between adjacent liver cells, passing easily between the so-called "stud-like" processes which extend from one hepatocyte into the surface of the adjacent cell, but did not permeate the tight junction on both sides of the bile capillaries. Five minutes after injection peroxidase was present in numerous small and large vesicles and occasional tubular structures in liver cells in the area adjacent and beneath the hepatic microvilli. In sinusoids peroxidase was taken up by Kupffer cells and by other cells lining the sinusoids. Thirty minutes after injection, peroxidase was present in large vacuoles of Kupffer cells and cytoplasmic extensions of sinusoidal lining cells overlying the hepatic microvilli; in liver cells peroxidase appeared to be localized either in large vacuoles resembling multivesicular bodies, or in vesicular and tubular structures at the biliary pole. The mode of transport from the vascular pole to multivesicular bodies and to the biliary pole are under study.

---

* Supported by GRS grant from the Boston City Hospital, Boston, Mass.

Measurements of Hormone Content of Individual Pituitary Cells

FAND, SALLY B. (Veterans Administration Hospital and Wayne State University School of Medicine, Detroit, Mich.)

Thin (2μ) sections of formalin-fixed, paraffin-embedded human pituitary glands were stained in the periodic acid Schiff (PAS) reaction to demonstrate the glycoprotein trophins thyroid-stimulating hormone (TSH), follicle-stimulating hormone (FSH), and luteinizing hormone (LH), in anterior lobe cells, intra-acinar colloid, and intermediate lobe colloid cysts. Using an integrating microdensitometer as a simple voltmeter, the cytoplasmic light absorption at 555mμ of the stained material was measured in cylindrical plugs of diameter less than 15μ. Plugs from positively stained cells of several types and colloid follicles were measured; neighboring unstained cells were used as controls. Values were recorded in arbitrary units of light absorption direct from the voltmeter. Results demonstrate that, per unit area, colloid contains, on the average, more than 5 times the PAS positive material of most basophils. Also discrete PAS positive vesicles vary both in the amount of positive material per vesicle and in their number per cell. There thus appears to be considerable capacity for altered storage of hormone within basophilic cells. This may provide for quantitative flexibility in the storage-release mechanism. Estimates of pituitary hormone content are calculated from the absorption measurements in single sections. The underlying assumption is that the PAS positive material measured is primarily hormone.

Lysosome Function in the Regulation of the Secretory Process in Cells of the Anterior Pituitary Gland

FARQUHAR, MARILYN G. (Department of Pathology, University of California School of Medicine, San Francisco, Calif.)

(No abstract submitted)

Characterization of Soluble Tissue Proteins by Microelectrophoresis on Polyacrylamide Gel

FELGENHAUER, K., G. G. GLENNER, and A. STAMMLER (Universitäts-Nervenklinik, Cologne, Germany and National Institutes of Health, Bethesda, Md.

A micro technique was developed which allows the electrophoretic separation of the soluble proteins of a single cryostat section. If homogenization is necessary a microhomogenizer can be used. Electrophoresis is performed in capillaries on polyacrylamide gel. The technical details for this method are described.
Quantitative evaluation of the resolved protein bands are possible by microdensitometry at appropriate wave lengths. Immunological identification of resolved bands can be performed after diffusion into agar gel or by precipitation in the separation gel itself. Applying the indirect technique of antigen detection 0.003 μg of a single protein can be detected.

Two applications of the method are given as example:
1) The protein spectrum of the cat cortex layers of gyrus cinguli.
2) Studies on several hydrolytic enzymes of different tissues.

DNA Content and Chromosome Number of L-cells

FILKUKA, J., J. SVEJDA, and V. AUBRECHTOVA (Ist. Instit. of Path. Anat., Med. Fac., Univ. Brno, Czechoslovakia)

Quantitative measurements of the DNA content of malignant tumors have been carried out by many authors. It was shown that tumors contain increased amounts of DNA with certain DNA distribution designed as the "DNA stem line." Some authors have found a correlation between the relative amount of DNA in normal and tumor cells and their chromosomal number. Our experiments were undertaken to find out the possibility of such correlation in L-cells. The DNA content of L-cells was estimated by Feulgen microphotometry and Rigler's Acridine Orange microfluorometry. The karyotype of L-cells was performed according to Rothfels and Siminovitch. The measurements were carried out on 1000 L-cells and 500 thymocytes and liver cells as a control. Both cytophotometric methods have shown the DNA stem line at hypotetraploid range, but the chromosomal stem line (modal chromosome number $= 56$) at hypotriploid range. This discrepancy may be explained by the increased DNA content per chromosome of L-cells. Similar results in dog's venereal tumors were achieved by Makino and in some human tumors by Richards and Atkin.

References:

(1) Makino, S.: N.Y. Acad. Sci. *108*, 1106 (1963).
(2) Rigler, R.: Acta Physiol. Scand. *267*, 7 (1966).
(3) Richards, B. M. and N. B. Atkin: Acta Un. Int. Cancer. *16*, 124 (1960).
(4) Rothfels, K. H. and L. Siminovitch: Stain Technol. *33*, 73 (1958).

Métabolisme du Glycogene dans La Glande Vésiculaire du Taureau

FILOTTO, U. (Istituto di Anatomia degli Animali Domestici-Istologia ed Embriologia, Università di Milano, Italia).

Des recherches histochimiques ont démontré la présence de glycogène dans l'épithélium et dans les lumières de la glande vésiculaire de "Bos taurus". L'a., au fin d'éclaircir le mécanisme de formation du fructose séminal chez cette espèce, étudie, par de techniques histochimiques et biochimiques, les activités enzymatiques liées au métabolisme du glycogène.

Les investigations consignées dans cette note portent sur 30 animaux sexuellement mûrs. On a étudié les systèmes enzymatiques suivants. 1. Phosphatase acid (méthode de Burstone, 1958). 2. Phosphatase alcaline (méthode de copulation avec des colorants azoïques, Pearse 1961). 3. Phosphorylase (méthode de Takeuchi et Kuriaki, 1955, avec les modifications de Ghua et Wegmann, 1959, pour la phosphorylase totale et la phosphorylase active). 4. Enzyme branchante (méthode de Takeuchi 1958). 5. UDPG-glycogène transferase et UDPG-pyrophosphorylase (méthodes de Takeuchi, 1962). 6. Techniques biochimiques pour la phosphorylase totale et la phosphorylase active (méthode de Cori et Coll., 1943, et de Sutherland et Wosilant, 1956, avec la variante de la démonstration de l'activité enzymatique en dosant les phosphates inorganiques produits par le G1P avec la technique de Tuassky et Shorr, 1953).

Le glycogène est toujours présent dans la glande vésiculaire du Taureau, soit dans l'épithélium (cellules prismatiques), soit dans les lumières. Les réactions pour la phosphatase acide et pour la phosphatase alcaline sont positives. La première est présente dans l'épithélium (pôle apical des cellules prismatiques) et dans les lumières; la seconde réagit dans l'endothélium des capillaires et dans les cellules prismatiques.

La phosphorylase totale est nettement positive dans l'épithélium et dans les lumièrès glandulaires; l'activité de l'enzyme branchante est démontrable dans les mêmes sièges. La phosphorylase active et toujours faible et même parfois négative.

Les activités UDPG-glycogène transferasique et UDPG-pyrophosphorylasique sont démontrables dans les coupes incubées dans des substrats contenants des activateurs. Il faut signaler que ces activités sont présentes, surtout, dans l'épithélium ou elles jouent, probablement, un rôle prédominant dans les mécanismes de synthèse du glycogène.

Les techniques biochimiques pour la phosphorylase, appliquées à la sécrétion de la glande vésiculaire, confirment les données histochimiques: la phosphorylase active est négligeable, tandis que la phosphorylase inactive est abondante.

Electrophoretic Mobilities of Serum γ-Glutamyl Transpeptidase and its Application in Hepatobiliary Disease

FISCHBEIN, J. W. and A. M. RUTENBURG (Yamins Research Laboratories, Beth Israel Hospital and University Hospital, Department of Surgery, Boston University Medical Center, Boston, Mass.)

The study of γ-glutamyl transpetidase (GT) isoenzymes was undertaken in an effort to explore their specificity in the detection and differential diagnosis of hepatobiliary-pancreatic disease. The electrophoresis of normal serum on paper separated GT into two components, a minor one in the albumin to α1-globulin region (GT_1) and a major one in the α2-globulin region (GT_2). The electrophoregrams of 52 patients with disorders effecting hepatobiliary and pancreatic systems showed a relatively large increase in GT_1 compared to GT_2 and frequently an abnormal peak in the β-globulin region (GT_3). The liver appeared to be the source of the abnormal serum GT components in these patients.

The presence of a dominant GT_3 peak was observed in patients with biliary cirrhosis secondary to choledocholithiasis or biliary atresia, and in some patients with liver metastases and severe toxic hepatitis. When the GT_3 was dominant, the prognosis was usually poor. In patients with hepatobiliary pancreatic disease the major value of GT electrophoresis was in evaluating the severity and progress of disease, prognosis, and response to therapy rather than in differential diagnosis. Electrophoretic and histochemical studies of tissue GT are in progress since demonstration of organ specific GT isoenzymes may prove useful in clinical diagnosis.

Localization of RNA Polymerase in Frozen Sections of Plant Tissues

FISHER, DONALD B. (Department of Botany, University of California, Berkeley, Calif.)

Endogenous RNA polymerase can be localized in fresh frozen sections of plant tissues (corn and onion root tips, and cotton embryos) by incubation with the nucleoside triphosphates ATP, CTP, GTP and ^{3}H-UTP, followed by autoradiography to localize the incorporated ^{3}H-uridine. Most of the incorporated radioactivity is present in the nucleolus, with additional activity in the remainder of the nucleus and possibly the cytoplasm. Little or no incorporation occurs in the presence of actinomycin D, in the absence of one of the nucleoside triphosphates, or if the sections are pretreated with DNase. The incorporated activity is not displaced by post-incubation in unlabelled nucleoside triphosphates, but it is removed by RNase.

The procedure should prove useful in studying RNA synthesis in instances where labelling of the living material is difficult, or in cases where patterns of RNA synthesis may be rapidly changing.

Some Histo- and Cytochemical Observations on Human Vaginal Epithelium

FORNI, ALESSANDRA, TORQUATO NENCIONI, and GIAN-FRANCO BALLARE' (Istituto di Istologia ed Embriologia Generale, Università di Milano, Milano, Italy)

The stratified squamous epithelium of human vagina in different functional conditions (follicular phase, luteal phase, menopause) was studied by histochemical methods. Particular attention was paid to the lipid content and to the process of keratinization. The study was carried out on 23 biopsy or surgical specimens and on smears obtained from 39 women. Whenever possible a comparison was made between sections and smears taken from the same woman.

At the end of the *follicular phase* the surface layers of the vaginal epithelium were much flattened and contained small droplets which stained with Sudan black B and were fluorescent with 3,4-benzpyrene. Starting from the basal layers up to the superficial ones, the peripheral part of the cells showed an increasing positive reaction to PAS, Sudan black B, Nile blue sulphate, PFAS (weak), to methods for SH groups, and fluoresced after staining with 3,4-benzpyrene, primuline, rhodamine B, phosphine 3R and Titan yellow. Between the superficial and the intermediate layers, a non-continuous stratum strongly positive for Baker's test and similar in its histochemical reactions to the stratum lucidum of the epidermis was present.

In the specimens obtained during the luteal phase, the superficial layers were less flattened and the layer similar to the stratum lucidum of the epidermis was lacking.

In menopause the patterns observed were very variable. In some cases the hypotrophic epithelium showed at the surface some layers of flattened cells which were strongly positive for both Baker's test and rhodamine B, and to Sudan black B, Nile blue sulphate, and 3,4-benzpyrene.

From a comparison between smears and sections and between histological and histochemical methods, cells with pycnotic nuclei, which stain pink with the Papanicolaou method (so called superficial keratinized cells) resulted in a positive Baker's test for lipine, while the cyanophilic cells with pycnotic nuclei contained sudanophilic droplets which fluoresced with 3,4-benzpyrene. The former seemed therefore to belong to the layer similar to the stratum lucidium, and the latter to the most superficial layers.

From our observations, the human vaginal epithelium contains various lipid components. In fact, unsaturated lipids, probably protein-bound, were present in the peripheral parts of the cells starting from the parabasal layers and in the layer similar to the stratum lucidum, and seemed to represent a specific differentiation. On the contrary, the lipid droplets of the cells of the superficial layers with pycnotic nuclei consisted of neutral fats and might have originated by lipophanerosis from the break-down of cell organelles.

62

Contribution to the Study of Cholinergic Innervation of the Heart

FOTIN, LUDMILA and MARIA POPESCU (Catedra de Histologie, Facultatea de Medicina, Bucharest, Romania)

Cholinergic innervation has been investigated in the heart of man, rat, guinea pig, dog, and ox by the Koelle and Friedenwald method as modified by Lewis, and by the Karnovsky method.

The findings reveal the presence of numerous nervous cholinergic fibres in the subpericardial connective tissue, running along the side of the vessels. An impressive wealth of cholinergic fibres were also found in the myocardium. Further, it was found that large nervous plexuses were located at the level of the aortic insertion, in the atrio-ventricular septum and at the level of the heart pillars. In the ventricle, the nervous cholinergic fibres run parallel forming close contacts to the muscular fibres.

Numerous microganglia, with a predominant localization in the auricle near emergence of the large vessels, are described.

On Two Phases in the Presynthetic Period of the Cell Cycle

FRANKFURT, O. S. (Institute of Chemical Physics, Moscow, USSR)

Important information on sequential events in the presynthetic period that lead to initiation of DNA synthesis may be obtained in studying changes in transition to the DNA synthesis phase induced by actions applied to different parts of presynthetic period. Effects of hydrocortisone, adrenalin, and actinomycin D on transition of squamous epithelial cells of mouse forestomach to the DNA synthesis phase was studied using autoradiography with H^3-thymidine.

Injection of hydrocortisone in physiologic doses or injection of low doses of actinomycin D 9 hours before transition of cells to the S-phase inhibited initiation of DNA synthesis. When these substances were injected 3 or 6 hours before transition of cells to the S-phase, there was no effect on initiation of DNA synthesis. Injection of adrenalin stimulated transition of cells to the S-phase, the stimulating effect appearing 9 hours after injection of adrenalin.

The results obtained show that to the end of the presynthetic period, there is a phase which differs from the preceding part of the presynthetic period in reaction to factors affecting the transition of cells to the S-phase. Evidently, within the presynthetic period, 6-9 hours prior to initiation of DNA synthesis, there is a critical transition which corresponds to start of processes inevitably leading to initiation of DNA synthesis.

The part of the presynthetic period which is characterized by resistance to hydrocortisone and actinomycin D was named as the R-phase (resistant phase). Experiments using actinomycin D show that transition to the R-phase is connected with synthesis of RNA necessary for synthesis of DNA. Diurnal variations in percentage of cells synthesizing DNA may be explained, on the basis of our experiments, on the action of hydrocortisone on transition of cells to the R-phase and information on diurnal variations in glucocorticoid level.

Cytochemical Staining of Multivesicular Body and Golgi Vesicles *

FRIEND, DANIEL S. (University of California School of Medicine, San Francisco, Calif.)

The origin of vesicles found within multivesicular bodies (MVB's) remains unknown. Cytochemical staining has not yet resolved this problem. In the rat epididymis, as in the vas deferens (1), horseradish peroxidase transported to MVB's from the cell surface and enzymes (ACPase, TPPase) transported to MVB's from the inner, concave surface of the Golgi complex are present only in the MVB matrix, not in the contained vesicles. Based on the findings of earlier studies (2), we employed the localization of reduced OsO_4 as an additional technique for investigating the origin of MVB vesicles.

Epididymides of rats hypophysectomized 17-21 days before sacrifice were minced and immersed in 2% OsO_4 for 40-48 hours at 40° C., stained in block with uranyl acetate, dehydrated, and embedded in Epon. In virtually all cells, dense deposits of reduced OsO_4 are found in: 1) several cisternae along the convex surface of the Golgi complex, 2) clusters of vesicles located at the ends of Golgi cisternae, and 3) most MVB vesicles. In a few cells, more limited deposits of reduced OsO_4 are found in cisternae of both the rough- and smooth-surfaced ER and in one or two cisternae along the inner, concave surface of the Golgi complex. No deposits are found in the MVB matrix, in vesicles along the inner, concave Golgi surface, or in coated vesicles.

The findings demonstrate that MVB vesicles have different cytochemical staining properties than either coated vesicles or smooth-surfaced vesicles found along the inner, concave Golgi surface; they do, however, resemble vesicles found at the ends of Golgi cisternae and vesicles along the convex Golgi surface, suggesting that they may be derived from the outer Golgi cisternae. In addition, the similarity of limited OsO_4 reduction in the ER and the inner one or two Golgi cisternae support the concept of a close functional relationship between these two compartments as proposed by Novikoff (3).

References:

(1) Friend, D. S. and M. G. Farquhar: J. Cell Biol. *35*, 357 (1967).
(2) Friend, D. S. and M. J. Murray: Am. J. Anat. *117*, 135 (1965).
(3) Novikoff, A. B.: Biol. Bull. *127*, 358 (1964).

* Research supported by U.S.P.H.S. research grant AM-09090 and research career program award 1-K3-GM-35, 313.

Synthesis of Nucleic Acids and Proteins, and Enzyme Localization in Antibody-Forming Cells

FUCHS, B. B. (Institute of Human Morphology, USSR Academy of Medical Sciences, Moscow I-110, USSR)

Late stages of immunodifferentiation were studied cytochemically. Metabolic peculiarities of different lymphoid cells participating and non-participating in antibody formation were investigated. Various research techniques were applied in different combinations: autoradiographic methods, electron microscopy, Coons and Jerne immunocytochemical methods, biochemical and immunological methods for the study of antibody and protein synthesis, and histochemical methods for nucleic acid and enzyme determination.

Two or three hours after second contact with the antigen, many lymphocytes and reticular cells show increased synthesis of ribosomal (actinomycin D sensitive) and messenger (actinomycin D resistant) RNAs. Later, the stimulation and synchronization of DNA synthesis in the cell population occurs. The synthesis rate of total RNA in small, medium and large lymphocytes differs. However, the ratio of synthesis rate of two RNA fractions in different groups of lymphocytes is similar, distinguishing them from that in reticular cells. This correlates with a high injurability of lymphocytes as compared to that of reticular cells during the RNA synthesis inhibition. When treated with actinomycin D, lymphocytes showed a rapid disappearance of nucleolar pars amorpha, nuclear euchromatin zones, and perished quickly. Under similar conditions, reticular cells exhibited changes of nucleoli only. During the inhibition of ribosomal and messenger RNAs (in the latent phase), antibody synthesis suffered to a greater extent than that of other proteins. This appears to be related to formation of new polysomes providing the antibody synthesis.

The phenomenon of a sharp stimulation of the synthesis of antibodies and some other proteins by small doses of ribonuclease was found. This can be associated with the stimulatory effect of RNA degradation products.

Certain metabolic peculiarities of antibody-forming cells were established with the aid of a modified Jerne method (antibody synthesis in a thin agar layer) combined with autoradiographic and enzymohistochemical techniques.

Cytochemistry of the Nucleolus in Myocardial Hypertrophy

FUCHS, B. B., V. D. ARUTYUNOV, and A. L. SHNAPER (Institute of Human Morphology, USSR Academy of Medical Sciences, Moscow, USSR)

The phenomenon of true nucleolar hypertrophy observed in cases of acute overloading of the left ventricle was described by V. D. Arutyunov in 1964. We performed a detailed cytochemical study of this phenomenon in rabbits and white rats suffering from stenosis of the thoracic aorta.

We investigated changes of the nucleolar volume, the RNA content and synthesis rate in nucleoli (as determined by Brachet's and uridine-H^3 incorporation methods), the heart size and weight in relation to increased loading upon the myocardium with and without an inhibition of ribosomal RNA synthesis (by actinomycin D). The possibility of influencing the rate of nucleolar hypertrophy with the aid of a mixture of nucleotides was also studied. All the processes were evaluated in time (from the moment of the stenosis induction) separately in the internal longitudinal, circular and external longitudinal layers of the myocardium. It was shown that, on the 3rd day after induction of thoracic aorta stenosis, an enlargement of nucleoli and RNA accumulation in them began in the internal longitudinal layer of the myocardium. Later on a progressive enlargement of nucleoli was observed in all myocardial layers (on the 5th day in the circular and on the 7th day in the external layer). The maximal sizes of nucleoli were found in all the layers of the myocardium on the 7th day (within the period from the 3rd to 7th day, the size of the nucleoli increased 3 to 4 times). From the 7th to the 13th day of the experiment the nucleoli gradually diminished in volume and became undetectable. From the first day of the stenosis induction a regular increase of the heart weight (20 to 40 per cent as compared with the control) took place. Although the increase continued, no hypertrophic nucleoli were detected.

Actinomycin D administered to animals with aortic coarctation in doses inhibiting the synthesis of ribosomal RNA arrested the enlargement of nucleoli and their RNA accumulation, simultaneously preventing an increase of the heart size and weight.

Parenteral administration of a nucleotide mixture accelerated and prolonged development of nucleolar hypertrophy. This can be considered as an indication of increase of compensatory myocardial capacity.

Data on the quantitative autoradiographic evaluation of ribosomal RNA synthesis and accumulation in nucleoli of the hypertrophic myocardium are presented.

Autoradiographic and Histochemical Indices of Cancerous Cell Involvement in Fibrillation

GABUNIA, U. A. and N. N. SHIUKASHVILI (Institute of Experimental Morphology: AS Georgia SSR, Tbilisi)

Presented are autoradiographic data on the involvement and distribution of radioactive isotopes of sodium sulphate S^{35} and proline C^{14} as well as the results of a histochemical analysis of mucopolysaccharides and proteins of argyrophilous and collagenous nature obtained on the basis of a study of mammary cancerous tumors of humans and mice of the highly cancerous line "A" and "C_3H."

The dynamics of the inclusion of sodium sulphate S^{35} and proline C^{14} has shown that, in the parenchyma of the tumor in question, there occurs an intensive metabolism of sulphomucopolysaccharides and proline-containing proteins, with sodium sulphate S^{35} and proline C^{14} being absorbed in abundance.

The character of mark distribution in cancerous cells and in their surrounding intercellular substance shows that plastic metabolism of sulphomucopolysaccharides and proline-containing proteins is effected through intracellular mechanisms: sulfomucopolysaccharides and proteins synthesized in the cancerous cells are excreted into their surrounding intercellular substance.

Histochemical studies have shown that the above chemical compounds are found in cancerous cells as well as in their surrounding intercellular substance. In the latter, these compounds are found in the shape of an amorphous mass in which formation of filamentary structures begins to show.

The amorphous substance and filamentary structures (not infrequently of granular construction) are to be found close to parenchymatous cells; these are argyrophilous and are brought to light by methods used for the detection of acid mucopolysaccharides in tissues. As we move away from the parenchymatous cells, coarseness of filamentary structures sets in, reaction to acid mucopolysaccharides weakens, argyrophilia is lowered and, reversely, reaction to neutral mucopolysaccharides increases.

The evidence obtained points to the participation of parenchymatous cells of mammary gland cancer in the production of chemical compounds necessary for build-up of fibrous structures.

Behaviour of Heterologous DNA in Plants

GAHAN, P. B., P. ANKER, and M. STROUN (Dept. of Biology & Cell Science, Woolwich Polytechnic, London., Botany Dept., University of Geneva and Radiobiologie, C.E.N. Mol., Belgium)

[3]H-DNA was prepared from a thymine-less strain of *Escherichia coli* grown in the presence of ^{3}H-thymine and fed to either cut shoots of *Lycopersicum esculentum* or roots of *Vicia faba* seedlings at a concentration of 200 γ/ml of solution. Stripping film and liquid emulsion autoradiographs were prepared from tissues processed by 1) freezing and sectioning (1); or 2) fixation for 3 hours in acetic acid-ethanol followed by either paraffin wax embedding or the Feulgen squash procedure; or 3) fixation in 2.5% glutaraldehyde and embedding in araldite for electron microscopy.

As controls, plants fed with ^{3}H-thymidine of a similar specific activity to the ^{3}H-DNA were processed similarly.

The ^{3}H-DNA, which retains a molecular weight of more than one million (2) was found to be present in phloem, cambium, parenchyma, epidermis and collenchyma of *L. esculentum* after feeding for 6 hours. Feeding for 6 hours with ^{3}H-DNA followed by 48 hours water increased the percentage of cells labelled in each tissue. A similar pattern of labelling was found with ^{3}H-thymidine but a lower proportion of the cells were labelled in each tissue than was found with ^{3}H-DNA.

The labelling was predominantly nuclear in all tissues, but phloem, cambial and parenchmal cells also had a cytoplasmic label which was found to be associated with mitochondria and plastids. In root meristem cells, both nuclei and mitotic chromosomes were labelled and exogenous, heterologous DNA was found to induce chromosome damage. The possible implications of these findings will be discussed..

References:

(1) Gahan, P. B., J. McLean, M. Kalina, and W. Sharma: J. Exptl. Bot. *18*, 151 (1967).
(2) Stroun, M., P. Anker, P. Charles, and L. Ledoux: Nature *215* 975 (1967).

Hydrolase Activity in Dividing and Differentiating Plant Cells

GAHAN, P. B. and JEAN McLEAN (Department of Biology & Cell Science, Woolwich Polytechnic, London S.E. 18)

β-glycerophosphate, naphthol phosphatase, and naphthol esterase activities were localized at particulate sites in the cytoplasm of dividing and non-dividing cells of root and shoot meristems and shoot cambium.

No differences were observed either in localization or in behaviour of the hydrolase activities between the two types of cells. There was a higher esterase activity in root and shoot pro-cambial cells than in meristem cells, this high level being maintained during mitosis of the pro-cambial cells. Thus, in contrast to reports for animal cells (1), activity was related to the particular tissue rather than to a stage of cell division. Studies of phosphatase and esterase activities during root differentiation have indicated that whilst the response time for esterase activity is one minute for all tissues, the corresponding times for phosphatase activities varied between 4 and 16 minutes, depending upon the tissue examined. Root cap, epidermis, cortex, and differentiated vascular tissues were found to have similar levels of esterase and phosphatase activities. High phosphatase and low esterase activities were found in apical initial and procortex regions, and the converse in the pro-cambium.

In the developing sieve elements, particles containing hydrolases were seen to line the walls between adjacent sieve elements. Particulate sites of enzyme end-product appeared closely adpressed to the walls, apparently in opposite pairs across the developing sieve plate. In the mature sieve plate, enzyme end-products were seen to be continuous from cell to cell across the cell wall.

References:

(1) Gahan, P. B.: Intern. Rev. Cytol. *21*, 1 (1967).

Studies on the Regulation of Cell Proliferation and Differentiation in Intestinal Mucosa

GALJAARD, H. and D. BOOTSMA (Med. Biol. Laboratory of the National Defense Research Organisation T.N.O., Rijswijk and the Dept. of Cell Biology, Faculty of Medicine, Rotterdam, The Netherlands)

In the investigation of cell renewal systems, the use of intestinal mucosa has the advantage that the process of cell proliferation and differentiation can be related to microscopic localisation of the cells along crypt and villus; furthermore, there is a uniformity in the end product of differentiation.

To obtain information about the factors involved in maintaining steady state growth and in the transition from a proliferating cell to a non-proliferating specializing cell, an enzyme-histochemical study was carried out on various segments of rat intestinal tract. The results were compared with those obtained after low doses (50-400 rads) of X-irradiation which exclusively influence the pool of proliferating cells in the crypts. From a large group of enzymes tested, only the activity of the non-specific esterases in the crypts was found to be markedly decreased after irradiation.

The time of onset of the effect (36 hours after irradiation) is independent of the radiation dose; the number of crypt cells involved and the duration of the effect increase with higher radiation doses.

Autoradiographic studies using ³H-thymidine showed that the migration of crypt cells into the top of the villi is not influenced by irradiation. In another set of labeling experiments it could be demonstrated that after irradiation there is a marked increase in the pool of DNA synthesizing cells at the expense of the pool of nonproliferating specializing cells in the crypts.

Combined autoradiographic and histochemical studies on the same frozen sections revealed a remarkable correlation between the decrease in esterase activity and the increase in the pool of proliferating cells. These results suggest a role of non-specific esterase in the specialization of the crypt cells.

Using Lowry's technique and spectrophotometric analysis at the ultra-micro level, the esterase activity was quantitatively determined in various cell compartments in crypt and villus. The results in control animals indicate an increase in esterase activity in the non-proliferating upper part of the crypt. After irradiation there is a loss of esterase activity in this part of the crypt which is not recovered during migration of the cell along the villus.

Microelectrophoretic studies on isolated crypts and villi using various substrates and specific inhibitors allowed further identification of the esterases involved. Experiments on germ free animals are discussed regarding the role of the villi in the regulation of cell proliferation in the crypts.

Histochemical Study of Pigmentary Neoplasms in Man

GANINA, K. P. (Kiev Research Institute of Experimental and Clinical Oncology, Kiev, USSR)

The metabolic processes and genetic alterations in human pigmentary neoplasms have been inadequately studied, although they determine the intimate processes of malignisation to a great extent.

We examined 100 nevi and malignant melanomas removed from patients without any treatment or after X-ray treatment. After special fixation, in addition to staining, DNA, RNA, promelanin, glycogen and lipoids were determined. Feulgen's test made it possible to study the intranuclear structure of chromatin, including determination of the content of sex chromatin (SC) in tumor cells and the surrounding tissues,

70

and the mitotic index (MI). In addition, the dimensions of the nucleolus, nucleus and cell were determined.

A comparative study of nevi, nevi with malignisation, and malignant melanomas showed pronounced structural changes in the foci of malignisation against a background of nevi, and to a great degree in malignant melanomas. This took the form of a coarsening of the intranuclear chromatin, a reduction in SC content in the tumor cells in women, a rise in MI, an increase in the dimension of the cell, nucleus, and nucleolus, accumulation of RNA in the cytoplasm and nucleolus, and intensification of pigmentation.

After X-ray irradiation, pronounced polymorphism of structure and histochemical indices was observed in the malignant melanomas. Typical features are the acute increase in tumor cells, a low RNA content in the cytoplasm, disorganization of the chromatin structure, fall of SC content to the vanishing point, decrease in MI, pronounced condensation of pigment in the cells and stroma, fibrosis and sclerosis of the stroma.

The investigations showed considerable histochemical and genetic disturbances in malignised pigmentary tissue and in malignant melanomas after X-ray treatment. The results may be applied to determine the degree of malignancy of pigmentary neoplasms in man.

Feulgen-DNA Values in Leucocytes After Hypotonic Treatments **

GARCIA, ALFREDO MARIANO * (Dept. of Anatomy, State Univ. of New York, Upstate Med. Center, Syracuse, N.Y.)

Peripheral leucocytes of rabbit were incubated in the following hypotonic media: a) 3 parts of blood or plasma diluted with two parts of distilled water, b) 0.95% sodium citrate, c) 0.2M sucrose, and d) Ohnuki's medium (1). For a, b, and c, the incubation time was 15 minutes at 37°C; for d, it was 90 minutes at room temperature. The leucocytes were then smeared in clean slides, air dried and fixed in methanol-acetic-formalin (75:15:10) mixture. Control smears were done at the same time. The slides were then stained by means of the Feulgen reaction and the amount of nuclear dye content was assessed with the Barr & Stroud Integrating Microdensitometer. The results seem to indicate that a) distilled water, sodium citrate and Ohnuki's medium swell the nuclei more or less uniformly and that this increase in size is correlated with an increase in Feulgen-DNA values (around 6 to 10%); b) on the other hand, leucocytes treated with 0.2M sucrose show a bimodal distribution: about 40% are greatly swollen or disrupted while the rest of them are shrunk; again, the cells with swollen nuclei yield higher Feulgen-DNA values while the opposite occurs in those nuclei which underwent compaction.

The results briefly exposed above reinforce previous findings that the physical state of the deoxyribonucleoprotein molecules influences the reactivity of the dye-binding sites (2, 3, 4, 5, 6) and that any change in the chromatin from compact to diffuse state enhances the reactivity of the functional groups of the DNP complex, thus accounting for the higher Feulgen nuclear content found in the present series.

References:

(1) Ohnuki, Y.: Nature *208,* 916 (1965).
(2) Hale, A. J. and E. H. Cooper: *In:* Current Res. in Leukemia, pp. 95-107 (Hayoe, ed.) London: Cambridge Univ. Press (1965).
(3) Gledhill, B. L., M. P. Gledhill, R. Rigler, Jr., and N. R. Ringertz: Exp. Cell Res. *41,* 652 (1966).
(4) Mayall, B. H. and M. L. Mendelsohn: J. Cell Biol. *35,* 88A (1967).
(5) Garcia, A. M.: *In:* Conference on Data Extraction and Processing of Optical Images. Ann. New York Ac. Sc. *In Press.*

* NIH Development Career Awardee 1K3-GM-11, 790-03.
** Work supported by grant AM 10016-03 (HEM) of the Nat. Inst. Arthritis and Metab. Diseases.

Long Term Changes in Submandibular Sympathetic Nerve Trunks after Ablation of the Proximal Nerves

GARRETT, J. R. (The Dental School, King's College Hospital, London, England)

The sympathetic nerve trunks to the submandibular gland of the cat pass along the surface of the artery to the gland. Various histochemical methods have been studied microscopically in these nerve trunks. The effect of unilateral cervical sympathectomy plus excision of the carotid and its tributaries, to a point just proximal to the submandibular branch, has been examined after different time intervals. The results, in each case, were compared with those in normal nerve trunks on the contralateral artery.

Normal submandibular sympathetic trunks show moderate catecholamine fluorescence, weak acetylcholinesterase (AChE) activity in neural bands, with accentuation of some axons, and much stronger non-specific cholinesterase (ChE) activity in thicker bands. Non-specific esterase and acid phosphatase were only faintly detectable. After ablation of the proximal nerves the distal trunks did not disappear and could always be recognised. Catecholamine fluorescence disappeared within 24 hours. There was a slight release of AChE and ChE activity after 24 hours, then a patchy decrease and by 12 days both were virtually absent. Some time later, however, there was a re-emergence of these enzymes, both of which eventually became more active than in the control nerves. Non-specific esterase activity was evident in scattered cells within 24 hours; it increased in intensity until about the 8th day and then slowly disappeared. A less dramatic increase in acid phosphatase was detected from 8 to 16 days. These changes will be discussed in relation to the electron microscopical appearances of adjacent parts of the same nerve trunks.

Quantitative Histochemistry and Insulin Content of Microdissected Human Islets of Langerhans *

GEPTS, W., F. GREGOIRE, and H. OOMS (Hôpital Universitaire Brugmann and Fondation Médicale Reine Elisabeth, Brussels, Belgium)

Lowry quantitative microchemical technics were applied to the study of the enzymatic activity of human islets of Langerhans. Pancreatic tissue from non-diabetics and diabetics was obtained either from surgery or early after death. Two islet tumors were also studied. The following enzymes were assayed: lactic dehydrogenase (LDH), isocitric dehydrogenase (ICDH), glucose-6-phosphate dehydrogenase (G6PDH), glutamo-oxaloacetic transaminase (GOT), acid phosphatase (ac. Pase). The insulin content of microdissected islets was measured by the double antibody method of Morgan and Lazarow.

The activity of LDH is much higher in the acini than in the islets. Two cases which had received large doses of cortisone show an increased activity in the islets.

ICDH is also more active in the acinar tissue than in the islets. An increased activity of this enzyme is present in the islets of the corticone treated cases.

No significant difference exists in the activities of G6PDH in acinar and islet tissue. GOT is usually more active in the islets than in the acini. Both islet-cell tumors show a very high GOT activity. Ac. Pase activity is slightly, but not significantly higher in the islets than in the acini. One of the cortisone treated cases shows a strongly increased ac. Pase activity in the islets, whereas in the acinar tissue the activity is lower than in non-treated cases.

Except for LDH, the pancrease of neonates has levels of enzymatic activity approaching those of the adult pancreas. LDH is lower in the acinar tissue, and higher in the islets, than in adults.

Up to this time of the study, no significant difference in the enzymatic activity of the islets has been observed between non-diabetics and diabetics.

Several discrepancies were noticed between the results of biochemical measurements and those of histochemical methods. A satisfactory explanation of these discrepancies can be suggested by analysing the mechanisms of the biochemical and histochemical methods.

The insulin content of the islets is lower in diabetic than in non-diabetic adults.

* This study was supported by grants 711 and 872 from the "Fonds National de la Recherche scientifique Medicale", Belgium.

Cytochemical Signs of the Presence of Cells Probably Involved in Salt Excretion

GERZELI, GIUSEPPE (Institute of Comparative Anatomy, University of Pavia, Italy)

Several data are known on the morphological and functional convergence of the salt glands, never homologous organs in some species of reptiles. I will report chiefly on comparisons between homologous organs of the species lacking specialized salt glands: cytochemical data seem significant, even if they merely represent indirect signs of the existence and trend of the phenomena of water and ions transport across the epithelia. The cytochemical study (proteins and amino-acids, nucleic acids, polysaccharides, lipids, hydrolases, reductases, transglycosydases) applies to head glands of one marine turtle (*Thalassochelys caretta*), one tropical lizard (*Iguana iguana*) and one desert lizard (*Sauromalus obesus*): their salt glands open to the exterior respectively near the posterior corner of the eye and into the lateral nasal cavity. Other species from different habitats, lacking a salt gland, have been also studied (Chelonia: *Testudo graeca, Testudo ibera, Kinosternon scorpioides, Emys orbicularis, Clemmis leprosa, Pelusios sinuatus;* Lecertilia: *Hemidactylus flaviviridis, Agama agama, Basiliscus basiliscus, Chalcides chalcides, Anguis fragilis, Lacerta viridis;* Crocodilia: *Caiman sclerops*). Finally, in two species (*Testudo graeca, Lacerta viridis*), attempts to stimulate a secretion have been made by salt loading and pharmacologically.

The secretory cells of the salt glands, rich in lipoproteins, are characterized by the high activity of the $NADH_2$- and the $NADPH_2$-tetrazolium reductases and succinate dehydrogenase; acid phosphatase and ATPase (Na-K-activated) are present in the basal part of the cells; the enzymatic activities involved in glycogen metabolism seem to be of little importance. The broad and irregular intercellular spaces between adjacent cells contain neutral and acid mucosubstances which are also detectable along the apical cell surface.

In the comparative observations, evidence has been found for the presence of cells having similar features exclusively in the lachrymal glands of the Chelonia and in the nasal glands of the Lecertilia: these cells, not very numerous, are inserted in the adenomeres among other cells engaged in mucin secretion. The sole species of Crocodilia studied shows this feature in a gland situated in the anterior margin of the orbit.

Following the attempts of stimulation, salt secretion has not appeared; however, in addition to generic marks of gland hyperfunction, a significant increase of the metabolic rate, especially the oxidation-reduction systems, has been observed: this increase is very marked in the cell type pointed out above after sodium and potassium chloride loading.

Apart from the possibility of a substantial salt excretion from the glands as a whole, it is assumed that the facts observed at a cellular level may represent the morphological and metabolic basis of a potentially inactive ion transport: an ultrastructural study is also in progress. These cells, presumably capable of salt secretion, are present in different glands but are generally homologous to the typical salt glands within each systematic group.

Value and Limitations of Quantitative Chemical Studies in Individual Cells

GIACOBINI, EZIO (Department of Pharmacology, Karolinska Institutet, Stockholm, Sweden)

The extreme heterogeneity and functional complexity of the nervous system strongly motivate the use of quantitative chemical methods at the cellular level. Remarkable methodological advances in recent years makes it possible to study with high precision and accuracy the activities of most enzymes and the levels of most substrates and metabolites in individual nerve cells. It is important to evaluate critically the limits and value of these techniques. In this discussion the following points will be taken into account:

A. *Choice of cell preparation.*

1) Vertebrate or invertebrate neurons. 2) Fresh or lyophylized cells. 3) Dissected cell bodies or tissue layers.

B. *Purity of the cell preparation.*

Contamination with glial cells and prescence of synaptic knobs.

C. *Preservation of the functional integrity of the cell.* Damage to the cell preparation produced by dissection and its consequences on:

1) functional activity of the cell; 2) integrity of the physiological barriers of the cell (membranes).

In view of the results presented, it is felt that pure morphological criteria (electron micrographs) are neither reliable or practical and that other criteria such as recording of electrical activity or measurement of cation content should be applied.

D. *Preservation of the chemical components to be studied.*

1) Minimization of enzyme inactivation or of unspecific loss (diffusion etc.).

2) Prevention of changes in substrate levels due to metabolic or diffusional factors.

E. *Does the enzyme activity found reflect only the accessibility of the substrate to the enzyme molecule (due to more or less integrity of the cell preparation) or is it a measurement of the "true" level of enzyme activity?*

In favour of the latter statement are:

Direct evidence found determining enzyme activities (AChE, ChAc and MAO) in the following cellular preparations:

1) freshly dissected "intact" cells; 2) frozen-dried cells; 3) disrupted cells (hypotonic shock, freeze and thawing, homogenization, sonication and detergents); and the following conditions: 4) intracellular sampling of cytoplasm and nucleoplasm; 5) kinetic data from single cell determinations.

For AChE, ChAc and MAO no significant difference in enzyme activity could be detected in the first three conditions and the results of intracellular sampling were in excellent agreement with whole cell determinations (4). The kinetic data speak for a good accessibility of the substrate to the enzyme molecule (5).

Indirect evidence:

1) correlation between "histo-" and cytochemical results; 2) constant distribution of enzyme activity pattern in the same structure (e.g., ganglion) and in the same species; 3) constant number of active and inactive cells in different experiments with the same type of structure (ganglion, AChE, ChAc, MAO); 4) specific and constant distribution pattern in the different parts of the neuron for certain enzymes (AChE, spinal ganglion cells, motor neurons, etc.); 5) reproducibility of results in measurements of enzyme activity from aliquots of extracts of the same cell.

F. *Possibility of measuring several enzymes or substrates in the same cell* (e.g., AChE and ChAc in single ganglion cells).

As an illustration of the potentiality of the microchemical methods, a model of a sympathetic neuron will be presented, which has been studied with respect to six different enzymes (AChE, ChAc, MAO, COMT, dopa DC, BuChE) involved in the metabolism of transmittor substances.

Integrating Microflamephotometer for Simultaneous Determination of Sodium and Potassium in Single Cells

GIACOBINI, EZIO and STEFAN HOVMARK (Dept. of Pharmacology, Karolinska Institutet, Stockholm, Sweden)

In order to decide whether some of the results obtained with inhibitors or metabolites on the impulse activity and metabolism of the stretch receptor neuron of the crayfish could be related to active transport processes, it was felt necessary to determine Na and K in cells treated with these inhibitors or wider conditions of modified ionic environment. An integrating microflamephotometer, suitable for analysis of either individual cells weighing 0.5-0.005 μg or microsamples of biological material below the nanoliter level, was constructed. The instrument has a sensitivity of 1×10^{-15} M for Na and K. The material to be examined is placed on a wire under microscopic inspection and introduced into the flame where excitation takes place. The time integral of the intensity of the light emitted at the chosen analytical spectral wavelength is proportional to the amount of the element present.

The emitted light pulse is transmitted through selective interference filters to two photomultiplier tubes. Their output signals are electronically integrated. The integrated signals are fed into capacitor memories and the background is subtracted. The two signals (Na and K) are registered on a potentiometric recorder and visualized on an oscilloscope. With the help of this technique, it has been possible to follow changes in Na and K content occurring during the initial period following isolation of single invertebrate neurons or during incubation conditions of reduced movement of Na and K. After isolation of the SRN the ionic events can be divided into two periods. During the first (initial) period, a continuous recovery from the initial distortion of the ionic content is going on. The intracellular cationic content is influenced during this period by at least two processes acting simultaneously and counteracting each other; one depending on the manipulation of the cell, the other on the active transport.

If the active transport mechanism is inhibited (e.g., under the effect of ouabain or in K-free or Na-free medium) the recovery does not take place.

Tetrodotoxin, glucose and short-lasting physiological stimulation do not significantly change the intracellular Na or K.

The same method has been applied to the study of cation content of neuron and glial cells from vertebrate as well as invertebrate nervous structures.

Mycoplasma: A Cell Concept of Colony Organization Based on Histochemistry

GILKERSON, SETH W. (Veterans Administration Hospital, Oteen, N. C.)

In our search for better methods to visualize colony morphology and determine histochemical organization of mycoplasma we developed a slide culture technic and used fluorescent stains. An examination of many cultures of four species, *Mycoplasma pneumoniae, hominis, lai-*

dlawii and *pulmonis,* by these methods lead us to postulate that mycoplasma colonies are organized histochemically like mammalian cells where the deoxyribonucleic acid is concentrated in the center of the colony forming a nuclear-type structure. Initially this concept was based on the results of the fluorescent stains, phenol-auramine and thiazine red, which differentiated in color the two, major parts of a colony in the same way they differentiated between the nuclei and cytoplasmic parts of human leucocytes and epithelial cells (1). By blue-light fluorescent microscopy (2) the nuclei fluoresced yellow-green and the cytoplasmic areas red. Each of the four species was distinctive and identifiable by colony morphology; the fluorescent reactions were closely identical except for variations in size and densities of the DNA-reactive centers. *Additional experiments added in support:* (a) Heat-fixed colonies of each species treated with Feulgen reagents (3) confirmed dense concentrations of DNA in the centers of the colonies. (b) Colonies of each species incubated in deoxyribonuclease and then Feulgen stained revealed marked reduction of DNA or no reaction while companion non-enzymed cultures were strongly positive for DNA. (c) Concentrations of DNA in bacterial colonies could not be demonstrated. *Staphylococcus aureus* and *Escherichia coli* cultured and stained by the same technics used for mycoplasma showed no histochemical differentiation. The results of our study show that a mycoplasma colony concentrates DNA in the cells in the dense center with little evidence of DNA in the thin-appearing outer part. In these respects a colony resembles the morphology and histochemical organization of a typical cell. If they function in a manner similar to a cell then the bio-physics of the "multicellular-cell-like" colonies could account for their microscopic size. We suggest that this type of organization represents an evolutionary development by which these metabolically primitive and wall-less forms gained the capability of autonomous growth (4).

References:

(1) Gilkerson, S. W. and M. Moss: Bacteriol. Proc. Abstr. p. 50 (1964).
(2) Gilkerson, S. W. and O. Kanner: Jour. Bacteriol. *86,* 890 (1963).
(3) Merchant, D., J. Kahn, and W. H. Murphy: *Cell and Organ Culture,* pp. 139-141, Burgess Pub. Co. (1960).
(4) Symposium: *Biology of the Mycoplasma,* Ann. N. Y. Acad. Sci., *143* (Art. 1) 1-824, (1967).

Elemental Analysis by the Laser Microprobe

GLICK, DAVID (Division of Histochemistry, Dept. of Pathology, Stanford University Medical School, Palo Alto, Calif.)

A laser microprobe apparatus and the technique for the emission spectrographic analysis of elements with a sensitivity in the range of

10^{-16}-10^{-21} moles, for histo- and cytological samples has been developed with Drs. E. S. Beatrice and I. Harding-Barlow, and in the earlier stages with Dr. R. C. Rosan, in my laboratory, and with Dr. N. Peppers and his coworkers at Stanford Research Institute. Analyses can be performed on single cells or bodies such as individual nuclei. Samples 80-200 μ^3 (e.g. 5-10μ diameter, 1-3μ depth, in a microtome section, tissue imprint or cell smear) of air- or frozen -dried material have been employed. With present equipment, up to 6 elements simultaneously can be quantitatively analyzed. For larger samples and qualitative estimation, greater numbers of elements can be included in the simultaneous analysis. The sensitivity is great enough to permit analysis of about 20 "trace elements" in single cells. Certain technical problems, such as development of suitable analytical standards, are under investigation. Although feasibility and capability have been established and will be demonstrated, technical improvements are still required to provide a practical instrument for general use. Progress in this regard is being made.

Mouse Kidney as an Induced Accumulator of Electricity

GODLEWSKI, H. G., A. HUSZCZUK, and BARBARA PENAR (Pharmaceutic Institute, Warsaw, and Institute of Experim. Pathol., Polish Ac. Scs., Warsaw, Poland)

Kidneys of C_3H mice were subjected to ionophoresis *in vivo* according to the technique described in Folia Histochem. Cytochem. *3*, 329, 1965. The influence of 2mA current passed for 15 minutes through the kidney *in situ* was evaluated by means of planimetric measurements of areas of altered histochemical distribution of the activity of certain enzymes, particularly that of succinate and lactate dehydrogenases. The three areas distinguished were: the area of partly inhibited activity at the anodal zone, the completely inhibited activity at the cathodal zone, and the intermediate zone in which enzymic activity had been preserved. Using the "EICO" set the voltage of both normal and ionophoretically treated kidneys was determined as 0.34V and 1.67V respectively.

The discharging of the electrically charged kidney was performed by employing short-circuits between the electrodes contacting the kidney poles for ab.1 hour and the resulting potentials measured by the use of "EICO" set. The shape of the potential curve representing the discharging process indicates that the behaviour of a charged kidney is analogous to that of an accumulator. It has been established that a spontaneous discharging of the charged kidney takes place after the ionophoretic treatment within 24 and 48 hours, but no evident regression of the current-induced morphochemical changes occur.

The Changing Distribution of Hepatic Copper in Wilson's Disease: A Cytochemical Study

GOLDFISCHER, SIDNEY and IRMIN STERNLIEB (Albert Einstein College of Medicine, Bronx, N. Y.)

Earlier histochemical studies have shown copper to be present only in granules of hepatic parenchymal cells (1-3) and these graffules have been identified as lysosomes in liver biopsy speciments obtained from some patients with Wilson's disease (4). In other patients, however, with equally high levels of hepatic copper, attempts to demonstrate the metal histochemically have failed, particularly during early stages of the disease (5). The present study was undertaken in order to (a) find out why these latter specimens do not stain for copper and (b) correlate the localization of copper in hepatocytes with the evolution of the disease. Paraffin embedded sections of formalin-fixed liver biopsy specimens from patients ranging from 3 to 51 years of age and in different stages of Wilson's disease were stained for copper with rubeanic acid and with the silver-sulfide procedure of Timm (6). Only the silver-sulfide stain visualizes copper in specimens from young, asymptomatic patients. In such biopsies copper staining is diffuse within the cytoplasm of hepatocytes and is not seen in lysosomes. Although specimens with diffuse staining often have the highest *total* levels of copper, the *concentration* of metal is apparently too low to be detected with the rubeanic acid technique, but evident with the more sensitive silver-sulfide procedure. In contrast, copper concentrated in lysosomes is stained by both rubeanic acid and the silver-sulfide technique. Data obtained by electron probe microanalysis confirms these differences in the distribution of copper Diffuse staining is most marked when fatty change and necrosis are evident; regenerative nodules are free of copper or show slight diffuse staining. In biopsy specimens from older patients copper is found mainly in lysosomes. This occurs where the hepatic architecture is relatively intact and in cirrhotic nodules of specimens that are free of necrosis and inflammation.

Our observations suggest that when copper is segregated into lysosomes damage to the cell is minimal. An apparent failure to incorporate copper into lysosomes may be responsible for the toxic effects of the metal that occur during the active stages of Wilson's disease.

References:

(1) Okinaka, S., M. Yoshikawa, M. Toyoda, T. Mozai, Y. Toyokura, and M. Kameyama: Neurol. Psychiat. 72, 573 (1954).
(2) Uzman, L. L.: Arch. Path. 64, 464 (1957).
(3) Howell, J. S.: J. Path, Bact. 77, 473 (1959).
(4) Goldfischer, S. and J. Moskal: Amer. J. Path. 48, 305 (1966).
(5) Sternlieb, I. and I. H. Scheinberg: New England J. Med. 278, 352 (1968).
(6) Timm, F.: Histochemie 2, 332 (1961).

Biochemical and Histochemical Characterization of the Enzyme and Matrix Components of Renal Lysosomes [*]

GOLDSTONE, A., E. SZABO, and H. KOENIG (VA Research Hospital, Chicago, Ill.)

Lysosomes in various tissues exhibit distinctive histochemical features in addition to their enzymatic staining reactions. Lysosomes contain PAS-positive (1) and autofluorescent (2) substances which are extractable from fresh, but not fixed, tissue with chloroform-methanol. Lysosomes are stained preferentially (3) and metachromatically (4) by basic dyes *in vivo* and *in vitro* in the fresh, unfixed state. The basophilia and metachromasia of lysosomes are based on electrostatic binding, since both are inhibited or abolished *in vitro* by competing cations and low pH(5). The chromotrope seems to be an acidic lipo-protein, as lipid solvents, detergents, phospholipase C, and denaturant fixatives such as HCHO abolish the metachromatic staining of lysosomes *in vitro* (5).

We have now identified and partially characterized the lysosomal components responsible for these histochemical properties. Lysosomes are isolated from rat kidney by isopycnic centrifugation of a mitochondrial suspension over a density gradient consisting of 1.75 and 2.0 M sucrose. The lysosomal fraction (1.75-2.0 M) is 10-15 fold enriched with respect to acid phosphatase and β-glucuronidase, and consists almost wholly of lysosomes when examined in the electron microscope. The lipid components (mg/mg protein) are: phosphatides, 0.113; cholesterol, 0.032; and glycolipid, 0.010. The phosphatides (% total) are: sphingomyelin, 36.7%; lecithin, 34.8%; phosphatidyl ethanolamine, 10.7%; phosphatidyl inositol, 5.2%; cardiolipin, 1.4%. The neutral lipids are mainly saturated ($>70\%$), while in phosphatides, the unsaturated fatty acids (18:1, 18:2, 22:1, 26.2) constitute 38-60% of the total.

After brief sonication (5 minutes, 4° C, in 0.45 M sucrose), solubilized protein (60-80% of total) is dialyzed, lyophilized, and separated into 3 fractions by DEAE-cellulose chromatography. Acid phosphatase and aryl sulfatase occur in fraction A; additional acid phosphatase, β-glucuronidase, and cathepsin are in B. A and B contain hexosamine ($>1\%$), and hexose (5%), and are PAS-positive. On polyacrylamide gel electrophoresis (P.G.E.), A is resolved into at least 8, and B, into 3 cationic proteins. A fluorescent component occurs mainly in B which has the spectral characteristics of pyridine nucleotide (excitation maximum, 360 mμ; emission maximum, 460 mμ). This protein-associated pyridine nucleotide seems to be responsible for the autofluorescence of lysosomes (2).

The C fraction is without enzyme activity, contains substantial phosphatide (0.25 mg/mg protein), and is resolved by P.G.E. into 2 anionic lipoproteins. These lipoproteins are sudanophilic, PAS-negative, and basophilic, and stain metachromatically with acridine orange from pH 5 to pH 11 (maximum at pH 7-9). Comprising 30% of the solubilized protein, these lipoproteins are quantitatively important components of the finely granular, osmiophilic, electron-dense matrix. In the intact lysosome the acid hydrolases, which appear to be cationic glycoproteins, probably are ionically linked in an inactive state to the acidic lipoproteins, as postulated earlier (1, 6).

References:

(1) Koenig, H.: Nature *195*, 782 (1962).
(2) Koenig, H.: J. Histochem. Cytochem. *11,* 556 (1963).
(3) Koenig, H.: J. Histochem. Cytochem. *11,* 120 (1963).
(4) Koenig, H.: J. Cell Biol. *19,* 87A (1963).
(5) Koenig, H.: J. Histochem. Cytochem. *13,* 20 (1965).
(6) Koenig, H.: *In*: Response of the Nervous System to Ionizing Radiation, pp. 403-417 (Haley & Snider, edit.) Boston, Little, Brown & Co. (1964).

* Supported by grants from NIH, AEC, and the National Multiple Sclerosis Society.

The Histochemistry of Metabolic Processes of the Cardiac Conduction System in Health and in Certain Pathological States

GORNAK, K. A. (Institute of Human Morphology, Moscow, USSR)

Apart from their functional and morphological peculiarities, the cardiac conduction system fibers also have some tinctorial and histochemical features which suggest that the metabolic processes occurring in them proceed in a special way.

The Purkinje fibers contain more glycogen than the working cardiac musculature. This glycogen yields to hydrolysis poorly and is not identical to the glycogens of other organs as far as physical and chemical properties are concerned. Also, these fibers differ from the working ventricular musculature in the content of proteins, lipids, and in enzyme activity.

Metabolic processes in the conduction system fibers proceed not only in a special way, but also at a slower rate than in the working musculature of the heart. This can be explained readily in physiological and biochemical terms. Whereas the working musculature, which performs great mechanical work, requires a tremendous amount of energy and its formation occupies a central place in the metabolism, no such quantity of energy is needed for the conduction system, and production of the requisite amount of ATP is provided for by a lesser number of mitochondria. Hence, the lesser activity of succinic dehydrogenase and cytochrome oxidase in Purkinje fibers.

In experimental myocardial infarction in dogs, those conduction system fibers situated within the infarcted zone not only failed to necrotize, but, in a vast majority of cases, even retained their glycogen, enzymatic activity, and usual tinctorial properties. As myocardial infarction proceeds, intact Purkinje fibers showed increased amounts of glycogen, increased activity of oxidative-reductive enzymes, and signs of hypertrophy which is to be regarded as a compensatory-adaptive process.

Intact fibers of the Purkinje system with differing levels of metabolic processes were observed in the wall of an experimental cardiac aneurism (acute, subacute, and chronic).

In experiments on rabbits it was shown that the conduction system fibers are fairly resistant, not only to local but also to general hypoxia induced by acute or chronic blood loss.

Die Wirkung von Milzextrakten auf das Enzymmuster des infantilen Meerschweinchenhodens

GOSLAR, H. G., P. GRIGORIADIS, and K. H. JAEGER (Anatomisches Institut der Universität Bonn, Bonn, Deutschland)

Extrakte bzw. Dialysate von Kälberwilz bewirken an den LEYDIG-zellen des infantilen Meerschweinchenhodens neben einer Grössenzunahme eine Aktivitätserhöhung von NADH-Cytochrom-c-Reduktase, Lactatdehydrogenase, Glukose-6-phosphatdehydrogenase und Steroid-3β-ol-Dehydrogenase. Das Milzdialysat war debei wirksamer als der Milzextrakt. Die Aktivitätssteigerung des ersteren kommt etwa der nach Serumkonadotropin beobachteten gleich, während die Grössenzunahme des LEYDIGzellkomplexes hinter der durch Serumgonadotropin bewirkten zurückbleibt.

Histochemical Investigations on Hepatic Carcinogenesis *

GÖSSNER, W. and H. BENOIT (Gesellschaft für Strahlenforschung, Institut für Biologie, Abteilung für Allgemeine und Experimentelle Pathologie, München, Germany)

Histochemical changes in rat liver parenchyma during the early stages of carcinogenesis have been studied. Wistar female rats have been fed with diethylnitrosamine (5 mg/kg/day) and N-nitrosomorpholine (8 mg/kg/day). Already 3-6 weeks after beginning of the treatment with the carcinogen, islands of liver cells can be found in which glucose-6-phosphatase has completely disappeared (1). In addition, these cells show a marked vacuolisation of their cytoplasm due to accumulation of glycogen which cannot be mobilised by starvation. About 12 weeks after beginning of the treatment, these cells lose their glycogen and develop an increased basophilia of their cytoplasm. This stage probably represents the transformation into the final tumor. The focal disappearance of glucose-6-phosphatase in combination with glycogen storage is among the first histochemical modifications occurring in sites of neoplastic transformation. In addition, a variety of enzyme histochemical methods (oxidoreductases, hydrolases, phosphorylases etc.) have been applied in an attempt to characterize further the enzyme pattern of these preneoplastic islands in comparison with normal liver tissue and the final hepatocarcinoma.

References:

(1) Gössner, W. and H. Friedrich-Freksa: Z. Naturforschg. *19b,* 862 (1964).

---

* Part of this work was supported by the Association Contract European Atomic Energy Community (EURATOM) and the Gesellschaft für Strahlenforschung mbH (GSF).

Radioautography of DNA Synthesis in Estimation of Proliferative Activity of Rat Brain Subependymal Cells

GRACHEVA, NINA D. (Central Research Roentgeno-radiological Institute, Ministry of Health of the USSR, Leningrad, USSR)

The subependymal zone of the lateral ventricles is represented by cells which apparently take part in the renewal of the rat brain glial population. Using H^3-thymidine and radioautography, we determined the parameters of the subependymal cell generation cycle by the curve of labelled mitoses as a function of time (T = 21 hrs, t_s = 9.8 hrs t_{G2} = 3 hrs, $t_M + t_{G1}$ = 8.2 hrs; labelling index 8% and mitotic index 0.6%). Proliferative pools calculated from these data amounted to 17%. The "saturation" with H^3-thymidine (repeated injections) resulted in labelling of 18%-26% of the subependymal cell nuclei. This indicates that the subependymal cells present a heterogenous population. With a view of studying the biological properties of the subependymal zone, the experiments were carried out with total body x-rays irradiation in the dose 300 r. delivered 80 minutes after an injection of H^3-thymidine. At the moment of the irradiation, 90% of the labelled cells were in S-phase and only about 10% of the labelled population moved to the G_2-phase. The mortality of the labelled and unlabelled population was studied. Interphase death of the labelled cells began in 2 hours and reached its maximum (89%) by 7 hours postirradiation. As follows from the data,

84

subependymal cells irradiated in the S-phase were highly sensitive to injury. Only about 0.4% labelled cells from all the initial population can take part in the reparation directed to the restitution of the total balance of glial cells. The share of the lost cells in the unlabelled population was about 32%, i.e., apparently exceeded the proliferative pool, which testifies to the possible death of the cells that ceased reproduction. These results will be discussed in relation to the findings of other workers.

Chemodifferentiation and Onset of Function

GRILLO, T. ADESANYA IGE (University of Ibadan, Ibadan, Nigeria)

Endocrine glands afford the histochemist and developmental biologist a unique opportunity to study chemodifferentiation in relation to the onset of function. The cytological characteristics of the cells of endocrine glands, i.e., their characteristic granules which have distinct shapes and tintorial affinities for histological stains, offer markers for microscopic studies. The histochemistry of the constituent parts of hormones may indicate too the assembly of polypeptides, proteins and steroids of hormones. Fluorescent antibody histochemical methods have recently made the identification of specific hormones easier.

While the identification of the hormones in embryonic endocrine glands has become precise, it has not been easy to define the onset of secretion of some hormones. Sensitive radioisotope immunochemical assay methods now make possible the detection of minute quantities of some hormones in foetal blood; but these methods have only been employed in determining the secretion of relatively few hormones.

More commonly, biological changes in the great organs of hormones have been used to mark the onset of secretion. These biological changes may be physiological or chemical. The chemical changes within cells of target organs afford developmental biologists immense opportunities for studies in chemodifferentiation. For example, it is now clear that some enzymes may exist in more than one form in a tissue. An enzyme may remain in an inactive form only to be changed into its active form by a hormone (e.g., inactive phosphorylase being converted by glucagon, in the presence of cyclic $3',5'$ adenosine monophosphate, to active phosphorylase). Among the problems which still remain to be solved are whether the chemodifferentiation enzyme forms of cells of target organs always completely depend on the secretion of specific hormones.

These problems will be discussed in relation to secretion of endocrine pancreatic secretions and the development of phosphorylase and uridine diphosphate glucose glycogen synthetase in embryonic tissues.

References:

(1) Grillo, T. Adesanya I, *et al.:* Gen. and Comp. Endocrinol. *4*, 446 1964).
(2) Grillo, T. Adesanya I: J. Endocrin. *36*, 151 (1966).
(3) Ellis, E. T., *et al.:* J. Path. Bac. *92*, 179 (1966).
(4) Grillo, T. Adesanya I, *et al.:* J. Endocr. *39*, 307 (1967).

The Development of the Freemartin Gonad in Cattle as Revealed by the Use of the Alkaline Phosphatase Reaction

GROPP, A. (Pathologisches Institut der Universität Bonn/Rhein-Venusberg, Germany)

In heterosexual chimeric twins, the female twin shows, nearly invariably, a more or less pronounced masculinization—a phenomenon called freemartinism.

The histogenesis of the freemartin gonad was hitherto explained, either under the aspect of a failure of secondary sex cord formation (Willier), or of a regression of secondary sex cords, after their formation (Bissonnette).

In the cattle ovary, primordial germ cells and the gonadal blastema give a positive alkaline phosphatase reaction (Gropp and Ohno). The labelling of these constituents by the alkaline phosphatase technique allows analysis of their fate in a series of embryonic freemartin gonads. The results have to be discussed in terms of a regression and degeneration of already developed cortical ovarian areas. The regeneration of tubular structures proceeds from the rete and closely neighbouring undifferentiated central areas of the gonadal blastema.

References:

(1) Bissonnette, Th. H.: Amer. J. Anat. *42*, 29 (1928).
(2) Gropp, A. and S. Ohno: Z. Zellforsch. *74*, 505 (1966).
(3) Willier, B. H.: J. Exp. Zool. *33*, 63 (1921).

Naphthlacetate Hydrolyzing Microsomal Isoenzymes of Rat Liver After Application of Hydrocortisoneacetate and Phenobarbital

GROSS, U. M. (Pathologisches Institut der Freien Universität Berlin, Berlin)

Naphthylacetate hydrolyzing microsomal esterases of rat liver are augmented after a 7 days treatment of animals with hydrocortisoneacetate whereas the microsomal glucose-6-phosphatase is diminished. Animals that are treated daily with phenobarbital show augmentation of naphthylacetate esterases and diminution of glucose-6-phosphatase. Significant differences between controls and treated animals can not be seen in histochemical analysis of rat liver (same substrate, coupling agent hexazotized pararosaniline). On the other hand, in isoenzyme patterns (electrophoretical fractionation) of naphthylacetate hydrolyzing enzymes, qualitative and quantitative changes are observed. The details will be presented and discussed.

The Analogy of Amyloid and Keratin as Suggested by X-ray, Amino Acid and Ultrastructural Analysis: Redefinition of Amyloidosis as a Dysplastic Mesenchymal or Epithelial Keratosis

GUEFT, BORIS (Albert Einstein College of Medicine, Bronx, N. Y.)

The X-ray diffraction studies of amyloid by Ashkenazi, Gafni, Sohar and Heller (1) showing 4 Å and 9 Å haloes, 4.50 Å and other equatorial arcs and a fiber axis repeat of 46 Å have been confirmed. It is pointed out that this pattern resembles that of feather keratin and has some features of beta proteins. Both proteins resist digestion by pepsin and trypsin and are insoluble in water and organic solvents. The amino acid analyses (e.g., wool root protein (2) and amyloid) have neither hydroxyproline nor very high values of combined glycine, alanine and serine, facts which exclude fibroins and collagen. Negative contrast electron micro graphs of dispersed amyloid, feather keratin (3), and wool protein (4) show fibril diameters and long helical periods within 50% agreement Both contain twisting filaments and protofilaments of almost identical width with protofilament widths in the 10 Å range. The fibril of each protein seems to be a coiled coil, and the Mercer terminology for keratin-like proteins applies to the subdivisions of both without exception. Reconsideration of these data leads to the suggestion that amyloidosis may be considered a type of keratotic dysplasia of mesenchymal cells in which the excessively produced protein is combined with polysaccharide, resulting in a specific staining reaction that is not found in the usual keratins. Epithelial amyloid deposits may also be explained by this hypothesis.

References:

(1) Ashkenazi, Y., C. Hersko, J. Gafni, E. Sohar, and H. Heller: Israel J. Med. Sci. *3*, 569 (1967).
(2) Mercer, E. M.: Keratin and Keratinization, p. 237, Pergamon Press, N.Y., (1961).
(3) Filshie, B. K., R. D. B. Fraser, T. P. MacRae, and G. E. Rogers: Biochem. J. *92*, 19 (1964).
(4) Dobb, M. G.: Nature, *207*, 293 (1965).

Glycogen, Glycogen Synthetase and Phosphorylase in Hamster Testis

GUHA, S. and J. P. FOUQUET (Institut d'Histochimie, Faculté de Médecine et Laboratoire de Physiologie Cellulaire, Faculté des Sciences, Paris, France)

Hamster testis, during the period of normal spermatogenesis (April to October), contains a considerable amount of glycogen (150-200 mg/100 g. of testis) in the seminiferous tubules. Testicular glycogen is present mainly in the Sertoli cells and spermatogonia of the tubules in stages IV through VIII (stages defined according to Leblond and Clermont '52). Some glycogen grains are also present in the lumen of the tubules, particularly at stage VII.

Phosphorylase activity is present in the tubules containing glycogen and in another set of tubules (probably stages I through III), while glycogen synthetase is present only in the tubules containing glycogen. Glycogen grains in the lumen also show phosphorylase and glycogen synthetase activity.

In new born hamster, glycogen is present mainly in the tunica albu-ginea. Gross grains of glycogen are found in sex-cords (1-4 grains per cross section of a cord). Glycogen in sex-cords and in tunica albuginea disappears progressively. In two week old animals, glycogen totally disappears from this tissue. Glycogen reappears in the tubules with the formation of lumen and with apparition of spermatocytes I (usually 3-4 week old animals). Glycogen increases slowly during the formation of spermatocytes II and immature spermatids (4-5 week). Assay shows about 50 mg of glycogen per 100 g. of testicular weight. With the advent of maturity, glycogen distribution becomes similar to that described for adult testis.

Glycogen synthetase activity appears at about the fourth week of development and increases progressively with sexual maturation. This activity is present only in the tubules containing glycogen. In all the cases this activity is much weaker than that of phosphorylase.

Phosphorylase activity in new born hamster is present in the cytoplasmic mass occupying the center of the sex-cord and also in the tunica albuginea. This activity remains more or less constant for first two weeks of development but increases from the third week. Most intense reaction is obtained in young mature animals and declines rapidly in animals older than three months.

Rete testis could not show any reaction for glycogen and glycogen synthetase activity. Phosphorylase, though variable, yet is present in rete testis epithelium of immature and mature animals. Occasionally, in mature animals, phosphorylase activity is observed in the lumen of rete testis.

In spite of certain correlations between glycogen synthesis on spermatogenesis, these two phenomena appear to develop independantly of each other, since some animals showing abnormal delay in spermatogenesis, show normal glycogen and enzyme content.

A Tissue Model for the Histochemical Demonstration of Catecholamines, Indoleamines, and Histamine *

HÅKANSON, R. and CH. OWMAN (Departments of Pharmacology and Anatomy, University of Lund, Sweden)

The serotonin-containing enterochromaffin cells have for long been considered the only important amine-storing cell system in the gastric mucosa. With regard to the high activity of aromatic amino acid decarboxylase, histidine decarboxylase, and monoamine oxidase, the distribution of which in the rat stomach did not correspond with that of the enterochromaffin cells, there was strong reason to believe that other cell systems were involved in gastric amine metabolism (Håkanson and Oman, 1966). This was confirmed by the finding that administration of L-dopa or L-5-hydroxytryptophan resulted in the formation of large amounts of the respective amines in the stomach wall (Håkanson and Owman, 1966). Histochemically, the amines were found to be stored in a system of epithelial cells that were numerous in the mucosa of the parietal cell region of the rat stomach (Håkanson, Lilja, and Owman, 1967). The morphology of the cells resembled that of the serotonin-storing enterochromaffin cells, mainly located in the pylorus; the cells were for that reason designated as "enterochromaffin-like" cells. It should be noted, however, that these cells are devoid of histochemically demonstrable monoamines in non-treated animals. On the other hand, using a recently developed fluorescence microscopic method for visualization of cellular histamine (Ehinger and Thunberg, 1967; Håkanson and Owman, 1967), it could be established that the enterochromaffin-like cells are the major storage site of gastric histamine in the rat (Håkanson and Owman, 1967; Håkanson, Owman, and Sjöberg, 1967). Only a minor portion is found in mast cells.

The capacity to produce and store monoamines, which is characteristic of the enterochromaffin-like cells, may suggest that physiologically important monoaminergic mechanisms operate in these cells, and that the experimentally demonstrated formation of dopamine and serotonin in these cells "mimic" such a mechanism; this probably involves some other monoamine, which cannot be visualized with the histochemical techniques available at present.

The properties of this new cell system offer unique opportunities for studies on the requirements for the histochemical demonstration of a variety of biogenic amines. We are concerned with a tissue model in which the amines are produced from any exogenous precursor amino acid chosen and subsequently stored by physiological mechanisms in a specific cell system with a very rich distribution.

References:

(1) Ehinger, B. and R. Thunberg: Exp. Cell. Res. *47*, 116 (1967).
(2) Håkanson, R., B. Lilja, and Ch. Owman: Europ. J. Pharmacol. *1*, 188 (1967).
(3) Håkanson, R. and Ch. Owman: Biochem. Pharmacol. *15*, 489 (1966).
(4) Håkanson, R. and Ch. Owman: Life Sci. *6*, 759 (1967).
(5) Håkanson, R., Ch. Owman and N.-O. Sjöberg: Life Sci. *6*, 2535 (1967).

---

* Supported by the Swedish Medical Research Council (grant No. K68-12X-1007-03).

Histochemical Demonstration of Histamine in Tissues by Fluorescence Microscopy *

HÅKANSON, R., CH. OWMAN, and B. SPORRONG (Departs. of Pharmacology, Anatomy and Histology, University of Lund, Sweden)

Recently, a histochemical method was devised simultaneously by Ehinger and Thunberg (1967) and Håkanson and Owman (1967) allowing the cellular localization of histamine, not only in mast cells but also in certain other cell systems. The method is based on the well-known reaction between *o*-phthaldialdehyde (OPT) and histamine which results in the formation of a fluorescent conjugate, and which was devised for the fluorometric quantitation of histamine (Shore, Burkhalter, and Cohn, 1959).

In the histochemical procedure, freeze-dried material is exposed to gaseous OPT, and the fluorophore is subsequently made visible by slight humidification. The method has been successfully applied to fresh cryostat sections as well as to freeze-dried and paraffin-embedded material (Håkanson, Owman, and Sporrong, 1968). Under the conditions used, the condensation product emits an intense blue, or sometimes, yellow fluorescence.

Histamine has been demonstrated with the present fluorescence microscopic method in mast cells from a variety of tissues in several species, including man. The amine is also well demonstrable in pathological material rich in mast cells, such as keloids and canine mastocytomas. However, the methodological precision achieved even permits the visualization of histamine in cell systems other than mast cells. For example, a fluorescence can be induced with OPT in numerous epithelial cells in the acid secreting region of the rat stomach; these cells morphologically resemble enterochromaffin cells in the pylorus, but contain no 5-hydroxy-

90

tryptamine (Håkanson and Owman, 1967). Further, an OPT-induced fluorescence has been detected in hitherto unidentified cell systems in the anterior pituitary and in the endocrine pancreas.

Results of model experiments suggest a high degree of specificity for the histochemical OPT method, but until the chemical basis for the reaction has been fully explored, it is necessary to combine histochemical analysis of different tissues with parallel quantitative determination of histamine (Håkanson, Owman, and Sjöberg, 1967).

References:

(1) Ehinger, B. and R. Thunberg: Exp. Cell. Res. *47*, 116 (1967).
(2) Håkanson, R. and Ch. Owman: Life Sci. *6*, 759 (1967).
(3) Håkanson, R., Ch. Owman, and N.-O. Sjöberg: Life. *6*, 2535 (1967).
(4) Håkanson, R., Ch. Owman, and B. Sporrong: J. Histochem. Cytochem *To be published* (1968).
(5) Shore, P. A., A. Burkhalter, and V. H. Cohn: J. Pharmacol. Exp. Ther *127*, 182 (1959).

* Supported by grant No. K68-12X-1007-03 from the Swedish Medical Re search Council, and from the Association for the Aid of Crippled Children, New York.

Elemental Analysis by the Electron Microprobe

HALE, A. J. (Searle Research Laboratories, High Wycombe, Bucks., U.K.)

The scanning electron microprobe permits identification, microscopic localisation and relative quantitation of many elements in cells or tissues. The microscopic resolution is around one micron and the detection sensitivity about 10^{-15}g. With properly prepared tissues the analytical method is non-destructive and all elements down to magnesium in the periodic table can be readily demonstrated. (Cosslett, 1962)

Endogenous elements and those deposited in the tissue by "staining" reactions can be studied. Sequential or simultaneous detection of different elements in the same locus permits more accurate interpretation of the quantitative and microscopic data concerning these elements. (Hall, Hale & Switsur, 1966; Hale, Hall & Curran, 1966.)

Problems of the theory and practice of qualitative and quantitative analysis of sections of material, biological or otherwise, which can be considered to be infinitely thin from the point of view of electron penetration, have been considered and the general theory of the thin film method of quantitative analysis (Marshall & Hall, 1966) will be discussed. The significance of the results obtained is greater than a simple demonstration that particular elements exist at certain loci. Complex relations of the elementary composition at a given site can be correlated with molecular content as demonstrated by cytochemical reactions visualised in the light or electron microscopes.

References:
(1) Cosslett, V. E.: Ann. N.Y. Acad. Sci. *97*, 464 (1962).
(2) Hale, A. J., T. A. Hall, and R. C. Curran: *In:* X-ray Optics and Micro-analysis. pp. 686-690 (Castaing et. al, edit.) Paris: Hermann (1966).
(3) Hall, T. A., A. J. Hale, and V. R. Switsur: *In:* The Electron Microprobe, pp. 805-833, (McKinley et al., edit.) New York: Wiley (1966).
(4) Marshall, D. J. and T. A. Hall: *In:* X-ray Optics and Microanalysis, pp. 374-381 (Castaing et al.; edit.) Paris: Hermann (1966).

An Electron-probe X-ray Microanalyser for Biological Applications

HALE, A. J., D. J. MARSHALL and V. R. SWITSUR (G. D. Searle and Co., Ltd., High Wycombe, and The Cavendish Laboratory, Cambridge, England)

The requirements of the biologist using a microanalyser are somewhat different from those of other users. Because of this, we have constructed an instrument with features not generally found in commercial micro-analysers. It has fixed instead of variable-angle spectrometers, a counter for continuous X-rays and a specimen stage for thin tissue sections. These will be described, and some examples given of the work being done with this instrument.

Histochemical Demonstration of Induction of Mitochondrial Oxidative Enzymes by the Carcinogens 3, 4-Benzpyrene and 3-Methylcholanthrene

HANKER, JACOB S., NICOLAS ZENKER, YOSHIHISA MORI-ZONO, CHANDICHARAN DEB, and ARNOLD M. SELIGMAN (Departments of Surgery, Sinai Hospital of Baltimore, Johns Hopkins University School of Medicine, and Department of Pharmaceutical Chemistry, University of Maryland School of Pharmacy, Baltimore, Md.)

The carcinogens 3,4-benzpyrene and 3-methylcholanthrene have been shown to induce the activity of enzymes involved in rat liver *microsomal* electron transport. Since the mitochondrion has its own capability for the production of protein enzymes, we decided to investigate whether certain carcinogens stimulate *mitochondrial* oxidative enzymes. Liver sections from rats injected with 3,4-benzpyrene or 3-methylcholanthrene when incubated in media specific for the histochemical demonstration of mitochondrial dehydrogenases or cytochrome oxidase showed enhanced activity of several of these enzymes over sections from control rats. These observations were made on 6μ fresh frozen sections of rat liver treated with the appropriate substrate and nitro-BT for dehydrogenase activity and the polymerizing reagent, diaminobenzidine for cytochrome oxidase activity. The rats were injected with the carcinogens in sesame oil intraperitoneally for 2 days and killed 19 hours after the last injection. The biggest increases in activity were seen in β-hydroxybutyric dehydrogenase, DPNH diaphorase and cytochrome oxidase. Little, if any, increase was seen in α-glycerophosphate, malic, or succinic dehydrogenase activity. These observations were confirmed by comparing the staining of mitochondria isolated by differential centrifugation from control and treated rats. Further confirmation of the validity of the histochemically observed induction was obtained by comparison of the rates of oxygen consumption by mitochondria isolated from equal weights of livers of the control and carcinogen treated rats using β-hydroxybutyrate or succinate as substrate in the presence of phosphate and phosphate acceptor, adenosine diphosphate, as measured polarographically with a Clark oxygen electrode. An inhibitor of protein synthesis, actinomycin D, diminished the increase in activity caused by the simultaneous administration of benzpyrene or methylcholanthrene providing evidence that the stimulation of activity was due to *de novo* protein synthesis, generally called induction. A difference in profile of enzyme stimulation by various agents was observed; this indicates that the changes observed in mitochondrial oxidative enzymes are not due to a generalized toxicity phenomenon.

Some Enzyme Histochemical Observations on the Zona Glomerulosa of the Rat Adrenal Cortex Under Different Experimental Conditions

HARDONK, M. J., J. D. ELEMA and JOH. KOUDSTAAL (Pathological Anatomical Laboratory, Groninger, The Netherlands)

The behaviour of glucose-6-phosphate dehydrogenase (G6PDH), isocitrate dehydrogenase (IDH), succinate dehydrogenase (SDH) and 3β-ol hydroxysteroid dehydrogenase was investigated by histochemical techniques in some experiments in which alterations in electrolyte balance had been induced. The distribution of fat was investigated by the Oil-red O technique.

Acute hyponatraemia, induced by peritoneal dialysis with glucose 5% gave an increased activity of G6PDH and IDH, detectable after 24 hours and of 3β-ol hydroxysteroid dehydrogenase after 48 hours. There was complete fat depletion of the zona glomerulosa after this interval. The SDH did not show any reaction under these circumstances.

Bilateral nephrectomy induced principally the same changes in enzyme activity after 24 hours except the 3β-ol hydroxysteroid dehydrogenase which showed some increase after an interval of 36 hours. A combination of these two procedures largely inhibited these changes when investigated 24 hours after starting the experiment. Fat depletion of the zona glomerulosa did occur. Intraperitoneal injections of KCl induced an increase of activity of the G6PDH, IDH and 3β-ol hydroxysteroid dehydrogenase, clearly visible after 48 hours.

The possible significance of these enzymes in steroid synthesis and the possible way in which their increase in activity is induced is briefly discussed.

Steroid-Producing Cells in the Human Ovary and in Ovarian Tumours

HARDONK, M. J. and JOH. KOUDSTAAL (Department of Pathology, Groningen, The Netherlands)

Steroid-producing cells have been studied by a number of histochemical and enzyme-histochemical methods. The applied enzyme-histochemical reactions can be divided in three groups:

I. Enzymes directly related to steroidogenesis:
   3β-ol hydroxysteroid dehydrogenase
   Secondary alcohol dehydrogenase
II. Enzymes possibly related to steroidogenesis:
   NADH- and NADPH-tetrazolium reductases
   Glucose-6-phosphatase and isocitric acid dehydrogenase
   Non-specific esterases
III. Enzymes not related to steroidogenesis:
   Alkaline and Acid phosphatase; ATP-ase and 5-Nucleotidase; Aminopeptidase; Lactic acid and β-Hydroxybutyric acid dehydrodgenase; Succinic dehydrogenase and α-Glycerophosphate oxidase.

Theca cells of ovarian follicles, hilus cells, stromal theca cells, stroma theca tumor cells and the periepithelial endocrine active stroma thecal cells contain all enzymes of group I and III.

Generally the spindle stromal cells and the periepithelial endocrine active spindle stromal cells do not show activity for the enzymes of group I. In contrast to the other cells, thecal cells of ovarian follicles show a strong activity of the enzyme aminopeptidase. Before ovulation, granulosa cells do not contain enzymes of group I and II (in contrast to theca cells, a moderate activity of the enzyme α-glycerophosphate oxidase can be seen). After ovulation, a high activity of the enzyme 3β-ol hydroxysteroid dehydrogenase can be shown.

Further use of the enzyme histochemical methods will give a better understanding of the precise localization of steroidogenesis in the normal and abnormal ovary and in the various ovarian tumors. The mechanism of the induction of various steroid-producing cells will be discussed.

5′-Nucleotidase Isoenzymes in Rat and Mouse

HARDONK, M. J., JOH. KOUDSTAAL, and PH. J. HOEDEMAEKER (Pathological Anatomical Laboratory, Groningen, The Netherlands)

In several tissues of rat and mouse, two 5′-nucleotidases were found: one showing greatest activity at pH 5.0., while the other is most active at pH 7.0-7.5. The localization of these two enzymes is different. At the acid pH, the deoxyribonucleotides are dephosphorylated faster than the ribonucleotides; at the neutral pH, the ribonucleotides are hydrolysed more rapidly.

Electrophoretic analyses of homogenates and cell component fractions reveal the persence of 5′-nucleotidase isoenzymes in the investigated tissues. At the acid as well as the neutral pH, five isoenzymes were found. The different combinations of 5′-nucleotidase isoenzymes in the investigated tissues possibly indicate different functions of the isoenzymes.

In two different ways an approach was made to get information about the localization of the isoenzymes. In the first place, the postnatal development of some organs of the mouse was investigated. In liver, kidney and small intestine, a correlation could be found between the changes in the isoenzyme pattern and the alterations in the localization of 5′-nucleotidase activity. In the second place, attempts were made to isolate the two rat liver isoenzymes and to make antibodies against them. Results of the isolation and immunization procedures will be communicated. A report will be given about the preliminary results obtained with the fluorescent antibody technique. The possible signification of the localization of 5′-nucleotidase isoenzymes will be discussed in relation to the function of the isoenzymes.

An Azoindoxyl Method for the Cytochemical Demonstration of Acid Phosphatase Activity

HAYASHI, MASANDO (Beth Israel Hospital and Harvard Medical School, Boston, Mass.)

Unsubstituted indoxyl substrates have been considered useless for accurate cytochemical localization in the indigogenic method, and this has also been true of indoxyl phosphate in demonstrating acid phosphatase activity.

The present report will demonstrate that unsubstituted indoxyl phosphate is an excellent substrate for acid phosphatase activity when used in a simultaneous-coupling medium containing hexazonium pararosanilin. Frozen sections cut from rat tissues fixed in formal-calcium at 4° C for 24 hours were tested. Adequate staining for enzyme activity was obtained following incubation of sections at pH 5.0 in a medium containing 1 mM each of indoxyl phosphate and hexazonium pararosanilin and 0.05 M acetate buffer for 30 to 60 minutes at room temperature or for a shorter period at 37° C. Sodium fluoride caused almost complete inhibition at 1 mM. The technique appears to have several advantages over that based on the formation of indigo dye and provides more precise localization and faster deposition of reaction product. The reaction product is non-crystalline and has no appreciable affinity for lipids. Also, no significant diffusion of stain has been noted following prolonged incubation.

Leukocytenenzyme in Blutkonserven

HELLER, A. (Med. Universitätsklinik, Köln, Germany)

Gewöhnlich werden Blutkonserven am häufigsten mit einem Alter von 7-14 Tagen transfundiert, das Höchstalter liegt bei 21 Tagen. Untersucht wurden über die Zeit von 21 Tagen hinweg die Leukocyten von Zitratblutkonserven auf ihre Enzymaktivität und ihren Dehydrogenasengehalt hin. Dabei fand sich zunächst bei frisch entnommenem Zitratblut eine normale Fermentaktivität aller untersuchter Enzyme. Bereits aber nach 4 Tagen war schon eine deutliche Aktivitätsabnahme vor allem von Leucinaminopeptidase und Adenosin-5-Triphosphornucleotidase zu verzeichnen, während z.B. die Cytochromoxydase nur eine geringe Abschwächung zeigte. Nach 7 Tagen fand sich bei allen Fermenten ein erheblicher Schwund der Aktivität, so daß z.B. der Score-Index der Alkalischen Phosphatase zu diesem Zeitpunkt nur noch 3 betrug. Wiederum war die Cytochromoxydase noch am geringsten von einem Aktivitätsschwund betroffen. Nach 14 Tagen war eine starke Abnahme aller Fermentaktivitäten zu verzeichnen und zwar nun auch der bislang ziemlich stabilen Cytochromoxydase, daneben war die Alkalische Phosphatase nicht mehr nachweisbar. Nach 20 Tagen waren bis auf einige Dehydrogenasen, die aber teilweise wie z.B. die Succinodehydrogenase ebenfalls keine wesentliche Aktivität mehr zeigten, praktisch sämtliche übrigen Enzyme nicht mehr nachweisbar. Die Untersuchungen zeigen, daß der Ablauf der Glykolyse in den Leukocyten bis über das Verfallsdatum einer Blutkonserve hinaus bei einer gewissen Anzahl der Zellen zwar noch aufrecht erhalten wird, dies geschieht aber wenn überhaupt zeimlich bald unter anaeroben Bedingungen, was die Funktionstüchtigkeit dieser Zellen doch stark herabsetzen dürfte. Eine weitere Ursache dafür dürfte sicher im Sistieren des Zitronensäurezyklus zu sehen sein, was zwar zu einem etwas späteren Zeitpunkt eintritt als die anaerobe Glykolyse, dann allerdings bei fast allen Leukocyten zum gleichen Zeitpunkt. Die Fähigkeit der Granulocyten Ausbreitungsformen zu bilden wie wir sie beobachten konnten—etwa in Anlehnung an die bekannte Thrombocytenausbreitung—ging bei einer Alterung des Konservenblutes zuerst verloren, so daß diese Fähigkeit zur Ausbreitung als sehr empfindliches Kriterium zur Funktionstüchtigkeit der Granulocyten angesehen werden kann. Dieses Phaenomen zeigt darüber hinaus allerdings ein unterschiedliches Verhalten bei verschiedenen hämatologischen Erkrankungen. Die Ergebnisse dieser Untersuchung machen deutlich, daß Blutkonserven, deren Alter mehr als 3-4 Tage betragen, nicht mehr zu einer Transfusion verwandt werden sollten, wenn auf eine gewisse Funktionstüchtigkeit der Leukocyten wertegelegt wird, denn andernfalls werden, was die Leukocyten anbetrifft, nur noch Zell-Leichen transfundiert.

Histochemistry of Experimentally Induced Carcinogenesis in Rat Liver

HERNÁNDEZ, F. and J. MARTINEZ DE MORENTIN (Department of Histopathology (CIB Felix Huarte), Univ. of Navarra, Pamplona, Spain)

The carcinogen diethylnitrosamine (DENA) was administered orally to Wistar rats during 22 weeks (5 mg/Kg body wt./day). During this period 3 treated rats and one control were sacrificed each week. Two portions of each liver were removed. One was frozen in liquid N_2 and stored at $-30°$ C and the second portion was fixed for 24 hours at $40°$ C in formol-calcium and then stored in sucrose-gum acacia. 10 μ sections were made on the cryostat and a histochemical study of the following enzymes was made: alkaline phosphatase, acid phosphatase, ubiquinone, esterase, β-hydroxybutyrate dehydrogenase, succinate dehydrogenase, DPN diaphorase, TPN diaphorase, lactate dehydrogenase, glutamate dehydrogenase, malate dehydrogenase and α-glycerophosphate dehydrogenase. Nodules of cancerous cells formed around the 18th week. The most important histochemical observations were: from the 8th week on, acid phosphatase was diffuse and less intense in some areas; alkaline phosphatase was positive in the bile canaliculi from the 15th week on; and ubiquinone was less intense in the tumor nodules. Malate dehydrogenase was completely negative only during the first 14 weeks.

On the basis of these results the role of lysosomes in cancer pathogenesis is discussed.

Sectioning of Undecalcified Bone for Histochemical Investigations

HERRMANN, HANS-JÜRGEN (Institut für Vergleichende Pathologie der Deutschen Akademie der Wissenschaften zu Berlin, 1136 Berlin)

After examining 21 domestic animals of various species and ages it was found that, with the aid of Mikrotom K (Messrs. Jung, Heidelberg), it was possible to obtain good sections of the compacta in the native state a certain length of time after death without previously having to freeze them. The duration of this period is inversely proportional to the animal's age and directly proportional to the moisture content of the bone and depth of temperature at which the specimen had been preserved. Suitable sections are obtainable after $2\frac{1}{2}$ months if preserved at a temperature of $-18°$ C or kept on dry ice. Using this cutting technique, it is possible to subject bone to histochemical examination a few minutes after death. Further, it is possible to estimate the influence of decalcification on the results of procedures for some fermentative histochemical reactions. With regard to its native condition, the bone section corresponds to tissue as are used for biochemical examinations. It is possible, under certain conditions, to determine the thickness of these segments with an adequate degree of precision by using mechanical means. Experience gathered in the preparation of histological sections from undecalcified spongiosa embedded in paraffin or polyethylene glycol is discussed.

Quantitative Histochemistry of Human Skin

HERSHEY, FALLS B. (St. Louis, Mo.)

Fluorometric micro-methods of Lowry have permitted quantitative analysis of enzymes not previously detected in the skin and comparison of their activity in various layers of human skin and skin appendages. Skin is frozen with liquid N_2, sectioned in the frozen state and dried. 0.5 to 5.0 microgram samples of epidermis, dermis hair follicles, sweat and sebaceous glands are weighed and analyzed. In human epidermal homogenates various Krebs cycle and glycolytic enzymes have approximately $\frac{1}{5}$ the activity of brain homogenates. The enzymes and isozymes of epidermis have characteristics differing somewhat from other tissues. For example, starch gel electrophoresis reveals a sixth isozyme of LDH which is associated with epidermal protein. This migrates faster than the five LDH isozymes found in all other tissues and shows chemical and physical properties of isozymes comprised of both types of LDH subunits.

Beta hydroxyl-acyl CoA dehydrogenase, one of the key enzymes linking carbohydrate and fatty acid metabolism, has been found in human epidermis and skin appendages with similar distribution to ICDH and G6PDH. These three enzymes are most active in sebaceous glands, where the NADPH produced is most likely the hydrogen donor for fatty acid synthesis. Aspartate and alanine transaminases linking carbohydrate and amino-acid metabolism, are also most active in sebaceous glands. The relative activities of sebaceous gland and epidermis are 3.5:1, 3:1, 2:1 for transaminases ICDH and B hychroxyacyl CoA dehydrogenase. Alanine transaminase increases 100% in the external sheath of active hair follicles.

Glutamic dehydrogenase in skin and skin appendages resembled the distribution of LDH and MDH. Epidermal GDH differs from liver and brain GDH in its Michaelis constants and other characteristics. It is most active in epidermis and in sweat glands, and unless the urea cycle enzymes are found in skin, GDH appears to be most important in control of ammonium ion in skin.

Further refinements of microtechniques by Lowry, including enzyme cycling oil well reaction chambers, etc., permit substrate analysis and enzyme analysis of even smaller samples and more precise histological localization.

Quantitative Chemical Neurohistology *

HESS, H. H., A. POPE, and N. H. BASS (McLean Hospital, Belmont, Mass. and Harvard Medical School, Boston, Mass.)

Quantitative chemical neurohistology substitutes chemical for morpho-logical measurements of histological and cytological components in nervous tissue. At the cellular level, it attempts to enumerate the cells present as well as to identify and describe a cell species as a morpho-logical unit differing qualitatively and quantitatively from other cell species in chemical and enzymic constitution. Cell enumeration *in situ* is most readily accomplished by combining DNA assays as an estimate of total cells per unit weight with differential cell counts on analogous fixed, stained tissue to determine both relative and absolute numbers of neurons and glia.

At the subcellular level, quantitative chemical neurohistology is con-cerned with identification and evaluation of chemical components of ratios of components as indices of organelles and other membrane struc-tures of cells. Establishing and testing of indices is approached both by the study of isolated cells and cell fractions and at a higher level of complexity by quantitative analysis of microslices consisting of selected cell populations from brain regions having well-known differences in proportions of histological elements. Information from all these sources should form an intercorrelated matrix. Quantitation at the subcellular level is achieved with varying degrees of precision by use of biochemical indices correlated with cytological features such as: (a) myelin or myelinated fibers (cerebrosides); (b) neuron plasma membranes and synapses (ganglioside sialic acid); (c) mitochondria (cytochrome oxidase); and (d) rough endoplasmic reticulum (RNA). With extension of cellular and subcellular quantitation over a significant spatial interval in a brain region, a composite picture is constructed of the normal histology and metabolism for comparison with possible changes in functional and pathological states.

Quantitative chemical histology has been used to analyze normal neuro-anatomical regions and to make phylogenetic comparisons between rat and human cerebral cortex (1). (a) The myeloarchitecture of both cortices was well-outlined by the intracortical distribution of cerebrosides. In the lightly myelinated human frontal secondary association area (Brodmann area 9), cerebroside concentrations were low in the upper layers and rose 6-7 fold through the deeper layers to white matter (2). In rat somatosensory area—the most highly myelinated region of rat cortex—values were higher throughout the laminae, but not in white matter, as compared with human brain. (b) The intra-cortical distributions of neuronal membranes, as depicted by ganglioside sialic acid (3), showed higher concentrations per unit dry weight in upper but not lower layers of human as compared with rat cortex. Expressed in terms of numbers of neurons present, gangliosides were 2-6 times higher in human than in rat cortex. The results are consistent with a higher ratio of dendritic/neuron somal surface area and a higher proportion of axodendritic synapses in human cortex (c) Mitochondria, as indicated by the intracortical distribution of cytochrome oxidase activity were most abundant in the upper layers of both cortices, especially in layer III (4,5). The oxidative capacity per unit dry weight was ten fold higher in rat cortex, paralleling an approximately six fold difference in basal metabolism per unit body weight.

Future refinements in knowledge of the molecular architecture of membranes and membrane subunits, as well as nuclei and chromatin of nervous tissue cells may provide more precise chemical indices for ultrastructural components and ultimately for determining the number, volume, surface area and subcellular constituents of specific cell types.

References:

(1) Hess, H. H. and C. Thalheimer: J. Neurochem. *12*, 193 (1965).
(2) Lewin, E. and H. H. Hess: J. Neurochem. *12*, 213 (1965).
(3) Hess, H. H. and E. Rolde: J. Biol. Chem. *239*, 3215 (1964).
(4) Pope, A., H. H. Hess, J. R. Ware, and R. H. Thomson: J. Neurophysiol. *19*, 259 (1956).
(5) Hess, H. H. and A. Pope: J. Neurochem. *5*, 207 (1960).

* Aided by Grants from the National Institutes of Health: NB-00361 and in part by NB-007297.

β-Glucuronidase Activity in the Tumors of the Human Skin. A Histochemical and Biochemical Compared Study

HEWITT, J. M., M. GUIGON and J. BOLUBASZ (Laboratoire de Recherches Dermatologiques, Hopital Broca, Paris, France)

β-glucuronidase activity was studied in benign (warts, pigmented-cell naevi) and malignant (basal-cell epitheliomas, squamous-cell carcinomas and malignant melanomata) human cutaneous tumors by histochemical methods using as substrates 6-bromo-2-naphthol β-D-glucuronide (simultaneous coupling method) and naphthol as BI-β-D-glucuronide (Fishman and Goldman post-coupling method), the specificity of which was established by inhibition experiments.

The histochemical results showed (1) a rather strong activity in the stroma of epitheliomata, whereas the tumor itself is generally weakly active; (2) a quite variable activity in malignant melanomata contrasting with the rather constant and moderate activity of benign pigmented-cell naevi.

Parallel to the histochemical study, the enzymatic activity of β-glucuronidase was measured by spectrophotometric technique using a substrate phenolphtalein-β-D-glucuronide.

The biochemical values showed important variations of β-glucuronidase activity peculiarly in the groups of pigmented-cell tumors, but it did not appear such a significant difference as could be expected from the histochemical results between the benign naevi and the malignant melanomata groups.

Protein Content of "Active" and "Inactive" Nuclei

HIMES, M. H. and C. BURDICK (Biology Department, Brooklyn College, Brooklyn, N.Y.)

Nuclei of hepatocytes and erythrocytes of *Rana pipiens* liver may be considered relatively "active" and "inactive." The differences in morphology of the two types of nuclei, with dispersed and condensed chromatin, has been correlated with differences in nonhistone protein content. These nuclei were isolated in sucrose, separated by sucrose density gradient centrifugation, and subjected to differential extraction of histones and nonhistone proteins. The proteins were then analyzed electrophoretically for qualitative and quantitative differences. At each stage of extraction, the proteins remaining in the nuclei were estimated cytochemically using cytophotometric methods. Staining properties of isolated nuclei and nuclei frozen *in situ* were compared. Although the morphological differences characteristic of the two kinds of nuclei were unaltered by the isolation, their dye binding capacity was changed.

The isolated nuclei of both types possessed many more binding sites for dye (azure B and fast green). The way in which proteins are bound to DNA and RNA in the two types of nuclei were compared, by differential extraction of nucleic acids, in an effort to elucidate functional differences.

Oxidized Amine Dye and Hemoproteins

HIRAI, KEI-ICHI and HIDEO TAKAMATSU (Dept. of Cytochemistry, Chest Disease Res. Inst., Kyoto Univ., Kyoto, Japan)

In cytochemistry, 3,3'-diaminobenzidine (DAB) as an oxidizable substrate has been used to demonstrate peroxidase activity in tissue slices fixed with aldehydes (1). It has been found that enzymatic oxidation of DAB in the presence of H_2O_2 gives a reaction product of brown deposits in microbodies, and the product is sufficiently fixed with osmium tetroxide to a black substance which very electron-dense.

The present experiments on rat hepatic cells have revealed that imide-type DAB which has been oxidized with atmospheric oxygen gives sensitive microbody staining without H_2O_2. This reagent was also precipitated on mitochondrial cristae as well as in microbodies, when tissue slices were fixed with formaldehyde or glutaraldehyde and incubated for 3-6 hr in a saturated solution of oxidized DAB in 0.1 M Tris-HCl buffer (pH 7.2). This staining occurs anaerobically and even after dry heating for 30 min at $100°$ C, so it would seem to be not an enzymatic oxidation. Inhibitory studies showed that all reactions of hepatic cells were completely inhibited by 10^{-3} M KCN, 10^{-3} M NaN_3 or 10^{-4} M $Na_2S_2O_4$. When 3-amino,1,2,4-triazole, which is an irreversible inhibitor of catalase, was employed (2), the reaction in microbodies was counteracted specifically. Spectrophotometric examination of cell fractions showed bond formation between oxidized DAB and cytochromes or catalase. These results show that oxidized DAB can react sensitively with hemoproteins directly. Finally, oxidized DAB may react with catalase in microbodies corresponding to peroxisomes (3), and with cytochromes in mitochondria.

Erythrocytes, leucocytes and Kupffer cells showed affinity for oxidized DAB. Inhibition of cyanide was observed in erythrocytes but not in some granules of leucocytes.

References:

(1) Graham, R. C., Jr. and M. J. Karnovsky: J. Histochem. Cytochem. *14*, 291 (1965).
(2) Margoliash, E: *In*: Haematin Enzymes, pp 259 (Falk *et al.*, edit.) New York: Pergamon (1961).
(3) de Duve, C. and P. Baudhuin: Physiol. Rev. *46*, 323 (1966).

Phosphamidase in the Gastric Cancer

HIROSE, SHUNTA (Dept. of Surgery, Osaka Univ. Medical School, Osaka, Japan)

Although many histochemical reactions have been described in cell research, no specific reaction has been reported in cancer cells. The reaction of phosphamidase seems to be the most useful for determining malignancy of cells.

Since the histochemical demonstration of this enzyme was originally reported by Gomori, detailed studies have not been reported on it. In order to make this reaction perfect in technique, we have designed satisfactory technical methods by means of applying our own freeze-substitution method and also using acetone in the course of embedding and paraffining, and our own synthesized substrate for the reaction. In our experience two characteristics were found in this reaction. There was an increase in the reaction corresponding to the degree of undifferentiation of the cancer cells though a strong reaction could be found in cancer cells irrespective of differentiation or undifferentiation. There was a difference in localization of the reaction product between cancer and normal cells.

A moderate reaction was shown only in the cupshaped zones of the surface epithelium, but no or a slight reaction was shown in other cells of the gastric mucous membrane. Beneath the striated border of the small intestinal metaplasia, a more dominant reaction was seen than in the surface epithelium, but the goblet cells showed no reaction.

In gastric cancers, the reaction was dominant in whole cellbodies, but well differentiated cancer cells showed a stronger reaction in the apical portion than in other parts of cell, as in the reaction found in the cupshaped zones of surface epithelium or zones under the striated border of small intestinal metaplasia.

The signet ring cell carcinoma showed a strong reaction not only in the solid part of cellbodies but also in vacuoles.

Aryl Sulfatase Activity in Individual Neurons and Neuropil and in Layers of Cerebellum

HIRSCH, HILDE E. (UCLA Center for the Health Sciences, Los Angeles, Calif.)

The highly sensitive method of Sherman and Stanfield (1), employing 4-Methyl-Umbelliferone Sulfate (MUS) as a substrate, was used to measure the aryl sulfatase activity in samples of nervous tissue as small as individual neurons. An endogenous, dialysable, inorganic inhibitor present in brain (probably phosphate ion) presents difficulties, but these can be overcome by employing high dilutions.

The sulfatase activity of brain homogenate as measured with MUS is strongly inhibited by phosphate and sulfite, and to a smaller degree, by sulfate and fluoride, but not by chloride. This is in agreement with the partially purified aryl sulfatase from brain studied by Balasubramanian and Bachhawat (2) who concluded that it is of type A, which is a lyosomal enzyme; however, the pH optima differ.

Activities in anterior horn cells of human spinal cord were 10-18 times higher than those of adjacent neuropil (7 times in monkey). In monkey cerebellum the granular layer had 3-4 times the activity of the molecular, and 5 times that of the underlying white matter.

Thus, neuronal perikarya, in comparison to dendrites, axons and glia cells, are very rich in (MUS-) aryl sulfatase, as they are known to be in β-galactosidase, β-glucuronidase and (α-naphthyl) acid phosphatase (3,4).

References:

(1) Sherman, W. R. and E. F. Stanfield: Biochem. J. *102*, 905 (1967).
(2) Balasubramanian, A. S. and B. K. Bachhawat: J. Neurochem. *10*, 201 (1963).
(3) Hirsch, H. E. and E. Robins: Fed. Proc. *20*, 342 (1961).
(4) Hirsch, H. E.: J. Neurochem. (1968) *(in press)*.

Electrophoretic Separation of Hormones Associated with Secretory Granules from Rat Anterior Pituitary Glands

HODGES, DONALD R., ALLEN COSTOFF, and W. H. McSHAN (Department of Zoology, University of Wisconsin, Madison, Wisc.)

A simple and rapid method for the preparation of highly purified rat anterior pituitary secretory granules has been developed. Homogenates were first centrifuged to separate nuclei and red blood cells. The supernatant was then filtered through a millipore filter to remove further cellular debris. The filtered supernatant was layered on sucrose-diodrast continuous gradients and centrifuged at 100,000 g for two hours. Six main zones were separated by the gradient and these were removed and further purified by filtration. The filtrates were centrifuged at 40,000 g and 100,000 g to obtain pellets of highly purified large and small granules. The degree of purification was determined by electron microscopic examination of the pellets.

Granules obtained by this method were investigated using 7.5% cross-linked polyacrylamide gel electrophoresis at pH 9.5. The granule pellets were pooled, extracted by homogenization in a glass homogenizer and the extract was run for 1.5 hours at four milliamperes per tube. The gels were removed and one or two were stained with Coomassie blue.

The stained gels were used as guides to cut corresponding zones from the unstained gels. These were extracted by homogenizing in saline and the hormone containing areas identified by bioassay.

After the gels were stained, three major and eight minor discs were visible. The first major disc had growth hormone activity and the second contained prolactin activity. The third major component was the fastest migrating protein on the gel and apparently did not contain hormone. Luteinizing hormone activity was found in a broad area between growth hormone and the top of the separator gel and could not be associated with a single visible disc. Protein lability was suggested by rather high loss of luteinizing hormone activity following electrophoresis. Follicle stimulating hormone activity was detected in the second minor disc below growth hormone. Thyroid stimulating hormone and adrenocorticotrophic hormone activities have yet to be determined.

Cytochemical Studies of Intracellular Digestion in Cells of the Nervous System *

HOLTZMAN, ERIC (Department of Biological Sciences, Columbia University, New York, N.Y.)

Cytochemical studies, by light and electron microscopy, have been carried out on a variety of neurons and associated cells *in vivo* (dorsal root and nodosal ganglia of the rat, rat adrenal medulla neurons) and in culture (mouse and rat dorsal root ganglia). Tissue was aldehyde fixed, frozen and incubated for demonstration of "marker" enzymes such as acid phosphatase and aryl sulfatase for lysosomes and TPPase for Golgi apparatus. The responses of cells to mechanical injury and to irradiation have been investigated. In addition, uptake of exogenous horseradish peroxidase has been followed by Karnovsky's method.

In normal perikarya, acid phosphatase is demonstrated in Golgi saccules of some neurons and in a system of agranular membranes (GERL) at the inner surface of the Golgi apparatus in all neurons studied. Vesicles derived from GERL also contain demonstrable acid phosphatase; some of these are coated.

At both inner and outer aspects of the Golgi apparatus in perikarya, rough endoplasmic reticulum may be closely apposed to Golgi saccules. Often such endoplasmic reticulum shows ribosomes only on the cisterna surface facing away from the Golgi apparatus.

Lysosomes in normal perikarya include dense bodies, multivesicular bodies and autophagic vacuoles. In neurons whose axons have been cut (chromatolysis) and in irradiated neurons, acid phosphatase or aryl sulfatase containing lysosomes accumulate in the central region of the perikaryon. The frequency of autophagic vacuoles in such injured neurons is much greater than normal. This is true in perikarya and in axons; normal axons show autophagic vacuoles relatively infrequently. Schwann cells of peripheral nerves and capsule (sheath, satellite) cells surrounding perikarya show acid phosphatase in saccules and vesicles of the Golgi apparatus. Lysosomes seen in these cells include multivesicular bodies, autophagic vacuoles and dense bodies.

Peroxidase is taken up by perikarya, unmyelinated axons and, in the adrenal medulla, by axon endings. Tubules and coated vesicles are involved in the uptake. Both in perikarya and in axons, peroxidase accumulates in bodies resembling multivesicular bodies. It is also demonstrated in membrane-delimited "cup-like" structures tentatively identified as precursors of multivesicular bodies.

Peroxidase is also seen in tubules, coated vesicles, "cup-like" bodies, multivesicular bodies and dense bodies of Schwann cells and capsule cells.

---

* Portions of this work were done in collaboration with the laboratories of Dr. A. B. Novikoff and Mrs. E. Peterson, Albert Einstein College of Medicine, New York.

Enzyme-histochemistry of Metastatic Carcinoma to the Skin—its Diagnostic Significance in Regard to the Site of the Primary Neoplasm

HOLUBAR, K., J. TAPPEINER, and K. WOLFF (Dept. of Dermatology, Univ. Vienna, School of Medicine, Vienna, Austria)

Tumors of the skin exhibit patterns of enzymatic reactivity characteristic for the tissue of their origin. To give but a few examples, leiomyomas contain enzymes typical for smooth muscle; neurofibromas show enzymatic reactions of nervous tissue; and hidradenoams reveal the enzymatic spectrum of sweat glands.

These enzymatic reaction patterns permit an enzymehistochemical distinction between tumors of connective tissue, nervous, or epithelial origin and between neoplasms with an eccrine, apocrine, or hair follicle histogenesis. It was thought that carcinomas of different internal organs might show equally distinctive enzymatic features and that the enzymehistochemical behavior of their metastases to the skin might be of diagnostic help in the search for the site of the primary neoplasm.

Skin metastases of internal malignancies (carcinomas of the breast, uterus, stomach, kidney) and the respective primary neoplasms were examined enzymehistochemically (two oxidative and seven hydrolytic enzymes) and the enzymatic reaction patterns were compared. The following observations were made:

1.) The enzymatic reactivities of corresponding primary and secondary neoplastic growths were identical.

2.) Different internal malignancies showed different enzymatic reaction patterns and so did their metastases to the skin.

Thus, it was possible to distinguish, histochemically, between skin metastases of various internal neoplasms and between these and primary skin tumors. In some instances the enzymehistochemical behaviour of a secondary malignancy to the skin was so distinct and characteristic for a certain internal malignancy as to give important diagnostic clues regarding the site of the primary neoplasm.

Enzymes Hydrolyzing Some Dipeptide Naphthylamides

HOPSU-HAVU, VÄINÖ K. (Department of Anatomy, University of Turku, Finland)

Recent studies have revealed that alanyl-alanine, glycyl-proline, glycyl-phenylalanine and leucyl-leucine naphthylamides are split by several different types of enzymes in mammalian tissues:

1. By metal or sulfhydryl dependent amino(poly)peptidases (naphthylamidases) so that the N-terminal amino acid is first liberated followed by the hydrolysis of the remaining amino acid naphthylamide. Such metal dependent peptidases are found e.g., in the soluble fraction of the rat liver and in the microsomal fraction of the rat kidney. Several enzymes of this type are found in many tissues.

2. By enzymes which require an N-terminal free amino group but do not attack amino acid naphthylamides or dipeptides and split the substrate into a dipeptide and naphthylamine. Such an enzyme with glycyl-proline naphthylamide as the preferential substrate was purified from the microsomal fraction of hog kidney and of several rat tissues. The enzyme was shown to be optimally active at neutral pH and to be independent of sulfhydryl groups and reagents as well as in metal ions and chelators. It was inhibited by several mono- and divalent ions.

3. By a lysosomal sulfhydryl dependent enzyme with pH optimum at 5.5. Such an enzyme hydrolyzing preferentially alanyl-alanine naphthylamide and more slowly glycyl-phenylalanine naphthylamide was purified from hog kidney. This enzyme also liberated dipeptides from glycyl-tyrosine and from glycyl-phenylalanine amide and was shown to be activated by chloride ions. This enzyme resembles and is most likely identical to cathepsin C.

All of these enzymes except those which are sulfhydryl dependent are demonstrated simultaneously in histochemical assay conditions when using these dipeptide naphthylamides as substrate and diazonium salts as couplers. A differentiation between individual enzymes can be made only in tissue sites where no amino acid naphthylamidases are present.

Effect of Actinomycin D and Puromycin on the Proteolytic Enzyme Systems in Growing Rodent Tissues

HOSANNAH, YVONNE, CARLTON E. BLACKWOOD, and INES MANDL (Columbia University, College of Physicians & Surgeons, New York, N. Y.)

Previous communications from this laboratory (1) have established that complexes between proteolytic enzymes and their inhibitors can be detected in the earliest stages of embryonic development. As development proceeds changing patterns of proteolytic enzyme activity specific for individual organs can be demonstrated. To gain insight into the molecular events underlying the formation of these enzyme systems the effect of intraperitoneal injections of Puromycin and Actinomycin D into pregnant rats during various stages of the gestation period was investigated. If normal progression of enzymatic activity is blocked by Puromycin injection, de novo synthesis of the enzyme or inhibitor at that stage is indicated. Inhibition by Actinomycin D injection indicates the involvement of messenger RNA synthesis.

To differentiate between the effect on soluble and on tissue bound enzymes biochemical and histochemical assay procedures have been used. Chromogenic naphthylamide substrates that lend themselves to both determinations were selected: benzoyl-arginine-β-naphthylamide, (BANA), and glutaryl-phenylalanine-β-naphthylamide (GPNA) as endopeptidase substrates with trypsin and chymotrypsin-like specificities respectively, and leucine-β-naphthylamide (LNA), arginine-β naphthylamide (ANA), α-glutamyl-β naphthylamide (GNA) and cystine-di-β-naphthylamide (CNA) as exopeptidase substrates.

Placental and embryonic tissues obtained from pregnant rats injected with Actinomycin D and Puromycin showed less ANA-ase and GPNA-ase as well as less of the naturally occurring inhibitor activities than control tissues at analogous stages of development. When Puromycin was injected at early stages of development, the embryo did not develop. Whether the embryo continues to develop to the end of the gestation period depends on the time elapsed between fertilization and initiation of the injections. Even when the animals injected with these anti-metabolites go to term, changes occur in the proteolytic enzyme pattern. Less exopeptidase and endopeptidase activity as well as changes in enzyme patterns were also observed in the liver and kidney of pregnant rats injected with Actinomycin D and Puromycin.

Histochemical observations of exopeptidase development paralleled the biochemical findings. After treatment with Actinomycin, the activity of LNA-ase, GNA-ase and ANA-ase in 5, 6, and 7 day old embryos was much less than in controls while after Puromycin injections, activity against the same substrates was totally inhibited. By the 10th day of gestation, in contrast to normal development, no exopeptidase was observed in embryos exposed to Actinomycin D, though high activity could be detected in extra-embryonic membranes. The same pattern of localization was found in 12 day old embryonic membranes and placenta with no activity in the embryos. Both Actinomycin D and Puromycin affected the synthesis of proteolytic enzymes and their inhibitors, but the effect of Puromycin was more pronounced.

References:

(1) Blackwood, C. E., Y. Hosannah, and I. Mandl: Proteolytic Enzyme Systems in Developing Rat Tissues. J. of Reprod. and Fertil. *in press,* (1968).

Microspectrophotometric Studies of Micrurgically Isolated Parts of Pedigreed Cells [*]

HOSKINS, GODFREY C. (The University of Texas Southwestern Medical School at Dallas, Dallas, Texas)

Microspectrophotometric analysis of parts of cells requires preservation of the structural integrity of the part to be analyzed, and isolation of the part from the remainder of the cell. Several commonly used methods for isolation of cell parts employ forces such as surface tension or osmotic pressure, and subject the isolated part to these same disruptive forces.

The use of micrurgy to physically cut out cell parts allows the isolated part to enter an environment designed to preserve rather than disrupt cell structure, where semiquantitative, (1) and quantitative (2) measurements can be made. This method has been used with (a) living cells *in vitro,* (b) cells fixed with any of several fixatives including methanol, ethanol, acetic acid, formaldehyde, and gluteraldehyde, and (c) material prepared as routine tissue sections. Regardless of the source of the material, the purity of the isolated part can be established by photographic recording while all contaminants, visible to light microscopy, are cut away.

When tissue cultures are used, they may be photographed by time-lapse cinematography through the process of fixation, and the film played backward to establish the "age" of each cell since mitosis (3). Furthermore this technique allow determination of the pedigree relationships of "sibling" cells having the same parent, and "cousin" cells having the same grandparent.

In the experiments reported here, the mouse fibroblasts (L929) and the method of cultivation were those used by Killander and Zetterberg (3); and as expected, when the values for nucleolus, nucleus, and cytoplasm were summed to give whole cell values, the results were compatible with their results. Percent error of measurement was greatest (up to 15%) for nucleoli, less for nuclei, and least for cytoplasm (less than 5%). Variations of dry mass and 2650Å total absorption were greatest for cells in G-2 period, less for cells in S period, and least for cells in G-1 period of growth. Unrelated cells showed greater variation of measurements than cousin cells, which showed only slightly greater variation than sibling cells.

Measurements of micrurgically isolated nucleoli indicate that during post mitotic reformation of the nucleolus both dry mass and 2650Å absorbing material rapidly increase to a value which is standard for any given family of cells. This value is achieved by 15 minutes after telophase, and thereafter the 2650Å absorbing material increases at a slightly greater rate than dry mass, and both continue to increase until mitosis. The nucleus (without nucleolus or cytoplasm) begins with a post mitotic (G-1) value for dry mass and for 2650Å total extinction. These values remain fairly constant (for a given family of cells) until S period, when they double. During G-2 period these values increase very little if at all. The cytoplasm begins with post mitotic (G-1) values for dry mass and for 2650Å total extinction. These too are more constant within a given family of cells. Both values increase throughout interphase until the next mitosis.

An interesting fact noted in the time-lapse films was that each family of cells has its own characteristic pattern of motion—some resemble amoeboid motion, while others reach far forward with a pseudopod, then let go at the rear to lunge ahead like a measuring worm. Some of these forms of motion may be more compatible than others with simultaneous pinocytosis and phagocytosis. This would provide yet another mechanism (in addition to enzyme activities) whereby genetic differences influence rate of growth.

These studies suggest the L929 strain to be made of several subgroups having different hyperdiploid karyotypes as is found in most tissue culture adapted cell lines. The slight variations between sibling cells and between cousin cells may be due to factors such as initial mass variation (3). However, the post mitotic (G-1) variation between non-related cells is compatible with the probability that the amount of DNA in one family differs from that of another and that there may be minor variations from generation to generation. It is suggested that diploid cells would provide a uniform population more suitable for studies of cell growth.

110

References:

(1) Hoskins, G. E., and P. O'B. Montgomery: Exptl. Cell Res. *26,* 534 (1962).
(2) Edström, J. E.: Biochem. Biophys. Acta *80,* 399 (1964).
(3) Killander, D., and A. Zetterberg: Exptl. Cell Res. *38,* 272 (1965).

* Supported by Post-doctoral Fellowship BPD 17266 (Cl). NINDB, and Damon Runyon Grant 943.

Alkaline and Acid Phosphatase Activities in the Duodenum of Foetal and Newborn Mice

HUGON, J. S. and M. BORGERS (E.M. Lab., Radiobiology. C.E.N. Mol., Belgium. Present address: Pathology, Dept., Medical Center. Sherbrooke University. Quebec, Canada)

Extensive studies of Moog (1) have demonstrated that in intestines of foetal mouse, alkaline phosphatase (AlP) activity appeared about the 14th day and increased progressively to birth. In the newborn, a new surge of activity takes place in the first days, culminating at 18th day. Acid phosphatase (AcP) did not show a similar pattern. We have observed, in preceeding work (2), a fine structural localisation of AlP and AcP in lysosome-like bodies of the duodenal absorbing cells of adult mouse. We investigated whether an identical aspect was present in the foetus and newborn. At 14th day of foetal life, a faint activity of AlP is observed on the still atrophic microvilli, on a few profiles of smooth reticulum and on multivesicular bodies in the apical part of the cells. Already sparce activity is noted on very few dense bodies. The same repartition is noticed the 16th day. Cells with glycogen content are totally negative. At this moment, AcP is present in all the dense bodies and in some smooth reticulum profiles mainly in the middle and basal part of the cells. From the 18th day, an increasing number of homogenous dense AlP positive granules appear above the negative Golgi zone. These bodies seem derived from the heavily loaded smooth tubules which may arise from deep invaginations of the positive apical membrane. These granules and some tubules show the AcP reaction. However, AcP positive tubules are never found in the most apical part of the cells. After birth, the granules remain numerous and reactive for both enzymes. Some of them contain empty droplets of presumably nutritive origin extracted by the preparative process. They disappear at the 16th days of life when Golgi cisternae become positive for AlP. At this moment, bodies with a heterogenous content, similar to adult duodenal dense bodies, mainly situated in the apical part of the cells, show the two enzymatic activities. From our results, it seems that AlP activity appears in the dense bodies of the foetal duodenum when pinocytosis begins on the luminal membrane. Moreover, as long as protein absorption through the epithelium is present in the newborn (3), clusters of special homogenous granules containing the two hydrolases are encountered in the apical third of the absorbing cells. They disappear with loss of intact protein absorption around the 15th day. At this moment, Golgi activity for AlP appears.

References:

(1) Moog, F.: Feder. Proc. *21,* 51 (1962).
(2) Hugon, J. and M. J. Borgers: Histochem. Cytochem. *14,* 629 (1966); Histochemie. *12,* 42 (1968).
(3) Wilson, T. H.: "Intestinal Absorption," Philadelphia, Saunders, (1962).

A Quantitative Study of Tetrazolium Salt Reduction in Slide Histochemistry

HURWITZ, LAWRENCE S. and LUCIEN J. RUBINSTEIN (Dept. of Pathology, Stanford University School of Medicine, Palo Alto, Calif.)

Using seven substrate systems, tetrazolium reductase activity was studied by slide histochemistry in 13 experimental tumor types implanted in the mouse brain and liver.

Two major groups of tumors were found in which activity of the NAD-dependent (soluble) alpha-glycerophosphate dehydrogenase substrate system was inversely related to that of the NADH tetrazolium reductase system. In the first group, which included 3 sarcomas and 1 anaplastic mammary carcinoma—all rapidly growing and poorly differentiated—, a relatively high NADH tetrazolium reductase activity was accompanied by low alpha-glycerophosphate dehydrogenase activity. In the second group, which included 4 more slowly-growing and fairly well-differentiated mammary carcinomas, a relatively lower NADH tetrazolium reductase activity was accompanied by relatively higher alpha-glycerophosphate dehydrogenase activity. Activity of the glyceraldehyde-3-phosphate and lactate dehydogenase substrate systems was found to parallel closely activity of the NADH tetrazolium reductase system.

The validity of quantifying pyridine nucleotide diaphorase and the NAD-dependent dehydrogenase systems by oxidation-reduction procedures depending on tetrazolium salt reduction was then tested by comparing the data obtained by optical densitometric measurement of the formazan deposition in tissue sections of two selected tumors with results obtained in the same material by standard microquantitative biochemical techniques.

Four substrate systems were studied. Two of them, the NADH tetrazolium reductase and the NAD-dependent alpha-glycerophosphate dehydrogenase, gave comparatively similar results with either method. The two others, glyceraldehyde-3-phosphate and lactate dehydrogenases, gave contradictory results.

The biochemical data suggest that, in the histochemical demonstration of dehydrogenase activity using the glyceraldehyde-3-phosphate and lactate dehydrogenase systems, tetrazolium reductase acts as the rate-limiting enzyme. Thus, in the presence of excessive levels of dehydrogenase, the histochemical reaction will reflect activity of the reductase, and not that of the dehydrogenase. On the other hand, when the tetrazolium reductase enzyme is not rate-limiting, as in the presence of considerably lower levels of NAD-dependent alpha-glycerophosphate dehydrogenase, the tetrazolium salt is reduced to formazan stoichiometrically, and the histochemical reaction reflects the level of dehydrogenase.

The tetrazolium reduction enzyme therefore appears to act as a rate-limiting enzyme in the histochemical estimation of dehydrogenase substrate system activity in some circumstances, but presumably not in others. It is probably not rate-limiting for those dehydrogenases which have very low turnover numbers and/or are in relatively limited quantities compared to the amount of reductase available. In these circumstances, quantification of the histochemical reaction would seem feasible in terms of the dehydrogenase, provided such measurements are standardized against a biochemical estimation.

Histochemical Identification of Mast Cells in the Mammalian Brain

IBRAHIM, M. Z. M. (Mountain View, Calif.)

Mast cells are traditionally regarded as rare or absent in the mammalian brain. They have, nevertheless, been described in the brain of the hedgehog (1), hamster (2), rat and man (3). My studies show abundant perivascular cells in the mouse, rat, hamster, guinea pig, rabbit, cat, dog, monkey and man. In the first three animals two such types of cells are seen. One (Type I) has the same morphology and staining properties (metachromasia) as mast cells elsewhere in the body. The second (Type II) contains coarser granules that are metachromatic only in formalin-alcohol fixed frozen sections. Practically identical cells are seen in the leptomeninges. It is the identity of this Type II cell that is in question. All the other animals have only Type II cells.

In formalin-fixed frozen and paraffin sections these cells are autofluorescent (yellow or greenish yellow) and stain orthochromatically but are often inconspicuous when stained with basic aniline dyes. Some granules are lost, becoming swollen or vacuolated on fixation depending on the species. Tryptic digestion makes the granules in the paraffin sections most obvious. They now stain blue with basic dyes and also show other signs of their acidic constitution; they stain intense blue with colloidal iron and red with Wolman's Bi-Col method and their methylene blue extinction is at pH 2.5. They do not, however, stain with alcian blue at pH 1.0. Sulphation induces metachromasia in them and they are PAS positive. All these facts point to the presence of a weakly acidic muco-polysaccharide which is probably heparin sulphate; the presence of heparin sulphate has been reported in mast cells in other parts of the body (4). The dramatic clarity of the granules brought about by treatment with trypsin is probably due to digestive removal of surface basic proteins that had been masking the acidic radicles present on the surface. Lipid, mostly acidic (sphingomyelin), is also present in the granules but they contain little protein and no RNA. In all the animals the cells show all the characteristic reactions of enterochromaffin and give a positive indole reaction; they probably contain serotonin. In the cat (only animal studied) they may also contain histamine. They are rich in acid phosphatase and nonspecific esterase and contain other hydrolytic enzymes and some oxidoreductive enzymes. In the irradiated brain they become anisogranular or degranulated and may also proliferate as they do in other parts of the body; some may transform into pigment cells and calcify some months postirradiation. They are never phagocytic nor do they show signs of migration. Compound 48/80 causes the cells (studied only in rabbit and cat) to become abnormal or degranulated. Although similar cells were described in the rabbit and rat by Cammermeyer (5) and his initial conclusion was that they were mast cells, in his final conclusion he doubted this.

It is concluded that the Type II cells are mast cells somewhat modified by their presence in the brain. In view of the composition of these mast cells and their effects on vascular permeability elsewhere in the body, I would like to stress the possible roles these cells might play in brain functions and probably in neuropathology.

References:

(1) Krabbe, K.: Acta Path. Microbiol Scand. *5*, 37 (1928).
(2) Kelsall, M. A. and P. Lewis: Fed. Proc. *23*, 1107 (1964).
(3) McGovern, V. J.: Austr. J. Sci. *17*, 176 (1955).
(4) Jorpes, J. E., B. Werner, and B. Aberg: J. Biol. Chem. *176*, 277 (1948).
(5) Cammermeyer, J.: J. Neuropath. Exp. Neurol. *25*, 130 (1966).

Postmortem Changes in Active and Total Phosphorylase of the Rat Brain

IBRAHIM, M. Z. M. (Mountain View, Calif.)

In the course of studies dealing with the changes in glycogen and phosphorylase content of the rat brain caused by exposure of the animals to ionizing radiation or to various forms of hypoxia, rapid postmortem changes were believed to be instrumental in causing some false results. Referral to the work of Breckenridge and Norman (1) indicated how this could happen; they described biochemically a sequence of changes in phosphorylase activity (active and inactive) which commenced within a few seconds after death. The tissue anoxia was considered to be the immediate cause. This report represents a histochemical verification of these authors' findings.

Rats were nembutalized, their brains exposed and then quickly transected coronally and immediately dropped into liquid propane. The method was such as to keep the animal alive until sectioning of the brain and thus the latter suffered only a few seconds of anoxia before freezing. The cut surface was later utilized for sectioning because of the critical nature of the rapidity of freezing (1). The brains of other animals were removed and frozen at varying intervals after death up to 10 minutes. Cryostat sections were incubated for active and total phosphorylase using a slight modification of Eränkö and Palkama's adaptation (2) of the Takeuchi and Kuriaki method (3). For active phosphorylase, AMP was omitted from the incubating medium. Contrary to what Godlewski (4) advocated, no EDTA (Versene) was added since Hori (5) showed that it may have other effects especially on the glycogen content of the sections. This glycogen does act as a primer and can certainly influence the apparent phosphorylase reactions. For total phosphorylase, AMP was included in the medium; the use of ATP plus magnesium ions (6) did not give quite as strong a reaction as when AMP was employed.

Active phosphorylase, initially very low, quickly increased and then started to decrease once again. Inactive phosphorylase underwent the reverse changes, i.e., initially high, quickly decreased and then began to increase again. The changes in both forms of phosphorylase were such as to maintain the total reaction constant for about five minutes. Thereafter, it started to decrease. These results are in full agreement with the biochemical findings of Breckenridge and Norman (1) and they illustrate how remarkably sensitive phosphorylase activity is to the effects of anoxia. Taking this into account, it was found that for comparison of pathological and control brains the animals had to be handled exactly simultaneously until their brains were mounted side by side frozen and cut. This precaution gave reproducible results. It was found desirable to freeze the brains within approximately 2-3 minutes of death.

During this study, the importance of glycogen in the medium as a primer was also checked. The results demonstrated that although it is not vital for qualitative detection of phosphorylase, it is essential for assessing the full quantitative activity. The disparity between this finding and that of Guha and Wegmann (6) who do not use a primer, is perhaps due to the fact that their studies were done on the liver which is far more rich in glycogen than is the brain. Such large quantities of glycogen could have obviated the need for added glycogen.

References:

(1) Breckenridge, B. M. and J. H. Norman: J. Neurochem. *8*, 383 (1962).
(2) Eränkö, O. and A. Palkama: J. Histochem. Cytochem. *9*, 585 (1961).
(3) Takeuchi, T. and H. Kuriaki: J. Histochem. Cytochem. *3*, 153 (1955).

(4) Godlewski, H. G.: J. Histochem. Cytochem. *11*, 108 (1963).
(5) Hori, S. H.: J. Histochem. Cytochem. *14*, 501 (1966).
(6) Guha, S. and R. Wegmann: J. Histochem. Cytochem. *9*, 454 (1961).

Histochemical and Electron Microscopic Studies on the Giant Cell Tumors of Bone (Especially Cultured Cells)

IMURA, SHIN-ICHI and MASANORI TAKEDA (Department of Orthopedic Surgery, School of Medicine, Kanazawa University, Kanazawa, Japan)

We studied three cases of human giant cell tumors of bone and cell lines derived from these bone tumors. Alkaline phosphatase activity was low in tissues, but in cultured cells it was weak or negative. In contrast, acid phosphatase showed moderate activity both in tissues and cultured cells. The activity of DPN diaphorase was highest, that of TPN diaphorase and lactic dehydrogenase was moderate or low, and that of other DPN- and TPN linked dehydrogenases was low. In tissues, oxidative enzyme activities except for DPN- and TPN diaphorases were higher in multi-nuclear cells than in stroma cells. In cultured cells, the activities were higher in multinuclear cells than in mononuclear cells. The differences in activity of DPN- and TPN diaphorases between mononuclear and multinuclear cells were not observed. ATPase and aminopeptidase activities were moderate, but non-specific esterase was low or negative.

With the electron microscope, cultured cells were characterized by a poorly developed rough surfaced endoplasmic reticulum and a large amount of free ribosomes. The rough surfaced endoplasmic reticulum consisted of isolated profils containing materials of low density. Similarly, the Golgi zone were poorly developed. Mitochondria generally appeared swollen. Many multivesicular bodies, small amount of dense bodies, and intracellular fine fibrils were observed. Sometimes annulate lamellae were found. The cell surface occasionally demonstrated villiform. The intercellular space contained some cellular debris and in rare cases extracellular fibrils were observed, but collagen fibers were not identified. Multinuclear cells contained cistic profils of endoplasmic reticulum and swollen mitochondria.

A Combined Biochemical and Histochemical Study of the Lead Method for Histochemical Demonstration of Adenosine Triphosphatase

JACOBSEN, N. O. and P. LETH JØRGENSEN (Department of Pathology and Institute of Physiology, University of Aarhus, Denmark)

The value of the lead method for demonstration of ATP hydrolyzing enzymes and the possibility of relating histochemical ATPase activity to active transport of ions were investigated. For histochemistry we used frozen sections of sediments, made by ultracentrifugation of the microsomal fraction from rat kidney, and compared the rate of histochemical staining with biochemical measurements of the $(Na^+ + K^+)$-, Mg^{2+}-, and Ca^{2+}-activated ATP hydrolyzing enzymes under conditions of the histochemical medium. Measurements of ATP hydrolysis were corrected for lead-catalyzed splitting of ATP.

In the test tube, the high specific activity of $(Na^+ + K^+)$ ATPase in this microsomal fraction was abolished by 0.5 mM of lead or more, even in the presence of tris-maleate 8 mM. The Mg^{2+} ATPase and the Ca^{2+} ATPase were also inhibited by lead, but with 2 to 4 mM of lead these activities reached an equal level 64% and 35% below the activity of the controls. Accordingly, no activation by $Na^+ + K^+$ or inhibition by ouabain $(10^{-3}M)$ could be demonstrated histochemically in the presence of the 2,4-3,6 mM of lead necessary for the precise localization of the reaction product. In the presence of 2,4-3,6 mM of lead the enzymatic hydrolysis of ATP was stimulated to an equal extent by Ca^{2+} and Mg^{2+} under biochemical as well as histochemical conditions. Under both conditions addition of 0.2 to 2.0 mM Mg^{2+} or Ca^{2+} caused a steep increase in activity and a maximal level was reached at 3 mM.

On the other hand, lead-catalyzed ATP hydrolysis, as related to concentrations of Mg^{2+} or Ca^{2+}, showed a linear increase and was negligible in comparison with the enzymatic hydrolysis. Following fixation of thin pellets in 3% glutaraldehyde at $0°$ C, enzymatic hydrolysis in the presence of lead was decreased. After fixation for one hour, activity was reduced by 80-90% and the corresponding reduction of histochemical staining was estimated at 80%. Even then, activation by Ca^{2+} or Mg^{2+} followed the above mentioned course under both histochemical and biochemical conditions.

According to these results, the presence of lead prevents demonstration of $(Na^+ + K^+)$ ATPase even in unfixed tissues, but agreement between the histochemical staining and the measurements of enzyme activity in the test tube made it possible to study the properties of the ATPase that can be demonstrated.

Ribonuclease and Deoxyribonuclease in Human Skin

JARRETT, A. (University College Hospital Medical School, Dermatology Department, University Street, London, England)

Acid hydrolases, acid ribonuclease, and acid deoxyribonuclease have been detected in human epidermal cells and in elastic fibres of the dermis. In the epidermis it is thought these enzymes are related to hydrolysis of nuclear proteins prior to formation of the keratin layer. The enzymes are sensitive to relatively small changes in pH and varying the pH alters the localisation of the reaction within the epidermis. Thus, in the case of RNase, at pH 5.4 the reaction is predominantly in nuclei of the basal cells and in the cells of the dermis; at pH 5.7 these sites remain positive but nuclei of the upper epidermal cells also give a positive reaction; at pH 5.9 the basal layer nuclei and nuclei of the dermal cells are no longer reactive. DNase shows similar changes as a result of alteration of pH. Both these enzymes have also been localised in elastic tissue of human dermis. Other acid hydrolases, such as acid phosphatase, beta-galactosidase, and sulphatase, have also been detected in these fibres but they do not give the same diffuse reaction throughout the whole length of the structure as do acid nucleases. The possible function of these enzymes in elastic tissue is discussed and work related to attempts to liberate these enzymes with lysosomal liberators will be described.

References:

(1) Jarrett, A.: J. Invest. Derm. *49*, 443 (1967).
(2) Jarrett, A. and R. I. C. Spearman: Histochemistry of the Skin: Psoriasis. Eng. Univ. Press (1964).
(3) Daoust, R. and H. Amano: J. Histochem. Cytochem. *8*, 131 (1960).

Histochemical Differentiation of Ribonucleases in Adipose Tissue

JOANDREA-CASIAN, CLAUDIA and CORNELIA PRUNDEANU (Catedra de Histologie, Facultatea de Medicina, Bucharest, Romania)

The presence of intense ribonuclease activity demonstrated by the film-substrate and metallic precipitate methods is reported in lipocytes, mastocytes, and capillary endothelia of rat adipose tissue.
Lipocytes exhibit especially acid RNase activity; in the capillary endothelia, alkaline RNase activity is predominant while in mastocytes both acid and alkaline RNase activities are present.
The enzymatic activity in these elements further shows certain properties specific for each of them.

On the Histochemical Demonstration of Two Oxidative Enzymes in Fixed Brain *

JOHNSON, ANNE B. (Albert Einstein College of Medicine, Bronx, N. Y.)

Light microscopic studies of two oxidative enzymes were performed on fixed brain with the aim of facilitating later ultramicroscopic demonstration. The enzymes NADH-tetrazolium oxidoreductase (NADH-D)

118

and non-NAD dependent α-glycerophosphate dehydrogenase (αGP-D) were selected for study because both are insoluble, are demonstrable with tetrazolium salts without dependence on a second enzyme, and remain active after fixation. Excellent morphology and considerable activity of NADH-D were preserved in frozen or chopper sections of blocks fixed for two and a half hours in charcoal-treated 3% glutaraldehyde. Formazan was not present in control sections incubated without substrate, indicating that reaction between bound glutaraldehyde and tetrazolium salts was not a problem. Activity of αGP-D survived fixation in 4% formaldehyde and could be demonstrated if phenazine methosulfate (PMS) was used as an intermediate electron acceptor. Menadione could not replace PMS with formalin-fixed tissue as the menadione reacted with bound fixative and formazan was produced. Similarly, PMS could not be used with glutaralydehyde-fixed tissue. The formazan of tetra-NitroBT was not solubilized by routine embedding in Epon or Araldite when the sections were taken directly from propylene oxide to plastic. The localizations obtained were compared to those in fixed and unfixed cryostat sections. In all sections, formazan deposition was consistent with the known mitochondrial location of both enzymes.

* Supported by USPHS grants NB-07298, NB-02255, NB-03356.

Histochemistry of Experimental Cutaneous Calcinosis *

JOHNSON, WAINE C. and DAVID S. ALKEK (The Skin and Cancer Hospital of Philadelphia, Temple University, Philadelphia, Pa.)

Cutaneous calcinosis was induced in female rats by sensitization with oral dihydrotachysterol (DHT) and the use of subcutaneous iron dextran, subcutaneous egg white, manual depilation, or exposure to ultraviolet light (UVL) utilizing methods previously reported (1). Controls consisted of animals treated with DHT alone, iron dextran alone, egg white alone, depilation alone, UVL alone, and normal skin. Tissue specimens were taken at intervals between 8 hours and 6 days after treatment, and were prepared by appropriate methods to demonstrate calcium, iron, mast cells, acid mucosaccharides, alkaline phosphatase, acid phosphatase, adenosene triphosphatase (ATPase), aminopeptidase, indoxyl acetate esterase and lactic (LD), succinic (SD) and glucose-6-phosphate (G-6-PD) dehydrogenase. The fibrocytes present in normal skin gave reactions as follows: alkaline phosphatase—usually negative; acid phosphatase—moderately positive; ATPase—moderately positive; esterase—weakly positive; LD—strongly positive; and SD and G-6-PD—moderately positive. The DHT treated controls did not show significant abnormal changes except for a possible increase in the number of fibro-cytes and mast cells, and the appearance of a few alkaline phosphatase positive fibrocytes. Iron dextran and egg white treated controls showed a marked increase in fibrocytes; a moderate increase in mast cells; the appearance of inflammatory cells beneath the panniculus carnosus muscle (PCM) and to a lesser extent in the dermis above the PCM; a few to occasionally moderate numbers of alkaline phosphatase positive fibrocytes; and increased amounts of material above and below the PCM showing histochemical properties of hyaluronic acid. Iron was present in fibrocytic cells or histiocytes as well as extracellularly in the iron dextran treated controls but no calcium was found. In animals given DHT and iron dextran, egg white, UVL or depilated, initial calcification usually occurred within 24 to 72 hours. Animals with skin in the grow-ing (anlagen) hair phase usually show calcification earlier than animals with resting stage follicles. Initial sites of calcification appeared to be re-lated to areas rich in acid mucosaccharides, or in some instances, collagen fibers. Numerous alkaline phosphatase positive fibrocytic cells appeared prior to initial calcification but were not always located immediately adjacent to initially calcified sites. These cells were absent or present in few to moderate numbers in normal skin and control animals. The fre-quent close approximation of the alkaline phosphatase positive fibrocytic cells to capillary endothelial cells suggests the origin of these cells may be from capillary endothelium. Mast cells did not show alkaline phos-phatase activity. The fibrocytic cells also showed a marked increase in activity of acid phosphatase and a moderate increase in ATPase, LD, SD, G-6-PD, and indoxyl acetate esterase activity. The fibrocytic cells or the stroma to the cells adjacent showed a marked increase in amino-peptidase activity.

A cellular infiltration consisting mainly of fibrocytic cells appeared prior to calcification and these alkaline phosphatase producing cells seem to play a significant role in the pathogenesis of tissue calcification. Whether this role is to inactivate polyphosphate inhibitors normally present in the tissues as suggested by Fleish and Newman (2) remains to be de-termined. Hyaluronic acid in extracellular spaces, collagen fibers (per-haps the chondroitin sulfate component), and acid mucosaccharides asso-ciated with hair follicles (in depilated animals) appeared to act as matrix substances to bind the calcium.

References:

(1) Johnson, W. C., P. D. Forbes, J. H. Graham, H. R. Gray: J. Invest. Derm. *43*, 453 (1964).
(2) Fleisch, H. and W. F. Newman: Amer. J. Physiol. *200*, 1296 (1961).

* Supported by USPHS Grant No. ES-269.

Enzymatic Parameters for Neurosecretory Activity *

JONGKIND, J. F. and D. F. SWAAB (Netherlands Central Institute for Brain Research, Amsterdam (Havens Oost), Holland)

Electron microscopical studies have indicated that increased neurosecretory activity of the supraoptic nucleus of the rat (SON), is paralleled by changes in appearance and distribution of organelles in cell bodies of the SON (Zambrano and de Robertis, 1966). In order to trace enzymatic parameters for neuroscretory activity, we assayed with qualitative and quantitative histochemical techniques a number of enzymes considered to be specific for Golgi-apparatus, endoplasmic reticulum, or lysosomes, in the process of increasing synthetic activity of the SON.

A quantitative histochemical method for thiamine diphosphate-phosphohydrolase (TPPase) was developed to support previous qualitative findings of increased TPP-ase activity in the Golgi-zone of the SON-neurons (Jongkind and Swaab, 1967). After dehydration for 3 days, TPPase activity in SON increased 40% and after 6 days thirsting, an increase of 80% was found. No change was observed in the adjacent hypothalamic area (AH). Quantitative determination of acid phosphatase under similar conditions did not show any increase in activity in SON and AH. The morphological distribution of other diphosphatases (UDP-ase, IDP-ase) and glucose-6-phosphate phosphohydrolase will also be compared with TPP-ase, while data on biochemical activity of UDP-ase and IDP-ase in freeze dried SON will be presented. The possibilities of this histochemical approach for neuroendocrinological problems will be discussed.

References:

(1) Jongkind, J. F. and D. F. Swab: Histochemie *11*, 319 (1967).
(2) Zambrano, D. and E. De Robertis: Z. Zellforsch. *73*, 414 (1966).

---

* This study was supported by a grant from the Netherlands Organization for the Advancement of Pure Research no. L. 515-24. I am grateful to Dr. Ezio Giacobini, Department of Pharmacology, Karolinska Institute, Stockholm, for his advice and financial support (U.S. Public Health Service NB 04561-04 and 05) for a part of this investigation.

Remarques sur la Localisation Histochimique des Disaccharidases Intestinales

JOS, J. (Unité de Rech. de Génétique Médicale, Paris, France)

En modifiant la méthode histochimique de Dahlqvist et Brun (J. Histochem. Cytochem, 1962, *10:* 294-302) et en utilisant un nouveau procédé d'incubation (Ann. Histochim. 1967, *12:* 53-61), nous avons pu obtenir une meilleure localisation des activités disacchardsiques (Nature, 1967, *213*: 516-518). Dans la muqueuse duodénale normale, l'activité enzymatique est très forte avec le maltose, moyenne avec le saccharose et le tréhalose, faible avec le palatinose et le lactose. Le précipité est localisé avec précision dans la bordure en brosse des cellules épithéliales qui tapissent les parois des villosités, tandis que la réaction est faible ou nulle dans le cytoplasme de ces cellules, sauf avec le lactose. L'activité est absente dans les cryptes. Dans le chorion, le précipité est présent dans certaines cellules.

Malgré ces résultats encourageants, la localisation histochimique des glycosidases se heurte encore à des difficultés. La principale tient à l' impossibilité de trouver dans le commerce des préparations de glucose-oxydase suffisamment actives, et dépourvues d'activités disaccharasiques. La seconde réside dans le choix du papier filtre qui joue un rôle important dans le nouveau procédé d'incubation que nous avons préconisé; ce papier filtre doit être mince, relativement lisse et à pores larges.

Enfin la nature du sel de tétrazolium utilisé est fondamentale. Le nitrobleu de tétrazolium ne peut convenir; son formazan a en effet une très grande affinité pour le cytoplasme des cellules.

En revanche, le formazan du tétranitrobleu de tétrazolium (T. N. B. T.) a autant d'affinité pour le cytoplasme que pour les membranes cellulaires. Les réactions des deshydrogénases, qui donnent une coloration essentiellement intracytoplasmique, démontrent que le formazan du T. N. B. T. n'a pas une affinité exclusive pour la membrane en brosse des entérocytes.

Ce formazan a cependant un inconvénient: il se fixe de façon non spécifique sur les parois des vaisseaux. Cette dernière coloration est due probablement à la présence de groupements thiols.

Protein Reabsorption and Digestion in the Ophidian Terminal Intestinal Cells

JUNQUEIRA, L. C. and A. M. SOUZA TOLEDO (Faculdade de Medicina, Univ. S. Paulo, Brasil)

During the study of electrolyte transport in the ophidian terminal intestine (Junqueira, Malnic, and Monge: Physiol. Zool. *39*, 151, 1966) it became apparent that in previously fed animals the epithelial cells of this organ present abundant granular inclusions in their cytoplasm. Histochemical tests showed that these inclusions were not lipids or polysaccharides but they gave a positive reaction for tryptophan and tyrosine (Lison and Pinheiro's reaction and Millon's reagent). These granules might be the morphological expression of secretion occurring in this epithelium or due to reabsorption of proteins present in the organs lumen. A series of studies was performed in the Colubrid *Xenodon* and the following observations strongly suggest that we are dealing with re-absorption and digestion of proteins in this organ:

1. The granules occur only in fed animals.
2. After feeding the granules appear first at the cell apex progressing wave-like to the cell base.
3. The size of the granules decreases from the cell apex to its base.
4. When an excess of mammalian haemoglobin is fed or injected in the terminal intestine of these animals the granules after reabsorption give a positive reaction for ionic iron (Perls reaction).
5. Electron microscopic observations corroborate in more detail the above mentioned morphological aspects, suggesting that proteins are reabsorbed by pinocytosis and digested in the cells.
6. A concentration of a strong histochemical reaction for acid phospha-tase in the cell apical region further strengthens this hypothesis.

The Body Covering of Animals and Plants as a Promising Comparative Goal for Cytochemical Studies in Regard to Neoplastic Changes, Occurring at Short-term Intervals

KAISER, HANS E. (Dept. of Anatomy, School of Medicine, The George Washington University, and the National Aquarium, Washington, D. C.)

Neoplasms occur in both the plant and animal kingdoms. The compar-ison of the different stages and steps of the process of multiphase car-cinogenesis requires a type of tissue which may be common in its distribution among species and comparable in both kingdoms. It could be shown in our studies with several methods that the body covering should be used as the most promising medium also with respect to com-parative histo- and cytochemical studies. Such a model has to fulfill cer-tain requirements if it is to be of effective use:

1. For comparative purposes it has to be a type of tissue occurring in both kingdoms, to permit a comparison in regard to function as well as to morphology.
2. Its use in experimental studies has to be simple in its application of carcinogens of different types.
3. This type of tissue should permit rapid removal from the organism allow vital stains, phase-contrast, electron microscopy in combination with other methods. Of special advantage may be the use of histochem-ical and cytochemical investigations.

Comparative Studies of the Cytology (Cytochemistry and Histochemistry) of Developing Cancer Cells

KAISER, HANS E. (Dept. of Anatomy, School of Medicine, The George Washington University, and The National Aquarium, Washington, D. C.)

The ability to transform a normal cell into a neoplastic cell is, with some exceptions, a general phenomenon of multicellular life. This process should be investigated in several species and types of tissues to distinguish between common stages, common to all neoplasms and specific steps, peculiar to specific neoplasms. Only short-time interval investigations will permit us to register the continuous changes as shown in film sequence. It is also necessary to use a series of several different methods (Kaiser, 1966 and 1968). These methods should range from light microscopy to biochemical and biophysical approaches. Of the utmost importance is the observation of the initial changes in the cell structure. The observation made with light microscopy must be augmented by other methods such as electron microscopy and chemical means and should include the employment of cyto- and histochemistry. At this point we would like to report upon studies undertaken at first with the phase-contrast and combined fluorescence microscopy (large Zeiss fluorescence microscope) and subsequently with histochemical methods to determine changes of carcinogenesis in connection with cell lipids and proteins, in both animals and plants. Attention should be turned to a determination of the rate and extent of absorption of 3,4-benzopyrene. The use of combined phase-contrast with fluorescence enabled us to follow the rate of spread and absorption of 3,4-benzopyrene at hourly intervals in different species. This combination of methods of observation enabled us to draw conclusions concerning species specific metabolism and the spread of the carcinogen. With reference to the suggestions made above, investigations are under way to follow up these results obtained with combined histochemical and electronmicroscopical methods or by chemical investigations of isolated basal cells of mammalian skin and other epidermal layers of plants and animals. It has been possible to show characteristic changes in the cytology of developing neoplastic cells in tissues of several species of both kingdoms, observed simultaneously with fluorescence-phase-contrast methods on the one hand and treated with histochemical stains on the other hand. The species used were taken from the phyla of coelenterates, different lower worm phyla, annelids, echinoderms, fishes and mammals regarding animals; cyanophyceae, algae, gymnosperms and angiosperms in regard to plants. The sequence of events in the spread of the carcinogen and the time intervals where it occurred will be shown in special comparative graphs and lantern slides.

Histochemical Investigations in the Myocardium of Guinea-pigs Treated With Reserpine

KAKARI, SOPHIA (Pharmacol. Dept., Royal Free Hospital School of Medicine, London University, London, England)

Using nitro blue tetrazolium salt (NBT), the cytochemical localization of succinate dehydrogenase (SDH) in the ventricular muscle of the reserpine-treated guinea-pigs revealed a blue "granular" pattern instead of the purple "fibrillar" appearance of the controls. Quantitative histochemical assessment of SDH activity, using neotetrazolium chloride (NT) (2, 3), confirmed these qualitative changes. Three groups of guinea-pigs were injected daily with reserpine subcutaneously. The first group received 0.1 mg/kg for 7 days, the second 2.5-3 mg/kg for 2-3 days, and the third 2 mg/kg for periods varying from 1 to 4 days. Controls were injected with saline or reserpine solvent. There was good correlation between the total dose of reserpine administered, the duration of treatment, and the reduction in the level of SDH activity. In the animals treated with the small dose of reserpine, SDH activity remained almost unchanged whereas in the animals treated with the larger doses the SDH activity was reduced by approximately 40 to 50%. SDH activity was also determined from progress-curves constructed by plotting μg-NT formazan deposited per μg of total residual nitrogen against time (min). Furthermore, phenazine methosulfate (PMS) (4, 5) added to the succinate-NT medium enhanced the SDH activity of the controls, but had almost no effect in the reserpine-treated animals. Heart rate was measured at the beginning and at the end of the treatment; there was good correlation between the dose of reserpine administered, the decrease in heart rate and the increase in the number of animals with abnormal localization of SDH (NBT). Accumulation of intracellular lipid and glycogen, and an increase in phosphorylase activity were also found in the myocardium of the reserpine-treated animals.

References:

(1) Nachlas, M. M., K. C. Tsou, E. De Souza, C. S. Chang, and A. M. Seligman: J. Histochem. Cytochem. 6, 217 (1957).
(2) Jardetzky, C. D. and D. Glick: J. Biol. Chem. 218, 283 (1956).
(3) Jones, G. R. N.: Exp. Cell Res. 43, 268 (1966).
(4) Farber, E. and C. D. Louviere: J. Histochem. Cytochem. 4, 347 (1956).
(5) Singer, T. P., E. B. Kearney, and V. Massey: Advanc. Enzymol. 18, 65 (1957).

Lysosomal and Extra-lysosomal Localization of Acid Phosphatases in Neurons of Sensory Ganglia of Rats

KALINA, MOSHE and JOSE. J. BUBIS (Tel-Aviv University Medical School, Department of Cell Biology and Histology, Tel-Hashomer Government Hospital, Tel-Hashomer, Israel)

Different intracellular distributions of acid β-glycerophosphatases were found in different neurons of sensory ganglia using the Gomori technique. With the light microscope the histochemical reaction product appeared in large neurons, as a fine filamentous network resembling the Golgi apparatus and also as scattered small granules, whereas only a granular pattern was present in small neurons. No such distribution of β-glycerophosphatases was observed in neurons of the central nervous system.

Electron microscopic studies revealed that the activity in the small cells was localized in lysosomes. In the large neurons, however, activity was mainly associated with the Golgi apparatus and the endoplasmic reticulum.

Fixation in glutaraldehyde or formaldehyde completely inhibited the network pattern of activity in the large neurons but did not have an apparent effect on the reaction in the small neurons. Molybdate appeared to activate the β-glycerophosphatase in the large neurons but did not affect the enzyme in the small neurons; both enzymes were inhibited by fluoride and tartrate. The enzyme in the small cells had a broader pH range of activity than that in the large cells. Additional evidence suggests the existence of two distinct acid phosphatases in the two loci of the large neurons.

Improvements in a Rapid Cell Spectrophotometer for Higher Resolution Cell Identification

KAMENTSKY, L. A. (IBM Watson Laboratory, Columbia University, New York, N. Y.)

An instrument has been described previously for measuring and displaying certain optical properties of biological cells flowing in liquid suspension at rates up to 1000 cells per second. This instrument was designed to characterize each cell by the results of up to four simultaneous measurements of total absorption, scattering, or fluorescence at specific wavelengths in the ultraviolet or visible. In a later paper a method of switching the flow path to sort cells was also described.

In this paper, I will describe modifications in the flow channel, optics, and electronics of the instrument that have resulted in a considerable increase in the ability of the apparatus to discriminate among different cell types. A number of populations of interest are now resolvable with the spectrophotometer using absorption or fluorescent stains.

Relative Stability of Cellular DNA, Apurinic Acid, and the DNA-Dye Complex in the Feulgen Reaction

KASTEN, FREDERICK H. (Pasadena Foundation for Medical Research, Pasadena, Calif.)

In order to assess the relative stability of cellular DNA and its derivatives during the Feulgen reaction, comparative cytophotometric data were obtained from mouse liver and kidney imprints after various treatments. Following fixation, cells in separate experiments were exposed for three hours each to one of the following solvents during the course of an otherwise standard Feulgen reaction: absolute ethanol or water (before hydrolysis), water, acetone or xylene (after hydrolysis), water or acetone (after dye binding and sulfurous acid rinses), and absolute ethanol (before mounting). Control preparations treated in identical fashion, except for the extended time in the specific solvent, were employed in each experiment.

It was found that in terms of the Feulgen reaction DNA is stable to both absolute ethanol and water prior to acid hydrolysis. After partial acidic degradation to the apurinic acid residue, extended standing in water removes 45-65 percent of the original stainable complex. After hydrolysis, acetone and xylene removes 28 and 24 percent respectively of the apurinic acid. Binding of Schiff's reagent to apurinic acid restores stability of the complex to the solvent tested. Experiments are in progress to determine whether the labile apurinic acid in fixed cell nuclei is extracted as intact molecules of residual DNA or as more degraded fragments. The influence of solvent extraction after acid hydrolysis adds still another variable to the list of factors influencing Feulgen-DNA cytophotometry.

Morphologic Characterization of Cell Nuclei by a Cytophotometric Scanning Method

KIEFER, GUNTER G. and W. SANDRITTER (Institute of Pathology, 78 Freiburg-B., Germany)

A method for the objectivation of the morphology of Feulgen stained cell nuclei is presented. Cytophotometric measurements (scanning method) were used to obtain the optical density of chromatin particles. The particles were classified according to their optical density (extinction level separation method). This classification of the nuclear chromatin particles reveals characteristic frequency distribution in different types of cell nuclei. Three classes of the cell nuclei may be distinguished:

1. Small dense cell nuclei in which 70% of the extinction values fall into the highest extinction level class while 30% belong to lower extinction levels (thymus lymphocytes).

2. Large cell nuclei (liver cells, mouse ascites tumor cells) in which 90% of the optical density values of the particles are arranged around a medium value showing the presence of an unimodal frequency distribution pattern.

3. Medium-large cell nuclei (thymus lymphoblasts, Kupffer cells), in which two populations of chromatin particles are found; one with a low and one with high extinction values.

The possibility to use this method for quantitative estimation of eu- and heterochromatin is discussed.

Acridine Orange Microfluorimetric Studies on the Activation of the DNP-complex in PHA Stimulated Human Leukocytes

KILLANDER, D. and R. RIGLER (Institute for Medical Cell Research and Genetics, Karolinska Institutet, Stockholm 60, Sweden)

Fluorescence intensities of human leucocytes stained with the basic dye acridine orange (AO) were measured by microfluorimetry (1). In accordance with a previous observation (2), a pronounced increase in AO fluorescence of leukocytes was found already a few minutes after growth stimulation with phytohemagglutinin (PHA). The PHA induced increase in AO fluorescence—due to an increased AO binding to the deoxyribonucleoprotein (DNP) complex—was dependent on the concentration and type of PHA. Structural changes in the DNP-complex were found parallel to the augmented AO binding capacity in the stimulated cells (3).

It was concluded that the increased dye binding after PHA stimulation was due to an unscreening of negatively charged phosphate groups in DNA. The most likely explanation for this phenomenon is an alteration in the binding forces between DNA and nuclear proteins induced by PHA. This view was supported by the finding that acetylation of basic groups in nuclear protein resulted in a marked increase in AO binding to DNP in non-stimulated leukocytes having a low dye binding capacity. However, in PHA stimulated cells—exhibiting a high AO binding capacity—acetylation resulted only in a minute additional dye binding.

Leukocytes sampled from infected or newly vaccinated blood donors exhibited spontaneously a high fluorescence intensity after AO-staining. In these cases PHA treatment did not induce any further increase in AO-binding. It was also concluded that the observed changes in the DNP complex of PHA stimulated leukocytes might be one prerequisite for later cell growth.

128

References:

(1) Caspersson, T., G. Lomakka, and R. Rigler, Jr.: Acta Histochem. Suppl. VI, 123 (1965).
(2) Killander, D. and R. Rigler: Exptl. Cell Res. *39*, 701 (1965).
(3) Rigler, R. and D. Killander: *These abstracts.*

Differentiation of Noradrenalin (NA) and Adrenaline (A) Containing Cells of the Adrenal by Fluorescence Histochemistry *

KING, M. P. and E. T. ANGELAKOS (Boston University School of Medicine, Boston, Mass.)

The formaldehyde (HCOH) method of Falck and Hillarp and the newly developed trihydroxyindole (THI) method (Angelakos and King, Nature *213, 391, 1967) were applied to the differentiation of NA and A cells in the adrenal medulla of a number of species (hamster, rat, rabbit, cat, etc.). In frozen sections (followed by freeze-drying) both NA and A cells gave the THI reaction without the necessity of any pretreatment. A distinct difference in fluorescence colors between NA (blue) and A (green) cells could be detected visually with the THI method. The NA and A cells in the adrenals of different species were found to have the same pattern as has been described with other histochemical methods. After the THI reaction microspectrofluorometry of pure solutions ("models") showed fluorescence spectra which corresponded closely with similar spectra obtained by microspectrofluorometry from NA and A cells with peak emission at 480 mμ (NA) and 515 mμ (A). A small difference, NA (470 mμ), A (480 mμ), was obtained by microspectrofluorometry with the HCOH method which, however, was difficult to detect visually. These results show that although NA and A may be differentiated by microspectrofluorometry after HCOH, the THI method is much superior in this respect. The procedure for the THI method for tissue sections has been fully developed and will be described.

* Supported by National Science Foundation grant GB 4386 and USPHS Career Development Award K3-HE 15,457.

Histochemical Studies on Transaminases in Animal Tissues

KISHINO, YASUO (Department of Surgery, School of Medicine, Tokushima Univ., Japan)

The development and evaluation of histochemical methods for localization of transaminases, including aspartate transaminase (GOT) and ornithine transaminase (OTA) were performed. For the demonstration of GOT activity in tissue slices, 6-benzamido-4-methoxyl-m-toluidine dia-zonium zinc salt was used since this couples easily with oxalacetate pro-duced by this enzyme reaction and yields a red colored compound. After the tissues are fixed with cold hydroxyadipaldehyde, frozen sections (20μ) are incubated for one hour in a medium of the following com-position, which is freshly prepared before use: 2ml of 0.2 M phosphate buffer at pH 7.4, 1 ml of 0.2 M L-aspartic acid, 1 ml of 0.1 Mα-ketoglu-taric acid, 5 mg of diazonium salt and 0.5 ml of 0.1 M EDTA. The reac-tion is terminated by washing the slices with distilled water and the stain is fixed with cupric sulfate solution. The specificity of this method for GOT was proved in control experiments by tests in the presence of B_6-enzyme inhibitors, such as INAH or hydroxylamine and in the ab-sence of aspartic acid. The presence of the enzyme in heart and skeletal muscles, liver cells, brain (especially cerebellar Purkinje cells) and the convoluted tubules of the kidney of rats was clearly observed.

For the detection of OTA in the tissue slices, frozen sections were in-cubated in a mixture containing L-ornithine, α-keto-glutaric acid, DPN, glutamic dehydrogenase, nitro-BT and veronal buffer at pH 7.4 for 15 minutes. The exogenous glutamic dehydrogenase catalizes the oxidative deamination of glutamate produced by OTA reaction and thus forms α-ketoglutarate, DPN being concomitantly reduced. Reoxidation of DPNH is coupled with the reduction of nitro-BT to yield formazan which deposits in situ and serves as the indicator of enzyme activity. We also performed the control experiments to demonstrate the specificity of this reaction by tests in the presence of several inhabitors, such as L-valine, canalline or hydroxylamine, and without L-ornithine or glutamic dehydrogenase. High activity was found in the tubules of the kidney and periportal areas of the hepatic lobules of rats.

Acetylcholinesterase (AChE) and Monoamine Oxidase (MAO) in the Hypothalamic Neurosecretory System

KOBAYASHI, H., A. URANO, and K. YOKOYAMA (Zool. Inst., Fac-ulty of Science, Univ. of Tokyo, Japan)

The neurosecretory cell body of the Japanese quail, tree sparrow and rat showed the most conspicuous concentration of AChE. The neurose-cretory axonal tract did not show any AChE reaction. The median eminence generally gave a moderate reaction, and the neural lobe showed a weak AChE activity. Thus, the neurosecretory neurons are of cholinergic nature. MAO activity was very weak in the neurosecretory cell perikarya and their axonal tract of the Japanese quail and the mouse. However, MAO activity was strong around the neurosecretory cell body. This suggests that the monoaminergic axons terminate densely at the cell bodies. In the median eminence of the quail and the mouse, the external layer gave a high concentration of MAO activity, whereas the internal layer showed a weak activity. The neural lobe of these animals gave a weak activity. In these organs, however, a strong reaction was observed around the capillaries. Tetra-nitro-blue tetrazolium demonstrated sharper staining pictures than nitro-blue tetrazolium.

Approaches to the Electron Microscopic Localization of Acetylcholinesterase

KOELLE, GEORGE B. (Department of Pharmacology, University of Pennsylvania, Philadelphia, Pa.)

The acetylthiocholine procedure, introduced nearly twenty years ago, has retained the advantage of being the most specific histochemical method available for localization of acetylcholinesterase (AChE, acetylcholine hydrolase, E.C. 3.1.1.7) and non-specific cholinesterase (ChE, acylcholine acylhydrolase, E.C. 3.1.1.8). Although the method has undergone numerous modifications to improve accuracy of localization, it has certain limitations for optimal results in electron microscopy, in particular the quaternary ammonium (hence poorly penetrating) structure of the substrate, and the crystalline nature of the primary reaction product. The thiolacetic acid method, while extremely non-specific, is superior in both these respects. By employing the two methods, with aurous gold as the capturing agent, the advantages of both have been obtained. Recently, further improvement in the localization of AChE at the motor endplate of skeletal muscle has been achieved by the gold-dithiolacetic acid ($CH_3COSSCOCH_3$) and gold-dithiolacetate ($Au(CH_3COS)_2$) methods. These substrates exhibit unusual properties that suggest that they react simultaneously with two esteratic enzymatic sites. Accordingly, implications can be drawn regarding the distribution of the enzyme at the membrane surfaces.

Cytochemical Changes Produced in Spinal Neurons by Fluorocitrate [*]

KOENIG, HAROLD and CHARLES HUGHES (VA Research Hospital and Northwestern U. Medical School, Chicago, Ill.)

Fluorocitrate (FC) is a selective inhibitor of the Krebs cycle which competitively blocks the aconitase-catalyzed conversion of citrate to isocitrate. When injected into the lumbar cistern of cat in a dose of 6-30

μg, FC causes clonic-tonic seizures in hind-limbs and trunk after a latent period of 30-120 minutes. Biochemical and metabolic studies (Koenig and Patel, 1967) indicate that FC inhibits the Krebs cycle in cat spinal cord by 20-45%. Oxygen consumption is not depressed initially due to utilization of the free amino acids, glutamic acid, glutamine, and aspartic acid for energy metabolism. Later the ATP level falls due to the increased energy requirements of seizuring neurons. Morphologic and histochemical studies were performed on spinal cord fixed *in situ* by vascular perfusion with buffered 4% paraformaldehyde or 4% glutaraldehyde, and stored in a cacodylate-buffered sucrose medium (Sabatini *et al.*, 1963). Frozen and nonfrozen sections (Smith and Farquhar, 1963) were stained for acid phosphatase (APase) (Gomori, 1952) and thiamine pyrophosphatase (TPPase) (Novikoff and Goldfischer, 1961) activities and examined by light and electron microscopy. Control sections incubated without substrate gave no reaction.

During the latent period, the principal changes in gray matter are: (1) a swelling of mitochondria in the perkaryon and dendrites of neurons (and glia); and (2) early neuroaxonal dystrophy, shown by an accumulation of APase-positive lysosomes in axon hillocks, and increased numbers of lysosomes and mitochondria within axons. After convulsions commence, these changes become more severe. In addition, single or multiple axonal balloons develop which are stuffed with APase-positive dense and lamellated lysosomes, normal and degenerating mitochondria, neurofibrils, and other structures. These dystrophic changes result from a disgorgement of cytoplasmic constituents into the axon, and seem to be initiated by an expansion of the mitochondrial compartment.

TPPase, a reference enzyme for the Golgi apparatus (GA), undergoes striking dislocation in FC-poisoned spinal neurons. In the light microscope, the concentration of enzyme product deposited in the GA diminishes. Coincidentally, Nissl bodies become reactive, followed later by a more diffuse staining. In the electron microscope, enzyme product is lost from the cisternae of the GA, and appears in many cisternae of the RER. Later some enzyme deposits also are diffusely distributed in perikarya, axons and nerve endings. A comparable displacement of TPPase has not been previously reported to our knowledge. We conclude that the TPPase stored within cisternae of the GA is conveyed through an essentially closed system of interconnecting, membrane-limited channels into cisternae of the RER. This route is the reverse of that proposed by Palade and associates as the general pathway for the intracellular transport of secretory proteins in the exocrine pancreas. It is significant that the transport mechanism in guinea pig pancreas is blocked by inhibitors of respiration and oxidative phosphorylation (Jamieson and Palade). The retrograde flow of TPPase from the GA to the RER in

132

FC poisoning may be related to an interruption of this transport mechanism.

References:

(1) Koenig, H. and A. Patel: J. Cell Biol. *35*, 72A (1967).
(2) Jamieson, J. D. and G. E. Palade: J. Cell Biol. *35*, 62A (1967).

* Supported by grants from NIH (NB04493 and NB05509) and the National Multiple Sclerosis Society (No. 372).

Carbohydrate-rich Tissue Components in Lung Cancer

KORHONEN, L. KALEVI (IInd Department of Pathology, University of Helsinki and Department of Anatomy, University of Turku, Finland)

In the present investigation, a battery of histochemical methods was applied to characterize the carbohydrate-rich tissue components (muco-substances) in lung neoplasms. Glycogen was present in epidermoid and large clear cell carcinomas. A sialidase-labile compound was revealed in adeno- and mucoepidermoid carcinomas. In other neoplasms, the secretion of neoplastic cells was sialidase stable. Small cell anaplastic carcinomas and bronchial adenomas, although having some gland-like structures, did not show secretory activity in neoplastic cells.

In most of the malignant neoplasms, a peculiar stromal reaction was observed. It consisted of hyaluronidase-labile mucosubstances, the main part of which was interpreted to be hyaluronic acid. Some sulphate-containing compounds corresponding to chondroitin sulphates were also present. This stromal material surrounded the islands of the neoplastic cells. It is possibly related to the high levels of carbohydrate-rich compounds revealed in serum, urine, and serous effusions in patients with malignant neoplasms, including those types in which the neoplastic cells themselves do not produce these substances. The possible site of synthesis of these compounds is in the stroma of the neoplasms. Some hyaluronidase and/or sialidase-labile material was also revealed surrounding the neoplastic cells more intimately. The presence of a mucosubstance "coat" around the malignant cells has been proposed which acts as a barrier, preventing the immunological reaction of the host tissue against the neoplastic cells.

The role of the mucosubstances in the morphological typing of lung cancer is discussed, as well as their relation to the stromal response and the lymphocytic and mast cell reactions. Some comparisons of the stromal reaction in neoplasms with that in inflammatory conditions are also presented.

Protein Synthesis in the Pancreas of Rat and Ferret After Feeding

KRAMER, M. F. and C. POORT (Department of Histology, Faculty of Medicine, University of Utrecht, The Netherlands)

The rate of protein synthesis is determined biochemically by measuring the radioactivity per unit DNA of the TCA-insoluble fraction and in autoradiography by counting the number of grains per cell in the pancreas of animals killed 10 minutes after i.v. injection of leucine-^{3}H. These methods are to be preferred over the measurement of the specific radioactivity of the protein present since the protein content of the gland varies during the secretory cycle.

Previous work with the above mentioned techniques did not reveal an increase in the rate of protein synthesis in the pancreas of rats after stimulation of the secretion by pilocarpine (1). This result is in contrast with the commonly accepted view on the relation between synthetic and secretory activities of the pancreas.

In order to investigate the effect of a physiological stimulus we now have measured the rate of protein synthesis up to 4 hours after feeding of previously fasted rats. Again, no increase in the rate of protein synthesis has been found.

Considering that in rats digestion is a continuous process, we repeated the feeding experiment with a predatory animal, viz., the ferret. Because of its discontinuous digestion, we expected the pancreatic activity to vary more clearly with the digestive function than in the rat, but in ferrets, too, no increase in the rate of protein synthesis up to 4 hours after feeding has been found.

The problems which arise from the presence of a continuous protein production with regard to the fate of the protein during absence of secretory stimuli will be discussed.

References:
(1) Kramer, M. F. and C. Poort: Zeitschr. Zellforsch., *in press* (1968).

Histochemical Demonstration of a Colchicine-induced Blockage of Enzyme Transport in Axons of Peripheral Nerves *

KREUTZBERG, GEORG W. (The Rockefelller University, New York, N. Y.)

It is known that after any injury of a peripheral nerve, axoplasm will pile up in the axons of the proximal stump. Oxidative enzymes especially have been shown to increase dramatically in these axonal stumps during the first 48 hours after the operation. On the ultrastructural level, an accumulation of mitochondria was found as the morphological

134

equivalent of the enzyme histochemical findings. There is evidence that mitochondria are piling up due to the damning of the proximo-distally moving column of axoplasm. Consequently, the increasing appearance of oxidative enzymes demonstrable by histochemical techniques can be taken as an indicator for axoplasmic transport.

By local administration of colchicine solution at the site of the injury, the piling up of oxidative enzymes (DPN-diaphorase, succinic dehydrogenase) was enhanced. Normal enzyme activity in the vessels and the Schwann cells did not change. It is assumed that the enzymes are not directly affected but their transportation is reduced.

The mechanism of colchicine action on the protein constituents of the axoplasm is discussed and the possibility of interfering with the transport mechanism in the axon is considered.

* Supported by NIH Grant No. 1 RO1 NB 07348-01.

Histochemische Untersuchungen von Lipiden im Samenblasenepithel des Bullen.

KÜNZEL, ERICH and ATTILA TANYOLAC (Institut für Veterinär-Histologie und Embryologie der Freien Universität Berlin, 1 Berlin 33)

Als Besonderheit sind im zweistufigen Epithel der Samenblase des Bullen die Basalzellen als "Zwergfettzellen" ausgebildet. Es sind pluri- und univakuoläre Fettzellen mit 3-15 μm grossen Fetttröpfchen. Die in den Hauptzellen apikal vorkommenden zahlreichen Lipidtröpfchen haben eine Grösse von 0.2 bis 1 μm, im Durchschnitt messen sie um 0.5 μm. Die Untersuchung wurde nach Formol-Calcium-, Kaliumdichromat-, Kaliumdichromat-Sublimat-, Osmiumtetroxyd- und Bouin-Fixation und am unfixierten Material vorgenommen. Extrahiert wurde mit Methanol-Chloroform, Methanol-Chloroform-Aceton, Äther, Aceton und Pyridin. Folgende Färbungen bzw. Reaktionen wurden angewendet: Sudanschwarz B, kolloidales Sudan, Brillantkresylblau, Nilblau, Nilblausulfat, Luxol-Fast-Blue, Phosphin 3 R, Benzpyren-Coffein, PAS und Ultraviolett-Schiffreaktion. Für die Elektronenmikroskopie wurde mit Glutaraldehyd, Osmiumtetroxyd nach PALADE, Osmiumtetroxyd-Kaliumdichromat nach WOHLFARTH-BOTTERMANN und Kaliumpermanganat fixiert. Die Kontrastierung erfolgte mit Uranylacetat und Bleicitrat. Es wurden neutrale Lipide, ungesättigte Lipide und Phosphatide nachgewiesen. Die Fetttröpfchen der Basalzellen bestehen vor allem aus Neutralfetten. Phosphatide kommen vor allem im apikalen Bereich der Hauptzellen vor; sie sind aber auch in den basalen Fettvakuolen nachweisbar. Die Reaktion für ungesättigte Lipide war ebenfalls basal und apikal positiv. Die histochemischen oder elektronenmikroskopischen

Befunde werden verglichen und das Ergebnis diskutiert, insbesondere auch im Hinblick auf die Lipidsynthese und den Lipidtransport.

References:

(1) Hake, T.: Lab. Invest. *14*, 1208 (1965).
(2) Riva, A.: J. Anat. *102*, 71 (1967).

Biochemical Studies on the Storage and Secretion of Anterior Pituitary Hormones *

LABELLA, FRANK S. (Univ. of Manitoba, Faculty of Medicine, Dept. of Pharmacology & Therapeutics, Winnipeg 3, Canada)

Centrifugation of bovine anterior pituitary homogenates has permitted the partial separation of cytoplasmic storage granules, each granule species enriched in one or more of the trophic hormones. Centrifugation, by itself, is inadequate for the preparation of hormone granules in a high degree of purity; isolated fractions show considerable overlapping of hormonal activities, and bioassays are probably complicated by interaction of multiple hormones present in any given fraction. The findings, however, suggest that each hormone is contained in a specific intracellular granule: GH and LTH are associated with the large acidophilic granules and LH, FSH, TSH, and ACTH with small granules. Concentrated efforts have been applied to the isolation of TSH-containing granules by differential and density-gradient centrifugation and by Millipore and Celite-column filtration. Granules with up to 5 U TSH/mg of protein have been isolated by these procedures. These studies may indicate whether the storage granules consist entirely of hormone protein or contain some inactive carrier protein. Gel filtration on Sephadex of TSH released from pituitary slices and extracted from isolated granules indicated that circulating TSH exists in physiological fluids as a monomer with a molecular weight about 28,000 and that pituitary TSH is stored in cytoplasmic granules in the form of readily solubilized and dissociated aggregates of the monomer. A biologically active fragment with a molecular weight of 7000 was prepared from a papain digest of purified TSH. Examination of the hexosemonophosphate (HMP) shunt in endocrine tissues indicated this pathway to be concerned more with the storage and secretion than with the synthesis of hormones. Metabolism by this pathway is not immediately coupled to the secretion mechanism but appears to be concerned with long-term processes, perhaps with renewal of membranes which are rapidly turning over in secretory cells. Quantitative estimation of the HMP shunt showed that the percentage of utilized glucose which was metabolized by this pathway amounted to 5, 4, and 3% in bovine anterior pituitary, posterior pituitary, and pineal body, respectively; the percentage was very low in brain cortex.

Synthetic vasopressin and oxytocin promoted the release of TSH from anterior pituitary *in vitro* in concentrations as low as 10^{-12}M; the dose-response curves for hormone release induced by each of these two peptides were bell-shaped, increasingly higher concentrations becoming less effective, ineffective, and even inhibitory. Epinephrine also released TSH *in vitro* but was less potent than the peptide hormones. Phenoxybenzamine blocked the stimulatory effect of epinephrine, but not of vasopressin, on TSH release. ^{14}C-1 or ^{14}C-6-glucose oxidation to ^{14}CO$_2$ was unaffected by concentrations of all three hormones which were maximally effective in enhancing TSH release. Incubations carried out aerobically at $0°$ resulted in only a slight depression of the stimulated release by vasopressin, but glucose oxidation was consistently reduced to approximately 10% of the level at $37°$. There appears to be no immediate, direct coupling between glucose oxidation and spontaneous or stimulated release of TSH from cytoplasmic stores.

The effects of vasopressin and oxytocin on ACTH release *in vitro* were almost identical to those on TSH release. In several experiments the effects of cortisol on the release of both ACTH and TSH from the same tissue preparation were examined. Cortisol consistently depressed ACTH release below control levels; increasing concentrations of the steroid produced a progressive decrease in ACTH release or, in some cases, became less inhibitory. In each case, release of TSH showed an inverse relation to that of ACTH. Experiments in which cortisol or thyroxine were combined with vasopressin or oxytocin suggested possible mechanisms whereby pituitary secretion is regulated by brain mediators and target gland products.

* Supported by USPHS grant AMO 5896 and the Medical Research Council of Canada.

Two Techniques for the Light and Electron Histochemical Demonstration of Acidic Substances

LANGLEY, O. K. (Institute of Neurology, Queen Square, London, W.C. 1., England)

Electron microscopists have now achieved the ultrastructural localization of a variety of enzyme activities in biological tissue. Of equal importance is the histochemical localization of biochemically definable macromolecules at an ultrastructural level. Only when this is accomplished may a correlation between structure and physiological function of biological tissue elements be meaningful. The electron histochemist seeks electron-dense compounds which act specifically on chemical ligands of tissue components and which are resistant to removal both by organic solvents of preparative manoeuvres and by the electron beam. Having selected such a stain, he examines it's specificity by the application of histochemical blocking reactions and enzymic digestion.

This report will discuss the results of staining peripheral nerve with colloidal iron oxide at low pH, and some attempts which examine the specificity of the method (1). The use of various heavy metal cations as stains for acidic tissue components also will be reported. Both the Fe^{3+} and Ba^{2+} ions appear suitable for electron histochemical application and both show remarkable specificity for the nodes of Ranvier of peripheral nerve. The affinities of the cations for specific chemical groups will be discussed and experiments will be described that shed some light upon the nature of the bond between cation and tissue binding sites. Tentative conclusions are drawn as to the chemical nature of the stainable entity.

References:

(1) Langley, O. K. and D. N. Landon: J. Histochem. Cytochem., 1968 *in press.*

Thésaurismose Hystiocytaire Glyco-protidique dans la Maladie de Cooley

LANZA, GIOVANNI B. (Istituto di Anatomia e Istologia Patologica dell' Università di Ferrara, Italy)

L'étude systématique de la moelle osseuse et de la rate sur plusieurs sujets atteints de la maladie de Cooley a démontré une constante hyperplasie du SRI avec, en général, des accumulations de volumineuses cellules histiocytaires, dans la plupart des cas aussi nombreuses—notamment dans la moelle—qu'elles réalisent des images de thésaurismose. Les cellules histiocytaires proliférées n'ont pas apparence écumeuse et ne renferment pas substances soudanophiliques, mais elles accumulent dans le cytoplasme un matériel glycoproteique pas-positif. Les investigations cytochimiques ont démontré deux differentes fractions glyco-protiques: 1) l'une acide, métachromatique et alcianpositive à pH 4, sensible à la sialidase et à l'hydrolyse faible par l'HCl, mais qui est résistante à la digestion par les hyaluronidases; 2) l'autre neutre, seulment pas-positive, pas attachée par la diastase, sialidase, hyaluronidases et hydrolyse acide. Le phenomen, qui est charactéristique de la maladie de Cooley et pour celà utile pour le diagnostique, doit être attribué aux modifications pathologiques de l'érytropoïèse cooleyenne et, comme telle, considéré exprèssion du damage métabolique des sialo-glyco-proteines.

Histochemistry of "Multiple Pheromone" Secretions of the Army Ants (Dorylinae)

LAPPANO-COLLETTA, ELEANOR RITA (Manhattan College, School of Arts and Science, Riverdale, Bronx, N. Y.)

Histochemical analyses of immature Army Ant secretory tissues (cuticular, salivary and neurosecretory), indicate a strong correlation between the onset of secretory function in the developing brood, and the adult colony behavior pattern. Intergeneric comparisons of the Dorylinae (*Eciton* and *Neivamyrmex*), have shown marked histological changes of secretory tissues coincident with the time of onset of high level colony activity (1, 2).

The use of histochemical techniques has delineated certain changing patterns of neurosecretory and digestive substances in tissues of the developing brood, coincident with critical periods of change in the cyclic adult colony behavior pattern of these ants. In addition, multiple sites of cuticular secretory activity and moulting substances, possibly pheromones in nature, have been recorded using histochemical staining procedures. These substances are believed to have potentially significant afferent functions, which will be tested in the field.

A hypothesis of summation effectiveness of "multiple pheromones" from these secretions of developing brood (larvae, prepupae and pupae) is set forth. This hypothesis correlates well with the established theory of cyclic colony behavior of the Army Ants. (3, 4) The possible role of these substances as "multiple pheromones" and as chemical and biochemical basis to the behavior pattern of the Doryline, Army Ants, is considered.

References:

(1) Lappano-Colletta, E. R., U. Geduldig, and T C. Schneirls: Nature, *207*, 959 (1965).
(2) Lappano, E. R.: Insect Soc., *5*, 31 (1958).
(3) Schneirla, T. C.: J. Comp. Psychol. *25*, 51 (1938).
(4) Schneirla, T. C.: Proc. Amer. Philos. Soc. *101*, 106 (1957).

Cell Renewal in the Crypts of Lieberkuhn of the Small Intestine in Mice

LEBLOND, C. P., J. MERZEL, and HAZEL CHENG (McGill University, Department of Anatomy, Montreal, Canada)

At various times after a single injection of ^{3}H-thymidine, small intestines of adult female mice were fixed by perfusion of the animals with paraformaldehyde. Pieces of duodenum, jejunum, and ileum were embedded in Epon, serially sectioned at 1 μ and taken through iron hematoxylin to stain nuclei and safranin to stain mucus. The intestinal epithelium is known to be composed of columnar cells, as well as mucus-containing goblet cells, Paneth cells and argentaffin cells (found to number 376, 24, 25, and 2 respectively in serial sections of a jejunal crypt). By preparing radio-autographs, it was hoped to examine whether these various cell types proliferate and are renewed.

At one hour after ^{3}H-thymidine injection, about 35% of the crypt *columnar cells* were labeled. Reactions were lacking over typical goblet cells. However, they were present over cells containing a few mucous droplets but otherwise looking like columnar cells and mainly restricted to the lower half of the crypts. Whereas typical goblet cells do not divide, these mucous cells do. At 12 hours, reactions appeared over some typical goblet cells, the incidence increasing with time. The results support the hypothesis that the crypt cells with a few mucous droplets proliferate and become goblet cells. Since only 14% of all crypt *mucous cells* are labeled, their renewal rate would be less than of columnar cells.

At one hour after injection, 100 crypts in each of five animals were examined. A reaction was present over *argentaffin cells* in only three mice. The average incidence of label (for the five animals) was about 4%. At later intervals after injection, the frequencies observed indicate that argentaffin cells renew slowly.

At the one-hour interval, no reaction was observed over *Paneth cells* in the five animals examined. However, by 2-4 days, a reaction did appear over an average of 2% of these cells. Presumably, then, Paneth cells arise from undifferentiated cells and are renewed quite slowly.

Thus, all cell types arise in the crypts. Labeled columnar, goblet, and argentaffin cells are found on villi at later time intervals; and, therefore, the steady state can be maintained by their loss from the villus tips. Labeled Paneth cells, however, remain in the lower half of the crypts; and the mode of cell loss is not known. In conclusion, all cell types undergo renewal in the crypts, but at quite different rates.

The Role of Thyroid Follicle Structures in the Elaboration of Thyroid Hormones

LEBLOND, C. P., N. J. NADLER, and ANNETTE A. HERSCOVICS (McGill University, Department of Anatomy, Montreal, Canada)

Each thyroid follicle functions as a unit in elaborating the iodinated glycoprotein, thyroglobulin, and in breaking it down to release the thyroid hormones, tetraiodothyronine (thyroxine) and triiodothyronine. After ^{3}H-leucine injection into rats, EM radioautographs of follicular cells revealed that at ten minutes silver grains were over the ribosomes of endoplasmic reticulum, later over the cisternae, by one hour over the Golgi apparatus, still later over apical vesicles, and after two hours over the lumenal colloid. These results indicate that the protein moiety of thyroglobulin is synthesized in association with ribosomes, migrates through the cisternae to the Golgi, and is secreted via the apical vesicles into the lumenal colloid. Furthermore, in thyroids incubated with ^{3}H-leucine *in vitro,* protein was shown to be synthesized as subunits of thyroglobulin, which aggregate about thirty minutes later. This synthesis is greatly inhibited by puromycin. Presumably, subunits are produced on ribosomes and aggregate during their migration through cisternae and/or Golgi.

In thyroids incubated with labeled galactose, radioactivity was associated from the start with the protein aggregate and was not decreased by puromycin during the first hour of incubation. Presumably, galactose addition occurs on preformed protein during migration, possibly in the Golgi.

After injection of ^{125}I, radioautographs of frozen-dried thyroids appropriate to retain water-soluble iodide showed silver grains over both cells and colloid. However, using conventional fixations to retain only iodine attached to thyroglobulin, silver grains appeared over colloid only. Hence, iodide is taken up by cells, deposited into the colloid and incorporated there into thyroglobulin.

Follicular cells send out from their apices streamers of cytoplasm which pinocytose droplets of colloid. When the colloid was prelabeled with ^{125}I, radioautographs showed silver grains over the droplets. After iodinated thyroglobulin is thus taken up into cells, it is proteolyzed by lysosomes, thereby releasing the hormones, which then pass through the cell bases into the circulation.

In conclusion, thyroglobulin subunits are first formed on the ribosomes, whence they migrate through the ER and Golgi. During this migration they aggregate and acquire carbohydrate, and are then secreted into the colloid. Iodination in the colloid lumen completes the formation of thyroglobulin. This substance is then taken up into the cells where it is proteolyzed to release thyroid hormones.

The Role of DNA Dependent RNA Synthesis in the Initiation of DNA Replication **

LEDERER, B.* and C. MITTERMAYER (Dept. of Pathology, University of Innsbruck [Austria] and Dept. of Pathology [Ludwig Aschoff Haus], University of Freiburg, Germany)

The synthesis of DNA during the mitotic cycle of mammalian cells is one of the most important events as it is a prerequisite for mitosis.
There are two possible ways of initiating DNA synthesis: a) It can start without foregoing specific transcription during G-1 phase, or b) it is triggered off by appearence of specific (DNA dependent) RNA, or by induction of specific enzymes. To get some insight into these possible

mechanisms, Actinomycin D was applied suppressing about 95% of total RNA synthesis, and the effect of this drug at different times of the mitotic cycle was checked.

The experiments were performed on random growing tissue culture cells (L-cells) and on mechanically synchronized L-cells (1). The rate of DNA synthesis was evaluated by means of autoradiography or liquid scintillation counting after ^{3}H-Thymidine pulse labelling. The age of the random growing single cells was determined by measuring DNA content by cytophotometry. In random culture during treatment with Actinomycin D, a drop in the mitotic index, in the labelling index and in the mean grain count/nucleus was seen. This can be explained as follows: Actinomycin D inhibits the start of DNA synthesis when applied during G-1 phase and it reduces the rate of DNA synthesis when given during S phase.

These facts are supported by data from experiments with synchronized cells. The experiments suggest a specific temporally limited transcription during G-1 phase for the initiation of DNA synthesis; apparently the information for the synthesis of specific enzymes is produced during this period of the life cycle.

References:

(1) Mittermayer, C., P. Kaden, and W. Sandritter: Histochemie *17*, 67 (1968).

---

* Fellow of the Alexander v. Humboldt-Stiftung.
** Supported by the Deutsche Forschungsgemeinschaft, the U.S. National Institutes of Health (5-RO5-TW 0024-02) and the Stiftung Volkswagenwerk.

Ultrastructural Localization of Glutamic Oxalacetic Transaminase (GOT) Activity in Rat Hepatocyte

LEE, SIN HANG and RICHARD M. TORACK (Department of Pathology, Cornell University Medical College, New York, N. Y.)

Oxalacetate, a primary reaction product of glutamic-oxalacetic transamination, is precipitated by Pb^{++} at a slightly alkaline pH, even in the presence of a chelating agent, aspartate. Based on this observation, a new histochemical method for demonstration of GOT activity in cryostat tissue sections has been developed (1). In light micrographs, the GOT reaction product is largely present as fine particles evenly distributed in the cytoplasm of the hepatocyte. The present communication deals with the adaptation of this histochemical method for electron microscopic localization of GOT activity in liver cell.

A thin slice of normal rat liver is rapidly minced in a cold (0-4° C) fixative containing 1% glutaraldehyde and 4% formaldehyde in 0.3 M sucrose, fixed in the same fixative for another two minutes and transferred to cold 4% formaldehyde in 0.3 M sucrose for an additional 30 minutes. These tissue blocks are washed in three changes of cold 0.3 M sucrose for half an hour. The fixative and washing media are all buffered at pH 7.2-7.4 with 0.05 M imidazole-nitric acid. Incubation of the tissue blocks is carried out at room temperature for 30 minutes in a modified GOT medium consisting of 20 mM 1-aspartic acid, 2 mM α-keto-glutaric acid, 6 mM lead nitrate, 50 mM imidazole and 0.3 sucrose, pH 7.2-7.4. Thereafter, the reacted tissue blocks are washed with 20 mM 1-aspartate in 0.3 M sucrose buffered at pH 7.3, postfixed by 1% O_sO_4, dehydrated and embedded in Epon. The dehydrating ethanol series is prepared by diluting absolute alcohol in 0.05 M imidazole-nitric acid, pH 7.5 in order to avoid undesirable effects of acid pH on the reaction product during the procedure of dehydration. Thin sections are examined unstained or after staining in lead citrate. Control experiments are carried out by following the same procedure except that the fixed tissue blocks are incubated in media in which α-ketoglutaric acid is withdrawn or in which d-aspartic acid is used in stead of 1-aspartic acid.

The final reaction product, lead oxalacetate is found to be precipitated in the limiting membrane and cristae of mitochondria as well as at the surface of the microbodies. There is no lead precipitate in the tissue of control experiments. The membranous localization of the mitochondrial GOT activity is in agreement with the available biochemical data (2) whereas the presence of GOT activity at the surface of the microbodies has not been supported by fractionation studies (3). The latter discrepancy between cytochemical and biochemical findings may be due to a high solubility of the enzyme at the microbody surface, which dissolves readily in the aqueous medium during the fractionation procedure, or it may be due to an activation of a latent enzyme on the microbody surface by one of the components either of the fixative or of the incubation medium.

References:

(1) Lee, S. H.: Amer. J. Clin. Path. *In press* (1968).
(2) Eichel, H. J. and J. Bukovsky: Nature *191*, 243, (1961).
(3) de Duve, C. and P. Baudhuin: Physiol. Rev. *46*, 323 (1965).

Glucose Metabolism in Rat Cerebellum Tissue Cultures as a Function of Age [*]

LEHRER, GERARD M. and MURRAY B. BORNSTEIN (The Mount Sinai School of Medicine of the City University of New York and The Albert Einstein College of Medicine, Bronx, N. Y.)

Methods will be presented for the study of glucose consumption and lactate production in volumes of less than 100μl by single tissue cultures of rat cerebellum growing in a Maximov slide assembly. Each culture represents 75-150μg wet weight of metabolizing tissue. Earlier obstacles to measurements due to inconstancy of volumes of media were overcome by including with the culture medium a small amount of ^{3}H antipyrene and referring all volumes to the original 0 time volume. The antipyrene did not interfere with normal development and myelinization of the cultures. The method also allowed determinations of total tissue water. Glucose consumption increased from 20-25 mmoles per kilogram wet weight per hour (mMKH) in 8-day-old cultures to 45-50 mMKH in 15 and 22-day-old cultures. In all cases, lactate production remained less than 2mMKH.

It is significant that these findings agree extremely well with the estimates of Lowry in developing mouse brain *in vivo* (1) as well as with previous measurements in living human brain (2) and other nervous tissues in other species (3). The fact that lactate production was extremely low speaks for the marked health of these tissues and is to be contrasted with the considerably higher lactate production values found in experiments with brain cortex slices *in vitro* (4).

The findings will be compared with previous data in tissue culture (5) and their significance will be discussed.

References:

(1) Lowry, O. H., J. V. Passonneau, F. X. Hasselberger, and D. W. Schulz: J. Biol. Chem. *239*, 18 (1964).
(2) Reinmuth, O. H., P. Scheinberg, and B. Bourne: Arch. Neurol. *12*, 49 (1965).
(3) Larrabee, M. G. and J. D. Klingman: in Neurochemistry, 2nd Ed., p. 157 (Elliot, K.A.C. et al., eds.) Springfield, Ill. C. C. Thomas (1962).
(4) McIlwain, H.: Biochemistry of the Nervous System, 3rd Ed., London, J. and A. Churchill (1966).
(5) Lehrer, G. M. and M. B. Bornstein: Exerpta Medica. Int. Cong. Series. No. 94, p. E140 (1965).

* Supported by Grants No. 293 and 433 from the National Multiple Sclerosis Society and USPHS Grants No. NB-05368 and NB-06735.

The Measurement of Sucrose Space and Sodium in Subcellular Components of Single Puffer Fish Supramedullary Neuron Cell Bodies (SMN's) *

LEHRER, GERARD M. and ROBERT KATZMAN (The Mount Sinai School of Medicine of the City University of N. Y. and The Albert Einstein College of Medicine, Bronx, N. Y.)

Seventy μC of ^{14}C sucrose, s.a. 420mC/mM, were applied in three doses over a 30-minute period to the medulla of a conscious, well-aerated specimen of *Spheroides maculata*. The brain was then frozen *in situ* in liquid nitrogen-chilled Freon and 30μ brainstem sections were prepared in a cryostat and dried without thawing. 5μ sections were also cut for radioautography of diffusable materials (1). Fragments of nuclei and cytoplasm of SMN's were dissected from the dry sections. Their area was measured and they were weighed on quartz fiber balances. The radioactivity of each fragment was measured and sodium measurement was then performed with a special micro flame photometer with a precision of better than 3%.

Remarkably high concentrations of sodium were encountered, both in nuclei and cytoplasm. The radioactive sucrose was used as an extracellular marker to control for movements of soluble substances between compartments during freezing, handling, and sectioning. Clear differences in distribution were shown between radioactivity from sucrose and sodium concentrations. Chromatography showed all tissue ^{14}C to be in sucrose. Both radioautography and measurement showed markedly lower levels of radioactivity in nucleus and cytoplasm of SMN's than in other tissue areas. Lactate levels in the same sections ranged from 1.7-1.95mM, indicating little or no anoxia prior to freezing.

Details of the radioautographic technique as well as of techniques for volume and density determinations on tissue fragments will be presented. The comparison between data obtained by the microflame photometric method and that obtained by electron microprobe X-ray spectrometry and activation analysis will be discussed (2).

References:

(1) Weeden, A. P. and H. I. Jernow: Amer. J. Physiol. *In press,* (1968).
(2) Katzman, R., G. M. Lehrer, C. E. Wilson, and A. Yam: J. Cell Biol. *31*, 57A (1966).

---

* Supported by USPHS Grants No. NB-05368 and NB-01450.

Role of Changes in Enzyme Activity of Cells in Intracellular Propagation of the Agents of Some Enteric Infections

LEITES, F. L., JU. J. TENDETNIK, E. S. RUCHADSE and O. E. RJADNEVA (Central Institute of Epidemiology, Ministry of Health, Moscow, USSR)

The interactions between *S. typhi abdominalis* and cells of the reticuloendothelial system were studied on the macrophages of bone marrow in tissue culture. It was shown cytochemically that, beginning from the first hours after challenge, the contents of ribonucleic acid and the activity of the enzymes diffusely distributed in the cytoplasm was markedly reduced in infected cells when compared with cells free from the bacteria. Among these enzymes (activity of which falls even lower in the later stages of phagocytosis) can be noted aliesterase, lipase, lipoprotein lipase, leucinaminopeptidase and others. Contrary to this, the activity of lysosomal enzymes cathepsin C and acid phosphomonoesterase markedly increases during the first three days after challenge while the number of lysosomes in the cells increases too. Synchronous activation and the identical distribution of both lysosomal enzymes in the cells point out the role of lysosomes as the organelle of phagocytosis. While the activity of lysosomal enzymes remains high, propagation of bacteria in the cells is insignificant. In the later phases of phagocytosis, as the activity of lysosomal enzymes falls, there occurs massive propagation of the agent in the cells causing their gradual distrophy and decay.

Analogous facts were stated while studying the activity of lysosomal enzymes in the rabbits which, after having the paratyphoid infection, remained bacteria carriers for a long time—up to a year. We showed that in internal organs of carriers, the activity of lysosomal enzymes is much less than in noncarriers. This was proven by direct microphotometry of sections of the liver after histochemical reaction on cathepsin C (the optical density in the sections of the carriers $= 0.34 \pm 0.052$, in the noncarriers $= 0.56 \pm 0.047$). These data suggest the role of disturbances of function of lysosomal enzymes in the pathogenesis of bacterial carriage.

An Electron Microscopic Histochemical Study of Connective Tissue Ground Substance in the Uterine Cervix of the Mouse *

LEPPI,** T. JOHN, PATRICIA A. KINNISON, and SUSAN P. GAFFNEY (School of Medicine, The University of New Mexico, Albuquerque, New Mexico)

A histochemical identification of connective tissue ground substance will depend primarily on the polyanionic nature of the acid glycosaminoglycans which comprise a large part of some extracellular matrices. Another critical requirement is the adequate preservation of sufficient ground substance with most of its anionic character intact and available for interaction with a staining reagent that results in an electron-dense complex.

The connective tissue components of the uterine cervical wall in the ovariectomized mouse, when under the influence of certain exogenous hormones of pregnancy, undergo abrupt changes resembling those associated with pregnancy and post-partum in the intact animal. A previous light microscopic histochemical study (1) of this connective tissue site revealed hormonally-induced alterations in gross size that were directly correlated with the amount of stainable ground substance present in the connective tissue extracellular space. In general, the larger cervices contained more stainable acid carbohydrate. Since the identification of connective tissue ground substance is obscure as compared to the clearer visualization of cells and certain fibers, an electron microscopic histochemical study of uterine cervical stroma from mice in different hormonal states was undertaken.

Several adult Swiss mice were ovariectomized and two weeks later each mouse received 4 daily subcutaneous injections of either $2\mu g$ estradiol benzoate (EB) in sesame oil (SO), 150 units relaxin (R) in benzopurpurine (BP), EB and R, SO or BP. All cervices were removed 24 hours after the last injections and small pieces of cervix were fixed 2-4 hours at room temperature in Flickinger's modification (2) of Karnovsky's paraformaldehyde-glutaradehyde mixture. Tissues were then rinsed for 30 minutes in several changes of 0.15M cacodylate buffer prior to being sectioned at 40-80μ in a 7% agar mounting medium on the Smith-Farquhar tissue sectioner. The sections of uterine cervical stroma were subsequently washed briefly in the cacodylate buffer to remove excess agar and then exposed from 12-18 hours to Mowry's variant of the Hale iron stain without Prussian blue conversion. Sections were then post-fixed in phosphate-buffered OsO_4 for 45 minutes, dehydrated and then embedded in Epon. Thin sections were viewed with or without further staining. In addition, some sections were exposed to 0.05% testicular hyaluronidase for 4 hours 37°C in 0.1M phosphate buffer (pH 5.5) prior to iron staining. Control sections were exposed to the buffer solution for the same time and temperature.

The administration of the hormone vehicles, SO or BP, or R alone resulted in a cervical stroma that contained closely packed, inactive fibroblasts characterized by small cytoplasmic processes and negligible endoplasmic reticulum (ER) and Golgi network. Nuclei of these "idle" fibroblasts contained condensed chromatin masses. Staining of an extracellular material by the polyvalent cationic iron reagent was minimal in some sections, but more often totally absent. This lack of iron staining was particularly noticeable in sections that were not subsequently stained by uranium and lead since the iron particles (30-60Å) could have been obscured by such staining. In comparison, cervical stroma from mice that received EB or EB and R was composed of fibroblasts that appeared more operative. The rough ER of these cells was extensive and widely dilated and most contained a prominent Golgi. In addition, cytoplasmic processes of the cells were attenuated and extended for considerable distances. The chromatin in the nuclei of these cells was limited to a thin rim just inside the nuclear membrane. There was prominent iron staining of both extracellular matrix and areas between collagen fibers, particularly in cervical walls from mice receiving both EB and R. Interfiber iron staining was indistinct in cervical sections from mice which received EB alone. Testicular hyaluronidase digestion resulted in an obvious reduction of iron staining in extracellular areas although staining remained in sites immediately adjacent to collagen fiber bundles and cell surfaces. Iron staining was unaffected in buffer control sections.

References:

(1) Leppi, T. J.: Anat. Rec. *150*, 51 (1964).
(2) Flickinger, C. J.: Z. Zellforsch. *78*, 92 (1967).

* Supported by United States Public Health Service grant AM-11064-02.
** Recipient of a Lederle Medical Faculty award, 1968-71.

Histochemical and Biochemical Studies of Canine Gastric Mucus Secretion

LEV, ROBERT, ANDRE GERARD, JACQUES DE GRAEF, and GEORGE B. JERZY GLASS (Department of Pathology & Section of Gastroenterology, New York Medical College, New York, N.Y.)

The dog gastric mucosa was examined histochemically and for in vivo synthesis of sulfated mucosubstances using 1 mc/kg body weight of Na_2 $^{35}SO_4$ i.v. in the resting state and following urecholine and histamine administration. The findings were correlated with chemical data on gastric juice from canine Heidenhain pouches.

In the resting mucosa, a sulfated, sialic acid-containing glycoprotein was identified histochemically within the surface epithelial cells and a similar material was recovered from the "visible mucus" component of the gastric secretions. A chondroitin sulfate (CS)-like compounds was found in fundic chief cells and pyloric glands and an analagous substance in the gastric juice where it was characterized chemically and electrophoretically as CS-A or CS-C and further identified by infrared absorption spectrophotometric studies (1) as CS-A. A neutral glycoprotein of blood group substance type was found in surface epithelium and in mucous neck and parietal cells and felt to be the source of a similar compound remaining in the supernatant after the acidic components of the gastric juice had been precipitated with cetylpyridinium chloride. Following a four hour infusion of the vagomimetic drug urecholine

(1-2 mg/hr), discharge of sulfated mucosubstances was observed from the fundic epithelium, peptic cells and antral pyloric glands; later, accelerated synthesis of sulfated mucosubstance was found in fundic crypt cells and pyloric glands. Partial depletion of neutral glycoprotein from mucous neck cells was also noted. Simultaneously, increased amounts of pepsin, sulfated glycoprotein and CS were detected in the gastric juice. The coexistence of CS and pepsinogen granules in resting peptic cells and the parallel excretion of CS-A and pepsin following urecholine suggest that CS-A may prevent the in vivo digestion of peptic cells by activated pepsinogen; numerous sulfated polysaccharides are known to exert anti-peptic effects. The data indicate that urecholine acts directly on mucosubstance-producing gastric cells rather than in-directly by releasing acid or gastrin.

A three hour infusion of 3 mg/hr of histamine induced accumulation of intracellular mucus and increased synthesis of sulfate mucus in fundic and antral crypt cells, but no significant extrusion of mucus was observed. The output of all sulfated mucosubstances and of pepsin in the secretions was not increased by histamine and was much less than following urecholine. Histamine is thus not a true mucigogue, at least in the dog.

References:

(1) Woussen-Colle, M. C. and J. De Graef: To be presented at the 16th Colloquium on Protides in Biological Fluids, Brugge (1968).

The Dual Histochemical Approach for Glycogen *

LHOTKA, J. F., JR. and J. W. ANDERSON (University of Oklahoma School of Medicine, Oklahoma City, Okla.)

In recent years the technic of choice for glycogen localization has become, to the exclusion of all others, the enzymatically controlled, usually diastase, periodic acid-Schiff method (D-PAS) initially published by McManus (Nature, *158*:202). The technic is based on known mechanisms and has a scientific appeal lacking in the older and often empirical approaches. Unfortunately, interpretation is based on a negative staining pattern and herein lies the seed of difficulty. In the past 12 years, the authors have had occasion to process literally thousands of sections for glycogen using all of the standard technics, including the Arzac-Flores and Gomori silver methods, the classic 1906 Best carmine technic, and the strong organic oxidative procedures typified by the Bauer chromic acid method. An interesting pattern developed. In cases where glycogen was frankly present in great quantity, the D-PAS technic appeared quite valid but in borderline cases where the amount of glycogen was minimal, especially in the face of concurrent diastase resistant neutral polysacharide localizations, accurate evaluations became most difficult with this method alone. The physical effects of the rigors of processing (especially the sulfurous rinses) aided in clouding interpretation with loss of delicate staining. Consequently, the authors adopted a dual approach, using serial sections, and staining in sequence with an older positive method, diastase control, the PAS technic and finally its control section. The Best carmine was selected by trial and error as the positive localization approach since in our hands it gave the most consistent results. The value of this dual approach was clearly demonstrated in embryonic studies on the chick notochord, the developing pig and human livers (Anat. Rec., *151*:440; *154*:498; Texas Rep. Biol. Med., *23*:640), where the D-PAS method alone would have made accurate analysis questionable. The double approach with its confirming negative and positive patterns of localization eliminated all doubts. In the hands of the expert histochemist, the D-PAS in most cases is quite adequate for glycogen localization. However, in borderline states and especially in the hands of inexperienced or inadequately histochemically trained individuals, the authors feel that the dual approach, as suggested here, is to be recommended.

------

* Supported by NIH research grant, HO1907.

The Staining of Ferritin by the Perls Reaction

LIBER, AMOUR F. (Bronx Veterans Administration Hospital and Albert Einstein College of Medicine, Bronx, N. Y.)

In the early days of the study of ferritin this substance was regarded as decidedly different from hemosiderin. Although no controlled experiments were made, it was tacitly assumed that the Perls acid ferrocyanide reaction revealed only hemosiderin but not ferritin in fixed tissue. In 1957 the close kinship of ferritin and hemosiderin micellae was demonstrated electron microscopically (1). Prussian blue was found to be deposited on ferritin particles seen with the electron microscope (2). A problem remaining to be resolved was whether ferritin gives a Perls stain visible with light microscopy. A substance believed to be ferritin was demonstrated in formaldehyde-fixed paraffin sections of human tissue by the cadmium sulfate reaction (3). This material "blended with hemosiderin" demonstrated by the Perls reaction, but the cadmium reaction was inconclusive as "clear circular crystals like those described by Granick were not observed" (3). Other authors have compared the results of the Perls reaction with attempted immunofluorescent identification of ferritin. They found incomplete spatial superposition of the materials characterized by the two methods (4,5). It was suggested that this might be due to a common immunologic specificity of ferritin and the iron-free apoferritin (4).

It is apparent from the above that the solution to the problem of stainability of ferritin is likely to be elusive in tissue preparations. In the present investigation a commercially available ferritin solution prepared from horse spleen was used. The solution turned intensely blue when mixed with acidified ferrocyanide. $CdSO_4$-induced ferritin crystals were washed repeatedly with $CdSO_4$ solution. Some were spread on albuminized slides and fixed with formaldehyde vapor. Others were suspended in melted gelatin containing $CdSO_4$. The blocks, solidified by rapid chilling, were fixed in formalin, embedded in paraffin and sectioned. Characteristic ferritin crystals were seen microscopically in unstained sections and smears. They were insoluble to prolonged washing in water. The Perls method stained them intensely blue.

Sections from fixed blocks of gelatin containing ferritin without $CdSO_4$ gave a diffuse stain, undiminished after prolonged prior washing in water.

The Perls reaction was negative in formaldehyde treated albuminized slides, sections of formalin-fixed plain and $CdSO_4$-gelatin, and $CdSO_4$ solution. The last became turbid but not blue.

The EDTA-Perls method gave the same results as Perls alone—a feature in common with hemosiderin in contrast with some other forms of iron (6).

In conclusion, ferritin in artificial conditions reacts tinctorially like hemosiderin and, after fixation, resists solution by water. Further problems are (1) behavior of ferritin in tissue, (2) development of a method for differentiating ferritin from hemosiderin.

References:

(1) Richter, G. W.: J. Exp. Med. *106*, 203 (1957).
(2) Thiéry, J. P.: Proc. 5th Int. Congr. Electr. Micr. 2, L-9 (1962).
(3) Dastur, D. K. and G. Smith: Acta Neuropath 2, 161 (1962).
(4) Craig, J. M.: Arch. Path. *79*, 435 (1965).
(5) Lorber, M. and J. L. Nemes: Acta Haemat. *37*, 189 (1967).
(6) Liber, A. F.: J. Neuropath., and Exp. Neur. *24*, 675 (1965).

On the Mechanism of Chromation Hematoxylin Sequence Stains

LILLIE, R. D., and PHILIP PIZZOLATO (LSU School of Medicine and US Veterans Hospital, New Orleans, La.)

Impregnations with 2.5% $K_2Cr_2O_7$ 18 hr 60°, 10-14 d 24° and 14-60 d 3° C effectively induce acetic hematoxylin staining of myelin, erythrocytes, bile casts and certain other structures. At higher temperatures moderate to much background staining occurs and the traditional

Weigert (1884) borax ferricyanide differentiation is required for good histologic results, while at 3° C background staining, even with 6-8 weeks chromation, practically does not occur. Bromination for 10 min in 1/400 Br./2% KBr(Aq.) completely prevents the reaction when chromation is done at 3° C, while with 18 hour 60° C chromation stronger bromine solutions and longer bromination time is required to produce moderate or severe impairment of the stain. Sulfation and acetylation have completely prevented the chromation acetic hematoxylin stain, but methylation up to 6 hr 60° C in 0.1 M methysulfuric acid/methanol is without effect.

Chromic acid solutions are also effective in producing the acetic hematoxylin stain; lower concentrations can be used and shorter times and lower temperatures, but freedom from background staining was not attained even at 3° C.

Potassium chromate solutions are ineffective. Trivalent chromium salts at 60° C produce acetic hematoxylin staining with short incubation, and then destroy it again. Acetic hematoxylin stains poorly or not at all when chromic salts are used at 24° C in equivalent concentrations, but excellent staining of myelin, erythrocytes, etc., is attained with neutral hematoxylin after 24 hours in 4.2 or 0.2% chromic acetate solution.

When sections are chromated 18 hours at 60° in 25% $K_2Cr_2O_7$ at pH 3.5 or 24 hours in 0.25% chromic acetate and then reduced 4 hours in 0.1 M $Na_2S_2O_5$ or 1 hour in 0.2 M $Na_2S_2O_4$ acetic hematoxylin fails to stain myelin, erythrocytes, bile casts or background, but neutral 0.1% hematoxylin stains myelin and erythrocytes light orange to pink. Reoxidation in 1% H_2O_2 only partly restored neutral hematoxylin staining after bisulfite reduction.

It appears indicated that traditional myelin and erythrocyte staining cannot be attributed to hexavalent Cr as such. Neither does it appear to be due purely to trivalent Cr. It would seem probable that a reduction phase corresponding perhaps to tetravalent Cr, as exemplified by CrO_2, must be invoked to account for the findings. It appears that both hydroxyls and ethylene groups in tissue are involved. However, the difference between 24° C Cr uptake and Cr deposition from hot $K_2Cr_2O_7$ may be only quantitative.

Comparative Quantitative Autoradiographic and Radiochemical Investigations

LINDNER, J., K. GRASEDYCK, G. JOHANNES and G. FREYTAG
(Pathologisches Institut der Universität, Hamburg-Eppendorf)

Demonstration of two ways of quantitative autoradiography: 1. Two small parts of the same material, one for the quantitative radiochemical analysis in suitable counter places, one for the usual ways of quantitative autoradiography, which depend on the incorporated labelled precursor. In the case of ^{35}S-sulfate-incorporation we use reflex-photometric-microscope measurements of intra- and extracellular areas such as for the adequate quantitative analysis of ^{3}H-thymidine labelled nuclear areas. Of this first way of quantitative autoradiography there are some examples demonstrated: embryonic and postembryonic development of connective tissue, of the latter especially inflammation, woundhealing, callus, atherosclerosis, etc. In this and in the following second method, too, double labelled material is used with very good results.

2. The second way of quantitative autoradiography is the following: One tissue section is analyzed successively autoradiographically and radiochemically.

We distinguish the two main possibilities:

a) Tissue sections are analyzed after mounting on special slices in a liquid scintillation counter, after washing (in order to remove the scintillator liquid) film or emulsion mounting and following autoradiographic evaluation. Both methods are carried out on the same tissue section.

b) Often it is desirable and necessary to analyse radiochemically autoradiograms a short or longer time after the morphological investigation, the usual quantitative autoradiographic evaluations, partly mentioned above, included. This is possible, especially after incorporation of labelled precursors with a long half life time. The best way is the quantitative solution of the section together with the film or emulsion (after their mechanical removal around the sections) by hyamin from the slide, addition of PPO-POPOP-scintillator and counting of the impulse rates in a scintillation counter. Comparison of the thus counted average impulse rates with the before counted number of labelled nuclei (in the case of ^{3}H-thymidine-labelling) showed linear correlations. The margin of error is very small; Quench can be estimated by evaluation of the internal and external standard. Some results of this second method are demonstrated and show the practical use in the further development of quantitative autoradiography.

Comparative Histochemical, Autoradiographic, Radiochemical and Immunological Investigations of the Atherosclerotic Vessel Wall

LINDNER, J., G. GRIES, and G. FREYTAG (Pathologisches Institut der Universität, Hamburg-Eppendorf)

These comparative investigations are carried out on human arteries, autopsy and biopsy materials after classification in four arteriosclerotic stages in comparison with morphologically unchanged parts of the same vessel wall and classification in four age groups. Thus, we can show with the methods used that the atherosclerosis starts with an increase of anabolism. The increased synthesis of ground substance is one of the first anabolic processes, combined with an increased O_2-consumption. The fibre synthesis starts later, as in every connective tissue development. The increased anabolism is followed by an increased catabolism. The turnover rates of the connective tissue components are enhanced, therefore. Then both processes decrease, catabolic more than anabolic processes. Therefore, the content of the histochemically, and biochemically analysed connective tissue components is enhanced in the further stages of atherosclerosis. In detail: the autoradiographic and radiochemical analyses show that the synthetic rates of the ground-substance decrease with ageing, are higher in intima than in media of all atherosclerotic stages and have the highest values of all stages in oedemateous plaques (the O_2-consumption runs parallel). Quantitative immunological analyses in the magnitude of native tissue sections demonstrate that the serum protein content is higher in the intima than in the media, that more albumins than globulins are demonstrated in every atherosclerotic stage and that the highest protein content exists in oedemateous plaques (with the highest increase of ground substance synthesis in the same time, as mentioned before). The biochemical fat analyses agree with the histochemical results. The disturbances of the lipid metabolism follow the disturbances of the connective tissue metabolism. The development of atherosclerosis in plaques represent the typical reaction of the vessel wall, as we could show in this separated analyses of isolated plaques (4 stages) in comparison to surrounding unchanged parts of the same vessel wall (see above). The comparison of the methods used (partly quantitative micromethods in the magnitude of tissue sections) shows that each new atherosclerotic process starts with an increase of the vessel wall metabolism (synthesis, especially of ground substance, before breakdown). In older atherosclerotic processes the catabolic process decreases more than anabolic processes. It results in atherosclerotic insufficiency of the vessel wall metabolism.

Autoradiographic Investigation of Brain Tumors

LIPCHINA, L. P., M. S. AKSYUTINA, and L. YA. YABLONOVSKAYA (Institute of Chemical Physics, Academy of Sciences, Moscow, USSR)

Autoradiography with thymidine-H^3 finds an increasingly wider application in studies of proliferation and differentiation of tumor cells. Tumors of the nervous system are less investigated in this respect than other neoplasms. Autoradiographic studies of brain tumors were applied only in single works (1,2). According to these studies, human brain tumors show a low labelling index (0.2 to 7.4 per cent) and, as these authors calculated, the generation time of tumor cells varies from 48 hours to 6 months.

In present work, transplanted glioblastoma of mice and rats *in vivo* and *in vitro* and human medullablastoma *in vitro* were studied. The generation time in mice glioblastoma was determined from the curve for labelled mitoses and was found to be 12 hours. The duration of mitotic cycle phases was: $G_1 = 2.15$ hours, $S = 6.8$ hours, $G_2 = 3.05$ hours. The proliferative pool was 1.0.

With pulse labelling, the fraction of labelled cells was different for various regions of the tumor: for those with dense packed cells, it was 50 to 60 per cent; for loose packed cells, 15 to 20 per cent. The labelling index for mice glioblastoma *in vivo* and for fragments incubated *in vitro* showed close values.

The generation time for polymorphous cell glioblastoma of rats, calculated from the labelling index *in vitro* and under the assumption that the mean t_s value is 6 hours, appeared to be 20 hours.

For human medullablastoma of typical structure with densely packed cells containing small dark nuclei, the labelling index varied from 10 to 20 per cent and attained 50 per cent after an 8 hours incubation; in regions with loosely packed cells showing the occurrence of differentiation which appeared in large, light nuclei and distinct cytoplasma, the labelling index was only 3.4 per cent, attaining 19 per cent after 8 hours of incubation.

The generation time of tumor cells calculated according to the formula

$$T = \frac{t_s}{I_a} \text{ at } t_s = 6 \text{ hours}$$

was 30 hours for non-differentiated tumor regions, and 170 hours for those differentiated. This is evidence that the population of malignant cells is heterogeneous with respect to its proliferative activity. This heterogeneity may be due to different differentiation levels.

References:

(1) Johnson, H. A., H. E. Haymaker, J. R. Rubini, T. M. Fliedner, V. P. Bond, E. P. Cronkite, and W. L. Hughes: Cancer *13*, No. 3, (1960).
(2) Kury, G. and H. W. Carter: Arch. of Path. *80*, No. 1, 38 (1965).

Tannin Interference with the Histochemical Demonstration of Phosphatase Activity in Plant Cells *

LIPETZ, JACQUES (Boyce Thompson Institute For Plant Research, Inc., Yonkers, N.Y.)

The presence of tannins in plant cells interferes with the demonstration of phosphatase activity by the metal sulfide (Gomori) (1) and the naphthol AS (Burstone) (2) methods. Interference in the Gomori method is due to the formation of partially soluble inks as a result of interaction between the tannins and the metal salt used. Ammonium sulfide also reacts with tannins to form partially soluble materials which obscure the histochemical reaction. Diazonium salts such as Fast Red Violet L.B. salt also react with tannins.

Attempts to selectively precipitate the tannins in fresh tissue sections by treating them with known precipitants of tannins such as caffeine, colchicine, permanganate, and carbowaxes were all unsuccessful. Fixation in cold acetone, formalin or glutaraldehyde also had little effect on the interfering reactions.

Numerous authors (3) have also demonstrated that tannins inhibit a broad variety of enzymes *in vivo* and *in vitro*. These findings, and the wide spread distribution of tannins in plant tissue, make it necessary to consider proper controls to eliminate tannin vacuoles as the reactive sites in the demonstration of phosphatase activity.

References:

(1) Gomori, G.: Microscopic Histochemistry, Chicago: U. of Chicago Press (1952).
(2) Burstone, M.: Enzyme Histochemistry, New York: Academic Press (1962).
(3) Kuc, J.: Ann. Rev. Microbiol. *20*, 337 (1966).

---

* The work reported here was done at the laboratory of Plant Morphogenesis, Manhattan College, Bronx, New York and supported by USPHS grant CA06955.

U.V. Fluorescence Studies of C.N.S. Acid Phosphatases, ATP-ases, and DPNH (Lactic bound) *

LIU, J. C. and L. ROIZIN (N. Y. State Psychiatric Institute, New York, N.Y.)

Fresh, liquid nitrogen quenched human cerebral biopsies and various anatomotopographic regions of the C.N.S. of *Macacus rhesus* were processed in accordance with modified technics of Gomori (1), Allen and Slater (2), and Wachstein and Meisel (3,4).

Incubation time for the respective enzymes was as follows: for acid phosphatase, 30 minutes to 3½ hours; for ATP-ases, 1 to 4½ hours; for DPNH (lactic bound), 10 minutes to 2 hours. At various periods of incubation the sections were mounted and were examined with U.V. light (200 HBO mercury lamp, BG 12 filter and Zeiss 50). Fresh, liquid nitrogen quenched and 10% formalin fixed serial sections of the same C.N.S. tissues were used for comparative purposes.

The combined examinations revealed a distinct differentiation of the

156

distribution of enzyme reaction products from the material displaying various degrees of fluorescence. Consequently, it offers a more complete cytological visualization of the relationship between enzyme reaction patterns and the fluorescent material.

References:

(1) Gomori, G.: Stain Technology 25, 81 (1950).
(2) Allen, J. M. and J. J. Slater: J. Histochem. and Cytochem. 9, 221 (1961).
(3) Wachstein, M. and E. Meisel: Am. J. Clin. Path. 27, 13 (1957).
(4) Iyengar, V. K. S., G. DiVirgilio, E. H. Robinson, G. Wodraska, and L. Roizin: J. Cell Biol. 23, 45A (1964).

---

* Supported in part, by the General Purpose Research Grant, 5650-1.

Intereference of Time Lapse Cinematography

LODIN, Z., M. KAGE, J. HARTMAN, and J. SRAJER (Institute of Physiology, Czechoslovak Academy of Sciences, Prague and Institut für wissenschaftliche Fotographie & Kinematographie, Winnenden, West Germany)

It was our task to introduce a method of permanent observation of the optical path difference (o.p.d.) in living cells by means of the Interference Microscope. The question resulting from this task, is the following: Is it possible to determine quantitatively the dry mass of living cells and their changes, connected with the changes in composition of their environment? In the first phase of our experiments, we tried to make use of this method for the study of changes, induced by alterations in the electrolytic concentration of nervous and glial cells. The stability of the Interference Microscope and the automatic regulation of the temperature and illumination are described. The possibilities of calculating the dry mass, even under conditions of imperfect stability of the system, were solved mathematically. A chamber was developed by us insuring suitable living conditions for the cultures and enabling an application of exact doses to the medium. In our paper, the influence of errors caused by the medium over and below the cell will be discussed as well as errors caused by different R.I. of the medium.

Histochemistry of the Arterial Wall

LOJDA, ZDENEK (Laboratory of Angiology, Laboratory of Histochemistry of the 1st Institute of Pathology, Charles University, Prague, Czechoslovakia)

A detailed knowledge of the chemical composition and metabolism of individual structural elements of the arterial wall is a prerequisite for a better understanding of its physiology and pathophysiology. The mutual comparison of the biochemical *in vitro* and histochemical *in situ* analysis elminates to a great extent limitations of each approach if applied separately. The contribution of the analysis *in situ* and its significance were critically reviewed by us (1) and particularly by Adams (2). Therefore this communication will be concerned chiefly with new findings in this field and particularly with those obtained in our laboratory. In the *in situ* analysis of lipids the limitations of newly introduced methods (3) are greater than generally assumed. This will be demonstrated on the basis of a comparison of results obtained *in situ* and by microchemical analysis of consecutive sections of the aortic wall. In the *in situ* analysis of acid mucopolysaccharides the preservation of all mucopolysaccharides in sections and the methods used for their detection are still a major problem. In the *in situ* analysis of enzyme activities, improvement was achieved in the localization of alkaline phosphatase which was demonstrated by more sensitive azo-dye methods even in the muscle cells of rat aortae; β-glucuronidase and acetyl-β-glucosaminidase; dehydrogenases by gel media. Histochemical methods also proved to be of great value in demonstrating enzyme activities in electrophoretically separated homogenates of samples of the aortic wall. Such zymogrames are very useful in the studies of the heterogeneity of enzymes. Examples are shown. The significance of these findings is briefly discussed.

References:

(1) Lojda, Z. Histochemistry of the vascular wall. *In*: Int. Symp. Morphol. Histochem. of the Vascular Wall, pp. 364-398 (L. Laszt, edit.) Basel, New York: Karger (1966).
(2) Adams, C. W. M.: Vascular Histochemistry (London) Lloyd-Luke Ltd., (1967).
(3) Adams, C. W. M.: Neurohistochemistry, Amsterdam, London, New York, Elsevier Publ. Co. (1965).

Problems of Methodology Related to Chemodifferentiation

LOJDA, ZDENEK (Laboratory of Histochemistry of the 1st Institute of Pathology, Charles University, Prague, Czechoslovakia)

Chemodifferentiation can be studied either by the *in situ* histochemical and cytochemical approach or the *in vitro* chemical analysis of homogenates or isolated cells. The choice depends on the problem under investigation. In the process of differentiation the heterogeneity is even more marked than in fully differentiated tissues. The analysis *in situ* is therefore very useful. An appropriate application of histochemical methods is a prerequisite for a correct evaluation of findings. It has to be kept in mind that fixation procedures and histochemical methods as such have their limitations. Typical examples taken from the histo-chemical detection of proteins, polysaccharides, lipids and enzymes are presented.

Heterogeneity of Enzymes of the Aortic Wall

LOJDA, ZDENEK and PREMYSL FRIC (Laboratory of Histochemistry of the 1st Institute of Pathology and 2nd Research Unit of Gastro-enterology, Charles University, Prague, Czechoslovakia)

Heterogeneity of alkaline and acid phosphatases, β-glucuronidase, car-boxylic esterases, lactate dehydrogenase and malate dehydrogenase was studied in homogenates of intima-media samples in aortae of rat and rabbit, normal and treated with thyreoidine and methylthiouracile and fed with cholesterol and in separate layers of human aortae. The method of agar-gel electrophoresis was used for separation. Enzyme activities were detected with modified histochemical techniques. Dis-tinct species differences and in some enzymes, organ or even topical specificity were demonstrated. The analysis of zymograms is more sensitive than determination of total enzyme activities. Changes in zymograms usually precede changes of total enzyme activities and are apparent earlier than morphological changes. Zymograms of some en-zymes show typical postnatal development and changes in atherosclerosis. The implications of these findings are briefly discussed.

Histochemical Characterization of Epithelial Mucins in Human Ovarian Mucinous and Serous Tumors

LONG, MARGARET E. and SHELDON C. SOMMERS (College of Physicians and Surgeons, Columbia University, New York, N. Y.)

Specific histochemical methods for carbohydrate groups have distin-guished three epithelial mucin components in primary ovarian mucinous and serous tumors. Comparative methods identified sulfomucins, sialo-mucins and neutral or vic-glycol mucins, respectively, by the presence of sulfate esters, sialic acid or vicinal hydroxyl groups (1,2). The types of tumors histologically included cystadenomas and cystadenocarcinomas subdivided into four groups of increasing grades of malignancy desig-nated Borderline, Grades I, II and III. Serial sections have been reacted for the three types of epithelial mucosubstances which, within each histological group, were assessed separately in the apical regions, the cytoplasmic portions and secretions of the tumor cells. A numerical value was assigned to each estimate ($+$ to $+++$) of the relative amounts of each type present in each site. Summation of these observations indicated that all three mucosubstances were present in the histological groups of both tumor types. The relative amounts of mucin components varied moderately in a given tumor and within each graded group. Benign tumors of both types secreted epithelial mucins differing in amounts of their group components from those in their malignant derivatives.

Neutral-type mucin predominated in benign mucinous tumors which produced relatively small amounts of sialo- and sulfomucins, the former in slight excess. Mucinous cystadenocarcinomas, with increasing grade of malignancy, produced diminishing neutral-type mucin and increasing sulfomucin. Amounts of sialomucin remained relatively unchanged throughout the malignant mucinous spectrum. Serous cystadenomas secreted sialo- and neutral-type mucins in approximately equal amounts and gave evidence of less sulfomucin production than the other two types. Differentiated serous cystadenocarcinomas exhibited more sulfo-mucin than their benign counterparts but with ascending grade of malignancy, the sialomucin type increasingly predominated among the epithelial mucosubstances. It is suggested that the application of specific histochemical methods for epithelial mucin components may be useful diagnostically in distinguishing between undifferentiated serous and mucinous tumors since sialomucin has been shown to predominate in the serous form, whereas the major mucin component in the highly malignant mucin-type was sulfomucin.

References:

(1) Mowry, B. W.: *In*: Mucous Secretions, pp 402-423, Ann. N. Y. Acad. Sci. *106* (1963).
(2) Spicer, S. S.: J. Histochem. Cytochem. *13*, 211 (1965).

New Observations on the Substrate Specificity of Rat Liver Cathepsin C

McDONALD, J. KEN and STANLEY ELLIS (NASA, Ames Research Center, Moffet Field, Calif.)

Existing reports indicate that cathepsin C is relatively specific for the hydrolysis and transfer of N-terminal dipeptides having an aromatic amino acid adjacent to an unsubstituted terminal residue (1). In this report, evidence will be presented, firstly, to show that cathepsin C also cleaves and transfers several non-aromatic N-terminal dipeptides, with maximal rates of activity exhibited on basic dipeptides, and, secondly, that peptide hormones such as adrenocorticotropin (ACTH) and glucagon are extensively degraded from their N-termini.

For the present study, rat liver was used as the source of cathepsin C. The enzyme was obtained in a highly-purified state through the adoption of the purification procedure reported by Metrione, *et al.* (2) for the isolation of cathepsin C from bovine spleen. The progress of the purification was followed by a direct fluorescence analysis of dipeptidyl arylamidase activity (3) on Gly-Phe-β-naphthylamide at pH 6.0. The relative rates of hydrolysis, by both the purified enzyme and the starting material, of the amide and β-naphthylamide (βNA) derivatives of a variety of dipeptides were as follows: Gly-Arg (100%), Ser-Met (28%), Ala-Ala (13%), Gly-Phe (5%), and His-Ser (4%). The results of this comparison indicated that these activities were purified in parallel. The hydrolysis of this group of substrates is evidence of the rather non-specific character of cathepsin C since the susceptible peptide bond in each substrate contains a residue which represents a distinct class of amino acid. Both the sulfhydryl requirement and the more recently observed Cl^- requirement (4) were manifest for the hydrolysis of all these substrates.

Gly-Phe-NH_2, the traditional cathepsin C substrate (1), is a relatively poor substrate compared with Gly-Arg-NH_2 and Gly-Arg-βNA. The most rapid rates of deamidation and dipeptide transfer were observed on Gly-Arg-NH_2. Of all the arylamide substrates tested, Gly-Arg-βNA was hydrolyzed over the widest range of pH (4.0 to 8.0). The hydrolysis of Gly-Arg-βNA cannot be attributed to tryptic enzymes since relatively high concentrations of pancreatic trypsin inhibitor had no effect.

It has already been demonstrated by Gruber and his associates that cathepsin C cleaves Ser-Tyr from ACTH (5). The results of our studies show that subsequent dipeptides are cleaved in rapid sequence. As judged by the rate of Ser-Met-βNA hydrolysis, the second dipeptide (Ser-Met) should be cleaved even more readily than Ser-Tyr.

As suggested by the hydrolysis of His-Ser-βNA, crystalline glucagon (Lilly) was tested as a substrate for purified liver cathepsin C. This hormone was readily degraded. His-Ser was the first product to appear, followed in rapid sequence by Gln-Gly and Thr-Phe. A time-course analysis for new N-terminals revealed that the degradation proceeded beyond Thr-Phe. Characteristic of the halide requirement of cathepsin C, glucagon-degradation did not occur if Cl^- was omitted from the reaction mixture. Since the cathepsin C used in this study was derived from liver, and since the "glucagon-degrading enzyme" of liver is reportedly a Cl^- and SH-activated enzyme (6), it would appear that cathepsin C is the "glucagon-degrading enzyme." Through the kindness of Dr. William W. Bromer, reaction products were bioassayed at the Lilly Research Laboratories. It was learned that the hyperglycemic activity of the hormone is lost upon the cleavage from the hormone of the first N-terminal dipeptide (His-Ser). Earlier reports have indicated that glucagon is not hydrolyzed by cathepsin C, however, the Cl^-

requirement of the enzyme may not have been satisfied since this property was not recognized until recently (4). Since cathepsin C is a lysosomal enzyme, it is of interest to note, as reported by Berthet and by Deter and de Duve, that the injection of glucagon into rats results in an altered permeability of the liver lysosomes with a resultant leakage and increased accessibility of lysosomal enzymes. Such a response could conceivably facilitate the inactivation of glucagon by cathepsin C. Although its attack is in the nature of a dipeptidyl aminopeptidase, cathepsin C specificity seems much less restricted than heretofore realized. It may be physiologically significant that the biological activity of peptide hormones such as ACTH and glucagon is terminated so directly with the cleavage of a single peptide bond by cathepsin C.

References:

(1) Tallan, H. H., M. E. Jones, and J. S. Fruton: J. Biol. Chem. *194*, 793 (1952).
(2) Metrione, R. M., A. G. Neves, and J. S. Fruton: Biochemistry *5*, 1597 (1966).
(3) McDonald, J. K., S. Ellis, and T. J. Reilly: J. Biol. Chem. *241*, 1494 (1966).
(4) McDonald, J. K., et al.: Biochem. Biophys. Res. Commun. *24*, 771 (1966).
(5) Planta, R. J. and M. Gruber: Biochem. Biophys. Acta *53*, 443 (1961).
(6) Kakiuchi, S. and H. H. Tomizawa: J. Biol. Chem. *239*, 2160 (1964).

Studies with Fluorescein-Labelled Antisera to Pituitary Hormones

McGARRY, E. E., R. NAYAK, E. BIRCH, and J. C. BECK (McGill University Clinic, Royal Victoria Hospital, Montreal 2, Canada)

Previous studies have shown localization of antisera to human growth hormone and to ACTH to the acidophils (1) and basophils (2) respectively in the human pituitary. More recently Kracht, et al. (3) found localization of antisera to synthetic B 1-24 ACTH in the acidophils of the rat pituitary. Studies in our own laboratory with antisera to ACTH harvested from a number of rabbits showed that in the rat pituitary localization occurred most consistently in large chromophobes although some antisera also localized in basophils or acidophils. All antisera to ACTH localized in the cells of the intermediate lobe of the rat pituitary while an antiserum to MSH localized only in the cells of the intermediate lobe. These results suggested that the site of localization was in part dependent on the individual animal source of the antiserum. Such individual variation in antibody combining site for an antigen has been demonstrated by others in other immunochemical systems (4,5). Studies with antisera to various growth hormone preparations showed that a number of antisera to human growth hormone harvested from a number of individual animals showed that some but not all antisera localized in the acidophils of the rat pituitary as did antisera to bovine and ovine growth hormone. Cross inhibition studies showed a minimum of two antigenic sites in the rat acidophils on which antisera to various growth hormones localized. Antisera to bovine and ovine growth hormone also localized in the acidophils of the bovine and porcine pituitary.

References:

(1) Leznoff, A., J. Fishman, L. Goodfriend, E. McGarry, J. Beck, and B. Rose: Proc. Soc. Exp. Biol. & Med. *104*, 232 (1960).
(2) Leznoff, A., J. Fishman, M. Talbot, E. E. McGarry, J. C. Beck, and B. Rose: J. Clin. Invest. *41*, 1720 (1962).
(3) Kracht, J., U. Hachmeister, H.-J. Breustedt, and K. Fischer: Acta Endocrin. (Kbh) Suppl. *100*, 36 (1965b).
(4) Arquilla, E. R., B. Catz, and J. Finn: J. Immunol. *90*, 843 (1963).
(5) Parker, M. L. The 53rd Ross Conference on Pediatric Research, March 1965 *(In Press)*.

Oxygen Effects on Tetrazolium Systems Containing Phenazine Methosulfate (PMS) *

McMILLAN, PAUL J. and CHRISTOPHER 'SEINDE ADEOYE (Loma Linda University, Loma Linda, Calif.)

The fact that the oxidation of reduced PMS by atmospheric oxygen can be catalyzed by the cytochrome *c* oxidase system has been documented (1). Azide and cyanide have been used in histochemical systems to inhibit this undesirable side reaction (2, 3). Conklin (4), however, has presented evidence suggesting that these anions also inhibit the PMS mediated transfer of electrons from reduced NAD to tetrazole. To investigate the possible role of such an action by azide in histochemical systems, we have determined the influence of various parameters on the PMS mediated reduction of a tetrazole, 2-(p-iodophenyl)-3-p-nitrophenyl-5-phenyl tetrazolium chloride (INT), by reduced NAD. The parameters which have been studied are 1) atmospheric oxygen, 2) PMS concentration and 3) azide concentration.

It has been found that azide does indeed impair the PMS mediated transfer of electrons. This is indicated by a reduction in the amount of formazan formed from a given amount of $NADH_2^+$. However, this effect of azide disappears when oxygen is excluded from the system. This oxygen effect has been found to be greatest at low PMS concentrations and almost negligable at high PMS concentrations. Preliminary experiments indicate that a similar effect of atmospheric oxygen can be demonstrated in a test tube experiment in which the $NADH_2^+$ is generated enzymatically. Experiments are continuing to determine if and under what conditions atmospheric oxygen influences the histochemical demonstration of dehydrogenases. The results of these experiments will also be reported.

References:

(1) Low, H., B. Alm, and I. Vallin: Biochem. Biophys. Res. Commun. *14,* 347 (1964).
(2) Brody, I. A. and W. K. Engel: J. Histochem. Cytochem. *12,* 928 (1964).
(3) Mathisen, J. S. and S. I. Mellgren: J. Histochem. Cytochem. *13,* 408 (1965).
(4) Conklin, J. L.: Stain Tech. *41,* 105 (1966).

* Supported in part by U. S. Public Health Service General Research Support Grants 5S01FR-5352-04 and 5S01FR-05352-06)

Leucocytic Lysosomal Hydrolases In the Process of Pinocytosis

MACOVSCHI, O. and C. BONA (Instit. Dr. I. Cantacuzino, Bucharest 35, Romania)

Since breakdown of materials taken up by pinocytosis, as well as phagocytosis, is obviously performed by lysosomes, the authors have studied the behaviour of some guinea-pig leucocytic hydrolases during the pinocytosis of two heterologous proteins: human gamma globulin (H.G.G.) an inert biological inducer and *Serratia mercescens* thermolabile endotoxin (T.L.E.) which possesses a well known biological activity upon cells.

The rate of pinocytosis, determined by fluorescent microscopic examination, histochemical evidence estimate by counting the enzyme-positive granules/cell and biochemical determination of the enzyme activity were performed.

The histochemical and biochemical methods used for the enzyme activity determination showed variable patterns of the enzyme activity depending upon the pinocytic inducers, time factor and rate of pinocytosis. The enzyme variations might be explained by an enzyme rearrangement between pinolysosomes and the cell cytoplasm, a cellular adaptability at the enzymic level or an inhibition of enzyme activities.

Quantitative estimate of the histochemical results performed by the authors revealed similar variation for: β-glucuronidase, cathepsin and acid phosphatase.

As revealed by our observations the leucocytic lysosomal hydrolases display rather important changes during the process of pinocytosis.

On the Modified Mercuric Nitrate Method for the Demonstration of Unsaturated C-C Bonds on Paraffin Sections—With Particular Reference to the Histochemical Characters of the Ceroid Pigment

MAEDA, RYUEI, NOBUO IHARA, and KOKICHI KANAZAWA
(Department of Pathology, Kansai Medical School, Moriguchi-shi, Osaka, Japan)

The principle of Okamoto's (1) mercuric nitrate method for phospholipids is based on the probable affinity of mercuric salt for unsaturated lipids (especially phospholipids). According to Pearse (2) this reaction may be due to combining of divalent mercury with unsaturated C-C bonds of the lipids. On the basis of this viewpoint, we applied the modified mercuric nitrate method to ceroid pigment-positive sections.

As to other histochemical methods for the detection of unsaturated C-C bonds, three have been developed: the first, the osmium tetroxide method, is unsatisfactory from the histochemical point of view; the second, the method based on halogenization cannot be applied to paraffin sections; the third, the PFAS (or PAAS) reaction method, has been developed by Lillie (3) as a method of research into the mechanism of unsaturated C-C bonds oxidation. However, it seems that the coloring of this reaction is rather faint on paraffin sections.

In the present investigation, we first modified the mercuric nitrate method (by greatly reducing the time required for section immersion in potassium iodide, for example) and then applied it to our paraffin sections, the consequence being that we were able to obtain fairly good results with respect to its specificity, reaction localization, and pronounced coloring.

The ceroid pigment, because of its unsaturated C-C bonds, gives a strongly positive reaction. This reaction is rendered negative by pretreatment with peracetic acid destroying unsaturated C-C bonds. On the other hand, a positive reaction can be obtained after pre-treatment with 2-hydroxy-3-naphthoic acid hydrazide. This fact reveals that the mercury-affinity reaction is not related to the carbonyl radicals. Furthermore, the present method reveals that so-called heart failure cells, in part, contain a yellow ceroid-like pigment showing a clear-cut red-purple coloring of the mercury-affine reaction rather than a blue-coloring of an iron-positive reaction.

Regarding substances or constituents which remain to be differentiated, the mercury-affine SH-radical is now under examination.

References:

(1) Okamoto, K.,M. Ueda, Y. Kusumoto, and D. Shibata: Bulletin of Kobe Medical College 3, 1 (1952).

(2) Pearse, A. G. E.: *In*: Histochemistry 2nd ed., pp 314-315, London: J. & A. Churchill (1961).
(3) Lillie, R. D.: Stain Technology, *27*, 37 (1952).

Steroid Histochemistry of Three Experimental Granulosa Cell Tumors

MAEIR, DAVID M. and LENORE WAGNER (Albert Einstein College of Medicine, New York, N. Y.)

The biosynthesis of almost all active steroid hormones from cholesterol appears to require the conversion of Δ_5 3β compounds to Δ_4 3-Keto-steroids (1). Methods to demonstrate the activity of enzymes required for this conversion were applied to the study of experimental ovarian tumors.

Three tumor lines were studied. One designated OS was produced in our laboratory by the transplantation of portions of ovary into the spleen of gonadectomized female rats. The inactivation by the liver of the hormones produced by the transplant allows for an unopposed effect of anterior pituitary gonadatrophins and results in a tumor (2).

The other two are granulosa cell tumors maintained at the Jackson Laboratory at Bar Harbor, Maine. The biochemical characteristics and hormone production of these tumors is well documented. Each of the Jax Lab tumors was studied in 15 animals and 132 OS tumors produced in our laboratory were examined for this report.

The enzymic activity of the tumors was visualized by incubation in a modification of the medium described by Rubin, *et al.*, (3) containing diphosphopyridine nucleotide (NAD), nitro-blue tetrazolium, and one of the following substrates: dehydroepiandrosterone (DHA), pregnenolone, pregnenolone-SO_4, DHA-SO_4, 4-Androsten-3β 17β-diol, and 4-Pregnen-3β 20β-diol. The degree of enzyme activity varied in the different cellular components of the tumors. Granulosa cells that had undergone luteinization contained more activity than those that had not. Enzyme activity was correlated with hormone production. The Jax tumor producing little or no estrogenic effect exhibited enzyme activity with only one of the substrates.

The activity of the tissue to the various substrates utilized varied. Certain tissues exhibited reactivity to all substrates while tumors with limited hormonal production exhibited reactivity only to certain substrates. Stromal elements in some tumors appeared to be the site of more enzymic activity than the masses of tumor cells.

These data corroborate evidence from the study of normal ovarian tissue of humans and animals that stromal elements play an important role in the elaboration of biologically active steroids. They also suggest that at

166

least in these ovarian tumors, histochemically demonstrable enzyme activity is a quantitative indication of hormone production.

References:

(1) Deane, H. W. and B. L. Rubin: Arch. Anat. Micro. Morph. Exp. *54*, 49 (1965).
(2) Deane, H. W. and D. W. Fawcett: J. Nat. Can. Inst. *17*, 541 (1956).
(3) Rubin, B. L., H. W. Deane, J. A. Hamilton, and E. C. Driks: Endocrinology *72*, 924 (1963).

Acid Phosphatases and Differentiation

MAGGI, VIVIANE, L. M. FRANKS, D. C. LIVINGSTON, and M. M. COOMBS (Tissue and Organ Culture Unit, and Division of Chemistry & Biochemistry, Imperial Cancer Research Fund, Lincoln's Inn Fields, London, W.C.2.)

Previous investigations have shown that tissues from normal, adult male C57 mice contain at least two acid phosphatases with histochemically different localization. One of these enzymes splits glycerophosphate and naphthol AS phosphate monoesters, and is inhibited by NaF or Na_2MoO_4. The other does not split glycerophosphate and is not inhibited or—in certain tissues—is activated by the two compounds (1). This second phosphatase is absent from tissues of 1 week-old mice and appears between the second and the fourth week of age depending on the tissue (2). In castrated adult male mice treated for 1 month or more with oestradiol an alteration of the epithelium of the seminal vesicles takes place which is accompanied by the disappearance of the second acid phosphatase (Maggi and Steggles, *in preparation*). In highly undifferentiated cells of very malignant and invasive cancers of both humans and mice the second phosphatase is absent. In well-differentiated non-invasive tumors both enzymes are present (3). The method of Gomori (4) as modified by Novikoff for electron microscopy, could not be used to study the ultrastructural localisation of the enzyme since the latter did not split glycerophosphate. The method of Davis and Ornstein (5) was found impracticable because of the considerable loss of pararosanilin-naphthol AS coupled end-product in ethanol or Durcupan used for dehydrating the tissue blocks. Some modifications of the method of Tice and Barrnett (6) were made to yield a completely reproducible method (Maggi, *et al.*, *in press*). These modifications, in part, consist in reduction of the tri-(4-nitro) lead phthalocyanin prior to solvent extraction rather than extraction of the nitro compound followed by reduction as described by Tice and Barrnett (6). This was found to be necessary to produce a tri-(4-amino) lead phthalocyanin having an acceptable lead composition. It was also found necessary to replace dehydration with ethanol by dehydration with Durcupan followed by embedding in Araldite. With this method, an electron microscopic study of the distribution of the enzyme has been undertaken and the results will be presented.

References:

(1) Maggi, V., L. M. Franks, and A. W. Carbonell: Histochemie *6*, 305 (1966).
(2) Maggi, V., L. M. Franks, and A. W. Carbonell: Biochem. J. *102*, 48P (1967).
(3) Maggi, V. and L. M. Franks: Biochem., J. (1967). *in press.*
(4) Gomori, J.: Stain Technol. *25*, 81 (1950).
(5) Davis, B. and L. Ornstein: J. Histochem. Cytochem. 7, 297 (1959).
(6) Tice, L. W. and R. J. Barrnett: J. Cell Biol. *25*, 23 (1965).

Localization of Fluorescent and Radioactive Insulin in Mouse Tissues

MAGGI, VIVIANE, L. M. FRANKS, PATRICIA D. WILSON, and A. W. CARBONELL (Tissue and Organ Culture Unit, Imperial Cancer Research Fund, Lincoln's Inn Fields, London, W.C.2.)

Crystalline beef insulin was lebelled with fluorescein isothiocyanate (FITC-insulin) or with I^{125} (I^{125}-insulin). 0.5 mg fluorescent hormone/mouse and trace amounts of I^{125}-insulin were injected intravenously and the animals were killed after 1 to 25 minutes. The localization of both compounds was ascertained by means of fluorescence microscopy, and by light and electron microscope autoradiography. FITC-proteins of low and high molecular weight were injected as controls and Trypan Blue was administered as a marker for lysosomes (1, 2). When FITC-insulin was used, the hormone was found mainly in the brush border area of the kidney proximal convoluted tubule cells only when the animals were starved for at least 18 hours. When I^{125}-insulin was injected, it was found in the same areas of the kidney cells of both starved and fed animals but in greater amount when the animals were starved for 18 hours. The radioactive hormone was found also, in decreasing quantities, in liver, heart, voluntary muscle, duodenum and small intestine, the capsule and trabeculae of the spleen, the basement membrane of seminal vesicles and ventral prostate, and in elastic tissue and cells of the lung. With high resolution autoradiography the radioactive hormone was found in the microvilli of the brush border of the proximal convoluted tubule cells of the kidney, in the apical mitochondria and in the nuclei, but never in the dense bodies or the Golgi apparatus. The hormone was found also to be localized in the liver and striated muscle mitochondria, in the sarcolemma and within the fibres of the muscle cells. The implications of these data with respect to the metabolic functions of insulin will be discussed.

168

References:

(1) Beck, F.: Exptl. Cell Res. *37*, 504 (1965).
(2) Maggi, V., L. M. Franks, P. D. Wilson, and A. W. Carbonell: *in press.*

The Mucopolysaccharide Production by Smooth Muscle Cells

MALYUK, V. I. (Institute of Zoology, Ukrainian SSR Academy of Sciences, Kiev, USSR)

In the intercellular substance of the media of blood-vascular trunks in mature rats, there were histochemically demonstrated sulphated mucopolysaccharides (SMPS) of the chondroitin sulphates A and C type and SMPS which were stable against the testicular hyaluronidase treatment (of the chondroitin sulphate B, heparitin-sulphate and heparin type). SMPS were not revealed cytochemically in differentiated smooth muscle cells. The use of sulphate-S^{35} provided evidence concerning the participation of smooth muscle cells of blood-vascular walls and intestines in the production of SMPS and allowed determination of the time parameters of their exchange within the intercellular substance. Sulphate-S^{35} was autoradiographically demonstrated in smooth muscle cells 15 minutes following intraperitoneal injection, and 30-60 minutes later it was observed in the intercellular substance. The concentration of label reached its maximum value in myocytes at 5-6 hours, and in the intercellular substance at 24 hours. The isotope content in the tissue gradually fell thereafter and only insignificant traces of radioactivity were registered by 10 days. The combined application of autoradiography and preliminary treatment with testicular hyaluronidase permitted the conclusion that the labeled sulphate in the smooth muscle cells of the aorta was incorporated in almost equal quantities into SMPS of the chondroitin sulphates A and C type and enzyme-stable SMPS. Having been released into the intercellular spaces, the substances of the first type were renewed very rapidly; their half-life period being 12-15 hours. The compounds of the second group appeared more stable; their half-life period amounted to 50 hours.

Enzymatic Activities in the Yolk Sac of Embriones and Alevins of Salmo irideus

MANFREDI ROMANINI, MARIA GABRIELLA, ANNUNZIA FRASCHINI, and FRANCA PORCELLI (Institute of Histology and Embryology, University of Pavia, Italy)

In our work (Manfredi Romanini and Fraschini, 1967) we have evidence that a relationship exists between the oxireductase enzymatic activities in the yolk sac of the chicken egg and in *Scyliorhinus stellaris* L. We have pointed out some differences in the histochemical descriptions of glycogen and alkaline phosphatase activities. In this note we describe researches on the yolk sac of *Salmo irideus,* an example of an oviparus fresh water fish hatching prematurely. Manfredi Romanini moreover has demonstrated, just for this type, that the beginning of digestive activity doesn't coincide with hatching, but with complete use of yolk reserves. Our researches regard the egg yolk sac before hatching and through complete use of the yolk reserves. At the same time we have also collected data about the degree of digestive activity of the various stages. We have studied the following groups of enzymatic activities: (a) Glycolytic enzymes: total phosphorylase according to Guha and Wegmann; branching enzyme according to Takeuchi; UDPG-transphorase glycogen, according to Takeuchi and Glenner. (b) Oxireductase activities: DPN and TPN diaphorases; succinate, lactate, malate, glutamate and β-hydroxybutyrate dehydrogenases, glycerophosphate, glucose-6-P dehydrogenase. (c) ATPase according to Wachstein and Meisel; carbonic anydrase according to Kurata. (d) Non-specific esterase, alkaline and acid phosphatase according to Burstone. (e) Activities really connected with the digestive processes such as lipase and leucine-amino peptidase.

Our descriptions of the glycogen metabolism state that in *Salmo irideus* there exists also a pathway of UDPG-glycogen transpherase which does not appear in the other yolk sacs studied till now. We also note a difference comparing the oxireductase activities with those studied till now. ATPase and carbonic anydrase activities present technical problems which need to be discussed in relation to various possible functional interpretations. The presence of weak lipase or peptidase activity shows digestive activity which undergoes a change during the various developmental stages that we have studied.

Histochemical Mapping of Monoamine Oxidase in the Brain of Squirrel Monkey

MANOCHA, SOHAN L., TOTADA R. SHANTHA,* and GEOFFREY H. BOURNE (Yerkes Regional Primate Research Center, Emory University, Atlanta, Ga.)

The activity of monoamine oxidase (MAO) has been mapped at the macroscopic level using 50μ thick frozen sections arranged in a caudo-cranial series of sections through the diencephalon and basal telencephalic centers of the squirrel monkey (*Saimiri sciureus*) brain. The histochemical technique of Glenner, Burtner and Brown (1957, J. Histochem. Cytochem.) was employed for the demonstration of MAO activity. The various thalamic nuclei show a mild MAO reaction except the nuclei periventricularis and para- and subfascicularis, which show moderate MAO reaction. The hypothalamus, in general, shows a stronger MAO reaction than the thalamus, because of its involvement in various autonomic functions. The nuclei paraventricularis, supra-optic and ventromedialis hypothalami are particularly prominent because of the stronger MAO activity they show in contrast to other hypothalamic areas. The nucleus caudatus, putamen and globus pallidus show mild to moderate MAO activity. The fiber bundles of fimbria hippocampi, fornix, anterior commissure and internal capsule show a moderate to moderately strong MAO reaction. The nuclei of the amygdaloid complex differ little from one another and show a mild MAO reaction. The nuclei dorsalis, medialis and lateralis septi show a mild MAO reaction compared to the moderate activity in the nuclei fasciculi diagonalis Brocae and triangularis septi.

---

* T. R. Shanthaveerappa in previous publications.

Cycle Secrétoire des Cellules Juxtaglomérulaires Rénales

MARIUZZI, GIANMARIO, ITALO NENCI, and GUIDALBERTO FABRIS (Istituto di Anatomia ed Istologia patologica, Cattedra di Istochimica Normale e Patologica dell'Università di Ferrara, Italy)

L'étude cytochimique et ultrastructural des cellules épithélioïdes granuleuses des appareils juxtaglomérulaires rénals, en differentes conditions expérimentales qui entraînent une activation fonctionelle (régime hyposodique, constriction d'une artère rénale, adrénalectomie bilatérale), a démontré une augmentation—variable mais constante—de la masse cellulaire fonctionnante, due à la transformation des cellules musculaires lisses de l'artèriole afférente (parfois même de l'éfférente) et à la proliferation des cellules granuleuses juxtaglomérulaires déjà différenciées. Leur hyperactivité se manifeste par une augmentation de leurs granulations spécifiques—très bien démontrables par la coloration selon Mariuzzi et Nenci—et de la basophilie cytoplasmique sensible à la ribonuclease. L'investigation ultrastructurale, outre qu'à documenter les phases de transformation des cellules musculaires lisses en cellules épithélioïdes granuleuses, démontre que le material de secretion est synthétisé dans le réticulum endoplasmique granulaire et après cela qu'il est transporté dans les vésicules du Golgi, où il s'accumule en granules électrondenses; l'excrétion du produit se réalise en grande partie directement dans l'espace intercellulaire à travers les pores de la membrane plasmatique et, en moindre partie, indirectement dans des vacuoles cytoplasmiques.

Microfluorometric Study of Ultraviolet Induced Fluorescence in Sarcoma 37 Tumor Cells (Ascitic Form) Of Mice Supravitally Stained by Thiazin Dyes *

MARQUES, DANTE, A. L. BASTOS, A. M. BAPTISTA, J. D. VIGARIO, J. M. NUNES, A. M. TERRINHA, and J. A. F. SILVA
(Institute Português de Oncologia "Francisco Gentil," Lisbon)

We described the phenomenon of primary inducible fluorescence (1) in secretory granules of tumor cells (2) when supravitally stained by a particular toluidine blue dye (British Drug Houses, Poole, England, C.I.52040, batch no. 651880; 2 drops of 1% alcoholic solution equally spread and dried on a microscope slide) and the cells are examined un fixed by fluorescence microscopy (3). Neither of the other batches of toluidine blue or thiazin dyes tested (thionin, methylene blue, azure I, azure II, azure A) showed this property but we observed a fluorescent reaction in cytoplasmic bodies after exposing supravitally stained cells to ultraviolet light. The fluorescent latent time reaction in those bodies was determined as: Thionin 26 seconds., 9 seconds SD, (10); azure II 1 minute 28 seconds., 9 seconds. SD, (11); toluidine blue 1 minute 35 seconds., 17 seconds SD (10); azure A 1 minute 40 seconds., 21 seconds SD (10); methylene blue 1 minute 57 seconds, 20 seconds SD, (13); and azure I 14 minutes 29 seconds, 2 minutes 21 seconds SD, (13). Azure II showed the lowest coefficient of variation (10%). Peritoneal cells of normal mice showed no fluorescence.

A high pressure mercury burner (HBO 200W) was used as the ultra violet light source with a MPV Leitz microfluorometer. Excitation filters were UG_1 (3mm) and BG 38, and the bright field condenser was 1.4 NA. Illumination of the microscopic field was according to the Köhler principle. Microscopic objective 0.85 NA, 63 x. Barrier filters K430, K460 K470, K490, K510 and K530. The area of the measuring field was 380 μ^2. The electric current from a photomultiplier (EMI 6094A), polarized by a high stabilized HT unit (NSHM BN 600 FNR. 244), amplified by a D.C. Amplifier (NE503B—Rank Nucleonics and Controls) was registered in a 1 mA recorder (VOM 5E Bausch and Lomb). Tumor cells with an estimated area ranging from 340 μ^2 to 380 μ^2 were selected for ultraviolet study of their fluorescent reaction.

The findings suggest that ultraviolet light reacts with thiazin dyes in this cell system to produce a photochemical fluorescent reaction which will be analysed. Cytochemical and cytological evidence for the lipoproteic nature of fluorescence in cell and cytoplasmic bodies induced by ultraviolet light will be presented and discussed, as will the lysosomal origin of secretory granules.

172

References:

(1) Bastos, A. L., Marques, V. D., Silva, J. F., Nunes, J. F. M., Correia, A. D., Vigário, J. D., and Terrinha, A. M., Z. Naturforschg. (B) *in press* (1968).
(2) Bastos, A. L., Marques, D. J., Nunes, J. F. M., Terrinha, A. M., Vigário, J. D., Correia, A. D., and Silva, J. F., III Intern. Congr. of Cytology (Rio de Janeiro), Abstr. no. 117 (1968).
(3) Bastos, A. L., Terrinha, A. M., Vigário, J. D., Nunes, J. F. M., and J. L. N. Petisca, Exp. Cell Res., *42*: 84 (1966).

* (Work supported by a Calouste Gulbenkian Foundation Grant).

Properties of Intranuclear Rodlets and Associated Bodies in Chick Sympathetic Neurons *

MASUROVSKY, E. B., H. H. BENITEZ, S-U. KIM, and M. R. MURRAY (Columbia University College of Physicians & Surgeons, New York, N. Y.)

Correlative light- and electron microscopic studies on the development of chick sympathetic ganglia in long-term, organized culture revealed unusual, sharply-demarcated intranuclear fibrillar bundles in ganglion chain neurons (1-3) which resemble similar formations reported in neurons and supporting cells of the central nervous system (4-6). The fibrillar bundle, which appears as a distinctive curvilinear structure in living and fixed neurons viewed in the light microscope, stains definitively with reduced silver reagents and basic ** dyes. Its general conformation and staining reactions closely resemble that of the intranucleaɪ rodlets described around the turn of the century by the classical histolo-gists. The 0.4-0.8μ spheroidal body, often seen associated with the fibrillar bundle in chick sympathetic neurons, stains in a similar manner, suggesting that certain types of constituents may be common to both structures.

This impression is bolstered at the electron microscope level in thin sections of glutaraldehyde-fixed, glycol methacrylate-embedded neurons subjected to DNase, RNase, and tryptic digestion. Preliminary results indicate that elements of both the fibrillar bundle and spheroidal body are extracted by trypsin. Some nucleic acid also appears to be associated with the spheroidal, granulofibrillar body. In glutaraldehype-fixed, post-osmicated neurons the *fibrillar bundle* appears as a highly ordered, compact array of 50-70 Å fibrils spaced 120-150 Å apart by intermittent fine cross-bridges. This degree of order is usually not seen in neurons fixed with buffered osmium tetroxide alone. Three dimensional reconstructions based upon serial sections through such intranuclear fibrillar

---

** i.e., acid-protein.

bundles indicate that they course through the nucleoplasm along arcing trajectories which may intersect a portion of the nucleolus, or run circumferentially subjacent to the inner nuclear envelope. Fibrils may sometimes be traced between these formations and the immediate environs of nuclear pores. Each 50-70 Å fibril appears to be made up of subunits in helical or cable-like arrays. Such fibrils sometimes may be seen extending between a closely juxtaposed 0.4-0.8μ *spheroidal, granulofibrillar* body and the fibrillar bundle in such a manner as to suggest that the spheroidal body may be synthesizing and/or orienting the fibrils along the fibrillar bundle. Supportive circumstantial evidence for this type of functional relationship is provided by tryptic digestion of certain fibrillar elements in both structures.

Clusters of 350-450 Å electron-dense granules, reminiscent of interchromatinic granules (or glycogen) in overall conformation, are occasionally found in the immediate vicinity of the fibrillar bundle. Studies are in progress to characterize further the nature of these granules, and the granular and/or fibrillar components of the spheroidal body and rodlet, with a view to establishing their role in neuronal metabolism and function.

References:

(1) Masurovsky, E. B., M. R. Murray, and H. H. Benitez: 2nd Intern. Biophys. Congr. (Vienna) Abstr. B12/08 p. 277 (1966).
(2) Masurovsky, E. B., H. H. Benitez, and M. R. Murray: J. Cell Biol. *31*, 73A (1966).
(3) Masurovsky, E. B., H. H. Benitez, and M. R. Murray: *In*: Proc. 25th Annual EMSA Meeting, pp. 188-189 Baton Rouge: Claitor's Book Store (1967).
(4) Siegesmund, K. A., C. R. Dutta, and C. J. Fox: J. Anat. (London) *98*, 93 (1964).
(5) Chandler, R. L., and R. J. Willis: J. Cell. Sci. *1*, 283 (1966).
(6) Mugnaini, E.: Sarsia *29*, 221 (1967).

* Supported by grants NINDB-00858 and NMSS 432 to M.R.M., and NIH GM-00256.

Design and Application of a New Fluorometric Cycling Method for DPN$^+$ and DPNH *

MATSCHINSKY, F. M., C. L. RUTHERFORD, and L. GUERRA (Washington University Medical School, St. Louis, Mo.)

Cycling methods for pyridine nucleotides are a prerequisite for the analysis of intermediates of glycolysis and citric acid cycle in quantitative histochemical studies (1). An efficient system for the measurement of DPN$^+$ and DPNH was designed, using glyceraldehyde-P-dehydrogenase and glutamate dehydrogenase in a coupled oxidation-reduction reaction. Because of their kinetic properties, these enzymes proved ideal analytical tools for this purpose. The assay mixture for the cycling step consisted of phosphate buffer, glyceraldehyde-P, α-ketoglutarate, NH_3, EDTA, ADP, arsenate, mercaptoethanol and the two enzymes. The useful assay range for DPN^+ or DPNH is 10^{-9} to 10^{-7} M. After cycling, the enzymes glyceraldehyde-P and α-ketoglutarate were destroyed by boiling in acidic H_2O_2 and the glutamate, which had accumulated during cycling, was then measured with an enzymatic-fluorometric assay. By alteration of the volume of the cycling step (from 1 to 50 μl) and of the concentration of the enzymes (to give amplifications from 100- to 30,000-fold) the sensitivity of the assay could be adjusted over a wide range. In combination with an oil-well technique (1,2) as little as 10^{-15} moles of the pyridine nucleotide was measureable. This new fluorometric-catalytic assay for DPN^+ and/or DPNH has, so far, been applied to the analysis of fructose diphosphate, lactate, citrate, and sorbitol in frozen-dried speciments of 0.1 μg obtained from retinal layers, epithelial structures of the endocochlear duct and the islets of Langerhans.

References:

(1) Lowry, O. H.: Harvey Lectures, Series 57, 1 (1963).
(2) Matschinsky, F. M., J. V. Passonneau, and O. H. Lowry: J. Histochem. Cytochem. (*in press.*)

* Supported by USPHS grant AM 10591.

Observation on the Chemical Composition of Human Brain Ribonucleic Acids

MATTURRI, L. and S. CURRI (Istituto di Anatomia e Istologia Pato-logica dell'Università Milano (Italy) and Centro di Biologia Molecolare, Padova, Italy)

Twenty human brains removed at autopsy were examined. These brains, free of any pathological lesion, belonged to subjects of both sexes, within the age range from 5 months to 80 years.
RNA was extracted by a guanidinium chloride technique: the final purified product, which was about 90% RNA, had a λ max of 258-260 mμ in 0.1 M NaCl. The phosphorous contained was determined by the Bachofer and Wagner method. Total nitrogen, ribose content and the mononucleotide composition of the RNA samples, were also determined. The percentage distribution of the purines and pyrimidines and the ribose content show some differences related to the age of the subjects. Topochemical analysis of samples of the same brains was also performed for comparison.

Variability of DNA Staining *

MAYALL, BRIAN H. (Dept. of Radiology, University of Pennsylvania, Philadelphia, Pa.)

Significant differences are found in cytophotometric analysis of DNA content of stained human leukocytes, with monocytes measuring about 10 percent higher than small lymphocytes and neutrophils. This effect is seen with preparations stained by either the Feulgen reaction or gallocyanin-chrome alum following ribonuclease treatment, and with measurements made on either an integrating scanning digital cytophotometer (CYDAC) (1) or an automatic two-wavelength cytophotometer (DATEM) (2). The measured differences are too large to be explained by any errors known to be associated with the instruments, but can be explained by differences in DNA content and by variations in efficiency of staining. Efficiency of staining is a function of both the stoichiometry and the optical properties of the DNA-stain complex.

Experimental results show that, in general, DNA staining increases whenever there is a decrease in the structural organization of the chromatin, be it due to the biological state of the cell or to physical and chemical denaturation. Cells in which the chromatin is highly structured, such as dense heterochromatic lymphocytes, show less staining than cells such as monocytes with diffuse, nonheterochromatic chromatin. If nuclear area is used as an index of chromatin compaction for leukocytes, it is found that a 33 percent increase in area is associated with a 10 percent increase in apparent stain content. A comparable relationship has been shown to exist for metaphase chromosomes (3). Similarly, it is possible to relate staining to cell flattening with any one class of leukocytes. Staining of DNA can also be modified by prior treatment of the cells. Agents which should denature chromatin (such as boiling water, formaldehyde, dimethylsulfoxide, or strong hydrochloric acid) increase staining of leukocytes and concomitantly decrease differences between cell types. Thus, variability in staining efficiency is of sufficient magnitude to explain the differences between leukocytes. It remains to be shown whether staining efficiency can be eliminated as a variable and to what extent the hypothesis of DNA constancy will then be upheld.

References:

(1) Mendelsohn, M. L., B. H. Mayall, J. M. S. Prewitt, R. C. Bostrom, and W. G. Holcomb: *In*: Advances in Optical and Electron Microscopy, pp. 77-150 (Cosseltt and Barer, edits.) New York: Academic Press (1968).
(2) Mayall, B. H., R. Q. Edwards, R. C. Bateson, J. R. Connolly, and M. L. Mendelsohn: Annals of the New York Academy of Sciences (1968). *in press.*

(3) Mendelsohn, M. L., D. A. Hungerford, B. H. Mayall, B. Perry, T. Conway, and J. M. S. Prewitt: Annals of the New York Academy of Sciences (1968). *in press.*

* Supported by Grant 5 RO1 CA03896 and Contract PH 43-62-432 from the National Cancer Institute, National Institutes of Health, Bethesda, Maryland, U.S.A.

Leukemic Leukocytes Examined with Histochemical Enzyme Technics

MELNICK, P. J. (Veterans Administration Hospital, Martinez, Calif. and University of California Medical Center, San Francisco, Calif.)

Cyto-histochemical enzyme technics suitable for tissue sections yield fragmentary or no results in blood and marrow formed elements, aside from alkaline phosphatase and a few other enzymes (1) because free-living cells have a structured cell periphery (2) and high electrical resistance (3) so that they are very impermeable. In contrast, the cells in tissue sections are cut across allowing direct access of substrate and capture reagent to their interiors. By prolonging incubation, allowing enough time for substrate to permeate, histochemical methods for 16 hydrolases and 15 dehydrogenases were applied successfully (4). This required finding a non-inhibitory fixative to retain the cells on slides, and also finding a protective additive to prevent leaching of soluble enzymes from cells. Cold acetone (4° C) for 60 seconds followed by 5% PVP (m.w. 40,000) for 60 seconds were found to be optimal. The most significant observations in the leukocytes were: a) enhancement of enzyme activity in leukocytosis, leukemoid reactions, florid stages of chronic and subacute leukemias, and in polycythemia vera; b) diminished or absent activity in undifferentiated cells; c) several enzyme defects, probably deletions; d) a number of primary mosaicisms; and e) diminished or absent activity, and a number of induced mosaicisms, developing in the course of antileukemic therapy. Erythrocyte enzymes were also demonstrated, including a number of hydrolases, the glycolytic and pentose shunt dehydrogenases, and isocitric and malic dehydrogenases (apparently non-mitochondrial fractions). No mitochondrial enzymes were present since erythrocytes contain no mitochondria. The technics were applied to the problem of viability testing of blood and marrow stored at low temperatures for clinical use (5). If certain enzyme defects can be correlated consistently with certain chromosome abnormalities observed with cytogenetic technics, basic information may be obtained about chromosome functions. The technics may also become applicable in other areas of hematology. The technics were also successfully applied to Detroit 6, HeLa and Chang liver cell lines (courtesy of Dr. John E.

Shannon, American Type Culture Collection), and may be useful in virology and oncology.

References:

(1) Hayhoe, F. G. J.: *In: Disorders of the Blood.* Edited by Whitby, L. E. H. and Britton, C. J. C. 9th Edition. New York: Grune & Stratton, Inc., 1963, pp. 131-158.
(2) Weiss, L. and E. Mayhew: New Engl. J. Med., *276,* 1354 (1967).
(3) Loewenstein, W. R.: *In: Biological Membranes: Recent Progress.* Edited by Loewenstein, W. R. and Meyer, E. M. New York: Ann. N. Y. Acad. Sci., *137,* 441 (1966).
(4) Melnick, P. J.: *In: Pathology of Leukemia,* edited by Amromin, G. D. New York: Hoeber Medical Division, Harper & Row, Publishers. (In press.)
(5) Melnick, P. J.: *In: Cryosurgery,* edited by Rand, R. W., Rinfret, A. P. and von Leden, H. Springfield: Charles C. Thomas, Publisher. (*In press.*)

Compaction and Boundaries in the Photometric Analysis of Human Chromosomes *

MENDELSOHN, M. L., T. J. CONWAY, B. H. MAYALL, B. PERRY, and J. M. S. PREWITT (Dept. of Radiology, University of Pennsylvania, Philadelphia, Pa.)

To be successful, an automated karyotype analysis must encompass the wide range of morphologic variability of normal and abnormal chromosome preparations. Compaction is a major source of such variability, and it has an obvious effect on the length, area and contrast of the chromosome, as well as a subtle effect on the staining of chromosomal DNA (1). In the presence of differential compaction, any attempt to standardize chromosomal measurements requires the definition of chromosomal boundaries by a procedure which is independent of compaction.

Boundaries of chromosomal images are characteristically unsharp due to the inherently fuzzy structure of the chromosome and the limited resolution of optical systems. Using CYDAC, a scanning digital cytophotometer, measurements of optical density across isolated stained chromatids indicate that the density profile is bell shaped, with both the height and width of the bell dependent on compaction. The inflection point of the bell is a convenient operational definition of the boundary, and we have developed a method which sets a grayness threshold at the inflection point of each chromosome individually. Preliminary tests have confirmed the feasibility of the method, and the procedure is now being applied to a population of chromosomes in order to study its effect on parameter extraction.

178

References:

(1) Mendelsohn, M. L., D. A. Hungerford, B. H. Mayall, B. Perry, T. Conway, and J. M. S. Prewitt: Annals of the New York Academy of Sciences (1968), *in press.*

---

* Supported by Grant 4 KO6 CA18540 and Contract PH 43-62-432 from the National Cancer Institute, National Institutes of Health, Bethesda, Maryland, U.S.A.

Cytochrome Oxidase and Succinic Dehydrogenase in the Liver of Mice Infected with Trypanosoma cruzi

MERCADO, TERESA I. (Laboratory of Parasitic Diseases, National Institutes of Health, Bethesda, Md.)

Using the method of Burstone (1) for the histochemical localization of cytochrome oxidase, histochemical tetrazolium methods for the detection of succinic dehydrogenase, and based on assays of liver homogenates employing the Warburg technique it was observed that mice heavily parasitized with the Tulahuen strain of *Trypanosoma cruzi* tended to have a lower enzyme activity than uninfected mice. Histochemically, normal mice exhibited uniform activity of both enzymes throughout the liver lobule with some accentuation around the periportal areas. In infected mice, activity was somewhat irregular and generally there was no periportal pattern. Biochemical determinations indicated that in mice infected eight or nine days with parasitemias up to 64 million trypanosomes per ml. of blood, activity was about one half that of the normal specimens. In younger infections, marked deviations from the normal were not observed. The functional significance of these findings cannot be evaluated at present, but the decrease in enzyme activity may be related to the marked fatty infiltration occurring in the infected livers. The disturbance seemed more extreme in those livers in which the fatty infiltration was most intense.

References:

(1) Burstone, M. S.: J. Histochem. Cytochem. *8*, 63 (1960).

DNA Variability in Gymnosperms

MIKSCHE, JEROME P. (Institute of Forest Genetics, North Central Forest Experiment Station, Rhinelander, Wis.)

Measureable DNA differences per meristematic cell, as determined by chemical and cytophotometric methods, exist among 2n = 24 diploid species of gymnosperm genera and within coniferous species. The two wavelength method and the diphenylamine reaction were used for determination of Feulgen absorption units and chemical estimation of DNA, respectively. Bulk root tip staining yielded Feulgen stain uniformity equal to that obtained by staining squashed cell preparations, thus eliminating the problem of tissue-depth penetration of stain. Alkaline fast green was used for histone estimation. The differences in DNA amounts from the lowest to the highest of the species studied varied by factors of 3.5 for chemical extraction and 2.9 for Fuelgen photometry; thus the two methods of analysis are in agreement. DNA variation within the species, *Picea glauca* and *Pinus banksiana* varied by factors of 1.6 and 1.5, respectively. These differences in DNA are probably related to changes in chromosome length and not chromosome strandedness. Our findings also indicate that within a species, the more northern sources possess more DNA than their southern latitude counterparts, and species with smaller amounts of DNA tend to have a wider distribution.

Lysosomes: Histochemical Demonstration of Latency Using Dimethyl Sulfoxide

MISCH, DONALD W. and MARGARET S. MISCH (Department of Zoology, University of North Carolina, Chapel Hill, N. C.)

Latent activity of acid phosphatase was tested in lysosomes of unfixed rat-liver tissue. Cryostat sections (8μ) were incubated in test media containing components of a Gomori reaction medium (1) at pH 5.0, or in a simultaneous-coupling azo-dye mixture (2) at pH 6.0 or 6.5, together with 0.6 M sucrose and 1.4 M dimethyl sulfoxide (DMSO).

Reactions made in test and control media at the same temperature (37°C, 23°C, or 15°C), same pH, and for the same length of time were compared for frequency and intensity of reaction sites within the parenchymal cells. An enhanced reaction was considered to be evidence for a more rapid penetration of substrate with a corresponding reduction in latency of the lysosomes so activated.

The inclusion of DMSO in the reaction media for both the Gomori and the azo-dye procedure resulted in enhancement of the reaction. The addition of 0.6 M sucrose to reaction media reduced diffusion artifact for the Gomori procedure as judged by the absence of nuclear staining, and the reduced size and increased intensity of enzymatic loci. Increasing concentrations of sucrose in the absence of DMSO, up to 2 M in control reaction media, resulted in only slight enhancement of the reaction, always much less than when DMSO was used with or without added sucrose. This effect of sucrose was less apparent with the azo-dye procedure.

Control incubations following heat inactivation, addition of 0.002 M d-tartaric acid, or omission of substrate were invariably negative.

The Gomori substrate (Na-β-glycerophosphate) was also employed for quantitative measurement of acid phosphatase activity in lysosome-containing cell fractions of rat-liver homogenates. The presence of 1.4 M DMSO during the enzyme reaction period resulted in an approximate 2-fold increase in phosphatase activity compared with fractions resuspended in sucrose alone (3). At the concentration employed, no direct effect of DMSO upon soluble acid phosphatase previously released from disrupted lysosomes was detected. Enhancement of acid phosphatase activity by DMSO occurred in the presence of intact lysosomes.

Naphthol AS-BI phosphate was used as substrate in the azo-dye procedure and was also employed for quantitative measurement of latency. Lysosome fractions assayed in DMSO showed an approximate 1.5-fold increase in activity compared with those in sucrose.

When glycerophosphate was used as substrate, the effect of DMSO on activation of lysosomes could be reversed for both the homogenate and the histochemical procedures by washing two times in sucrose.

Our observations indicate that latency of lysosomes can be preserved routinely in cryostat sections of unfixed rat liver, can be appreciably reduced in the presence of DMSO (thereby increasing enzyme activity), and can be re-established upon removal of DMSO. The use of dimethyl sulfoxide provides a method by which the intactness or permeability of lysosomes can be examined using a histochemical procedure which is closely parallel to a corresponding homogenate assay. The method is potentially applicable for detecting the onset of a wide variety of conditions within cells which may involve alteration in permeabilty and activity of lysosomes.

References:

(1) Holt, S. J.: Exp. Cell Research 7, (Suppl.) 1 (1959).
(2) Barka, T. and P. J. Anderson: *In:* Histochemistry, pp. 244-246, New York: Hoeber (1963).
(3) Misch, D. W. and M. S. Misch: Proc. Natl. Acad. Sci., 58, 2462 (1967).

Fixation of RNA in Plant Tissue

MITCHELL, J. P. (Department of Botany, University of Edinburgh, Scotland)

Estimation of total nucleic acid, essentially RNA, in explants from *Helianthus tuberosus* tubers fixed in methanol, 80% ethanol and ethanol

/acetic acid shows that no loss of nucleic acid takes place. Aqueous formaldehyde (4%) apparently extracts about fifteen per cent of the total nucleic acid. This discrepancy probably is due to the binding of nucleic acid components to other tissue constituents preventing their extraction for measurement, since transfer RNA—the most likely nucleic acid species to be removed during fixation—is found in extracts of undegraded RNA from formaldehyde fixed tissue.

Nucleic acids in tissues fixed in ethanol/acetic acid and in fixatives containing formaldehyde are resistant to aqueous extraction and to treatment with an EDTA solution. This latter treatment is desirable as a means of preparing suspensions of individual cells. Tissue treated with citrate buffer under conditions used for Azure B staining loses substantial quantities of nucleic acid and the presence of Azure B does not wholly prevent this loss.

Ribosomes and Polyribosomes during Mitosis in L-Cells *

MITTERMAYER, C., G. KIEFER, and W. SANDRITTER (Department of Pathology, University of Freiburg, Germany)

Protein synthesis is almost exclusively confined to ribosomes held together by messenger-RNA, thus forming polyribosomes. The work presented here is concerned with the question, whether the physiologic block of RNA-synthesis during mitosis has an effect on the polysome pattern.

L-Cells were synchronized by selective mechanical detachment (Terasima and Tolmach: Exp. Cell Res. *30*, 344 (1963); Mittermayer, Kaden, and Sandritter: Histochemie *12*, 67 (1968), Ribosomes and polyribosomes were prepared in the following way: 5×10^5 synchronous L-cells received a 10 minute pulse label with 100 μC/ml Leucine-H^3 and were then homogenized in a medium containing 0.025 M KCl, 0.005 M MgCl$_2$, 0.125 M sucrose and 0.05 M Tris-buffer, pH, 7.2. The homogenate was centrifuged 10 minutes at 15,000g and the supernatant mixed with Na-deoxycholate, final concentration 0.2%. 1.0 ml of this preparate is layered over a 5-20% sucrose density gradient and centrifuged 100 minutes at 125,000g at 4° C. The 260 mμ absorbance and the radioactivity of the effluent of the centrifuge tubes is monitored. During mitosis no or almost no polyribosomes can be observed. Immediately after completion of cytokinesis a complete pattern of polyribosomes appears with aggregates up to 20 ribosomes. The disappearance of polyribosomes during mitosis is not an artefact of the preparation procedure, since electron microscopy also shows lack of polyribosomes during mitosis. It is not yet clear whether the disappearance of polyribosomes during mitosis is due to lack of messenger-RNA or whether polyribosomes are degraded in the course of cell division.

* Supported by the Deutsche Forschungsgemeinschaft and the U.S. National Institutes of Health (5-RO5-TW0024-02).

The Detection of Thiamine Derivatives by Means of High-Resolution Electron Microscopic Autoradiography

MIZUHIRA, VINCI, KAZUKO UCHIDA, TAKANORI AMAKAWA, HIDEO SHINDO, JUNICHI TOTSU, and IKUO SUESADA (Department of Anatomy, Laboratory of Cell Biology, School of Medicine, Tokyo Medical and Dental Univ. Bunkyo-ku, Tokyo, Japan)

We have succeeded in fixation of tritiated thiamine and its derivatives in tissue blocks, as to make the insoluble precipitates with hydrogen platinum chloride (thiamine-^{3}H-PtCl$_6$). In the present report, the absorption mechanism of tritiated BTMP(s-benzoylthiamine-^{3}H o-monophosphate) in rat intestinal epithelium and the distributions in heart muscle, liver and kidney were studied with the aid of high resolution electron microscopic autoradiography.

88.5 μc of BTMP-^{3}H was injected into a loop of small intestine, or into tail vein. After 1, 10 and 30 minutes, small tissue blocks were taken from the body and prefixed with 2.5% glutaraldehyde, then postfixed with 1% OsO$_4$ both containing H$_2$PtCl$_6$. The tissue blocks were dehydrated with graded ethanols containing H$_2$PtCl$_6$ and embedded in Epon. Ultrathin sections were made by glass knives. The sections on a grid were coated with Sakura NR-2 nuclear research emulsion, exposed for 4 weeks and carefully developed with a gold latencificated Elon-Ascorbic-Acid method at 17° C for 14 minutes. Then removal of gelatin and electron staining proceeded with a lead monoxide solution for about 20 minutes.

The precipitates of thiamine-PtCl$_6$ are clearly observed as small dense granules directly, estimated to be 100 to 300 Å in diameter. Developed fine silver halid grains appear as closely overlapped on the precipitates. In the intestinal epithelium, precipitates and grains appear inside the cell membrane of microvilli, on the core of microvilli and in the organelle of epithelial cells, such as mitochondria, Golgi-membranes endoplasmic reticula and nuclei.

We also demonstrated the enzymatic activity of alkaline phosphatase which concerned the dephosphorylation of BTMP on the outer surface of cell membranes of microvilli when it was absorbed into the microvilli. After the dephosphorylation, it transforms to thiamine or s-benzoylthiamine (SBT).

In the heart muscle, precipitates of thiamine-^{3}H-PtCl$_6$ were observed in the glycogen area around mitochondria, in mitochondria, in nuclei and in red cells.

Histochemical Studies on Ornithine Carbamoyltransferase Activity

MIZUTANI, AKIRA (Dept. of Cytochemistry, Chest Disease Research Institute, Kyoto Univ., Kyoto, Japan)

The recent development of a Gomori-type method for the histochemical demonstration of ornithine carbamoyltransferase (OCT) activity has established that it is localized only in the mitochondria of hepatocytes in the rat and mouse (1). In this method, phosphate ions released through enzymatic action are captured by lead ions present in the incubation medium; it is one of the few histochemical methods involving two primary substrates (L-ornithine and carbamoylphosphate). If either substrate is lacking, the specific mitochondrial reaction does not occur. Further studies emphasized the following points: (1) Fixation is a critical factor. The best fixative was found to be 4% fresh formaldehyde solution obtained from depolymerization of paraformaldehyde, containing 1% CaCl$_2$, 8% sucrose and 0.04M cacodylate buffer (pH 7.2). Adequate fixation of liver occurred in 4-6 hours; the slices should be as thin as possible for rapid fixation. Formaldehyde perfusion for 30 minutes if possible, through the portal vein gave the most reliable results for ultrastructural observation. Although an intense reaction was also obtained by 3.13% glutaraldehyde perfusion for 2-3 minutes, the control experiment with the medium from which L-ornithine was omitted sometimes gave the similar, but less intense, findings. (2) Each 10 ml of the incubating medium contained L-ornithine, 5 mg; carbamoylphosphate Li salt, 3 mg; 0.05M tris-maleate buffer at pH 7.2, ml; 1% lead nitrate, 1 ml; and sucrose, 0.8 g. Carbamoylphosphate is unstable and, therefore, the incubation time should be limited to 15 minutes at 20° C or, if necessary, the medium should be renewed. (3) Quantitative assay of citrulline formation with fresh kidney homogenate confirmed that the moderate reaction in lysosomes and brush borders was nonspecific. This histochemical method demonstrated that OCT activity was localized in hepatic mitochondria, mostly in the matrix, in the mouse, rat, guinea pig, rabbit and frog. The degree of intensity was in the following order: rat, mouse > frog > guinea pig, rabbit.

References:

(1) Mizutani, A.: J. Histochem. Cytochem. *15*, 603 (1967), Ibid. *in press.*

184

Studies in Excretory System. Extraneous Coats in Transitional Epithelium of Urinary Tract

MONIS, BENITO and ALBERTO CANDIOTTI (Instituto de Biologia Celular. Univ. Nac. de Córdoba. Córdoba, Argentina)

A surface mucous coat was identified in transitional epithelium of urinary tract of man (1). Electron microscopic observations revealed a filamentous coat which was characterized by histochemical ultrastructural technic (2).
These observations indicated that an acid glycoprotein which contains sialic acid was present in the extraneous coat.
Chemical quantitative studies and histochemical staining procedures with enzymatic digestion tests will be reported on the glycocalyx of transitional epithelium of various species.
A distinct functional role for mucous coats of urinary tract is postulated.

References:

(1) Monis B. and H. Dorfman: J. Histochem. Cytochem. *15*, 475 (1967).
(2) Monis B. and D. Zambrano: Z. Zellforsch. 1968, *in press.*

The Behaviour of the Schiff Positivity in Fluorescence on Pig Gastric Mucosa Homogenate

MORI, G. and A. INGRAMI (Department of Pharmacology, Medical College, University of Milan, Italy)

An acriflavine solution at 0.5% in distilled water, saturated with bubbling SO_2 was used to follow at the spectrophotofluorimeter the course of the PAS positivity of three homogenates of mucosa: these were obtained from three different parts of the pig stomach: the fundus, the middle gastric and the antropyloris areas. The mucosa, separated by an automatic sliver, after the 3 various sectors were sectioned, was homogenated in distilled H_2O (1.1 by weight). Centrifugation was then effected at 15,000 r.p.m. for 15 minutes. The supernatant was then placed on a cellulose acetate strip of the Schleicher und Schuell Firm (Western Germany). The quantities in weight were maintained even through the determination of the dry residue. The oxidation was obtained through immersion of the strip in a periodic acid solution, first at 0.6%, this concentration exceeded slightly those suggested by the PAS techniques. After washing the strips in water, they were immersed again in a bath of the following solution: acriflavine-SO_2 solution at 0.5% in H_2O distillated for 15 minutes, and the colour was removed through repeated soakings in ethylic alcohol (at 70%). The cellulose acetate strip was divided into equal sections: one corresponding to the deposited stain, the other was used as a "blank" sample in the neighbouring section. These were eluted in 5 ml of a chloroform/ethanol solution (9/1) and reading was thus carried out in cuvettes at the spectrophotofluorimeter, at a wave length of 490 millimu for excitation, of 525 millimu for the emission. At the standard times of the Schiff's reaction we had alternate readings, while at 30 and 60 minutes the fluorescence data has a tendency to increase in percentage, but this was without a significant difference from the reading of the blank. This induced us to increase the concentration of the oxidizing agent. With periodic acid at 1% we already obtained after 10, 15 and 30 minutes a constant reading, showing a modest discrepancy from the fluorescence of the blank. After one hour of oxidation, a maximal fluorescence peak was obtained for all three samples, as compared with a constant reading of the blank. Increasing slightly the homogenate weight doses, and altering slightly the reading techniques, some difference was evidenced in the fluorescence of the three homogenates. This fluorescence shows a tendency to extinguish itself after 2 hours of oxidation. If we consider that the measurement of the completed oxidation, and therefore, of the Schiff's basis- aldehyde reaction, the stability and constant value of fluorescence of the Schiff's reagent used, we may assume oxidation of the homogenates is completed within one hour of oxidation with a concentration of 1% of the oxidizing medium. Since prolonged oxidations may provoke a PAS-positivity of complexes although PAS negative at the beginning, this fluorimetric evaluation of the Schiff's aldehyde reaction, seems to us a useful means of investigation to evaluate the Schiff positivity of tissular homogenates, whose mucous incretion is excitated differently in different pathological conditions.

Studies on Alkaline and Acid Ribonucleases in Mammalian Tissues. Immunohistochemical Localization and Immunochemical Properties

MORIKAWA, SHIGERU, MASAO YAMAMURA, TAKAYUKI HARADA, and YOSHIHIRO HAMASHIMA (Department of Pathology, Faculty of Medicine, Kyoto University, Kyoto, Japan)

The localization of alkaline and acid ribonucleases (RNase) were investigated in mammalian tissues by fluorescent antibody techniques. Alkaline RNase from bovine pancrease and acid RNase from bovine spleen were used as antigens, and the antibodies were obtained from rabbits.

Anti-alkaline RNase antibody reacted against alkaline RNase, and anti-acid RNase antibody reacted against two different acid RNases with no cross-reactivity between alkaline and acid RNases. Each kind of antisera inhibited the enzyme activities of the corresponding RNases from heterologous sources as well as bovine RNase to various extent. The localization of RNases in the pancreas, liver, small intestine, kidney, spleen, lymph node, thymus, adrenal and heart were investigated. Some differences in staining results were observed depending upon the tissues and fixatives employed. Two different patterns of distribution of alkaline RNase in the pancreas and kidney were observed depending upon the fixatives, and liver alkaline RNase was revealed only in frozen sections fixed with acetone. Acid RNase was proved in some of islet cells of the pancreas. The conjugates could stain the sections from some heterologous materials, such as human, rat, mouse, etc.

Light Scattering by Biological Cells and its Relation to Cell Size [*]

MULLANEY, P. F., P. N. DEAN, and M. A. VAN DILLA (Los Alamos Scientific Laboratory, University of California, Los Alamos, New Mexico)

Both theoretical and experimental investigations have been undertaken to determine whether the size distribution of cells in aqueous suspension can be measured by optical means. For the purposes of calculation, cells have been taken as homogeneous spheres (diameter $= d = 2$ to 20 microns) suspended in water (relative index of refraction range $= 1.03$ to 1.07). The light scattered at the wavelengths 488.0 and 632.8 nm has been calculated with a computer following the method of Hodkinson (1). Unlike the rigorous Mie theory treatment (2), scattering is treated as a combination of diffraction, reflection, and refraction. The size parameter measured depends on the range of scattering angles covered by the detector. The diffraction contribution, which is dominant at small angles, depends only on the external silhouette of the cell and is independent of refractive index and internal cellular details. Between 0.5 to 2.0°, our calculations predict a scatter signal closely proportional to d^3; over 80% is diffracted light. As the scattering angle increases beyond 5°, the refraction and reflection contributions increase and the diffraction contribution decreases; refractive index, membrane reflectivity, and other interior details become increasingly important. If these factors are independent of particle diameter, the predicted signal is approximately proportional to d^2 for the scattering angle range 5 to 25°. In this range of scattering angles, a 4% change in refractive index results in a 150% change in signal. A size determination derived from light scattering measured in this angle range would be difficult unless the refractive index is accurately known for all cells encountered in a given experiment, a difficult requirement to satisfy. If a range of scattering angles (3.0 to 6.0°) is chosen so that only the diffracted light from the smaller cells and only the reflected and refracted light from the larger cells is collected, there is no simple relationship between signal and cell size. The plot of signal versus d^2 is of an oscillatory nature. Thus, the theoretical indications are that measurement of forward scattering with small detector acceptance angle (0.5 to 2.0°) holds the most promise for reliable cell sizing.

Preliminary experimental work designed to test these theoretical results has been performed using equipment similar to that described at this meeting, by Van Dilla *et al.* for cellular fluorescence studies. The mercury light source has been replaced by a Spectra-Physics Model 130B He-Ne laser emitting at 632.8 nm. The small diameter (1.5 mm) and divergence of the light beam from the laser permit measurement of small angle scattering which would not be possible with conventional light sources. Light scattered by Chinese hamster ovary and other mammalian cells has been measured in the 5 to 25° angle range. In general, the scatter signal increases with increasing particle diameter, but the exact relationship between signal and cell size is not simple. This may be a result of refractive index differences as noted above. The apparatus is being modified to allow scattering measurements to be made between 0.5 to 2.0°, and these experimental results, along with a comparison with theory, will be presented.

References:

(1) Hodkinson, J. R., and J. Greenleaves: J. Opt. Soc. Am. *53*, 1336 (1963).
(2) Mie, G.: Ann. Phys. Lpz. *25*, 377 (1908).

* This work was performed under the auspices of the U.S. Atomic Energy Commission.

Die Toluidinblaumetachromasie im Nierengewebe, Urin und Speichel nach Applikation von sulfatiertem Polyanion

MÜLLER, GERHARD (Anatomisches Institut, Universität Mainz/ Rhein)

Sulfatiertes Polyanion findet als Medikament (SP 54) vielseitig Verwendung. In vitro reagiert es mit Toluidinblau metachromatisch. Im Tierversuch war nach intravenöser Injektion von SP 54 eine starke metachromatische Reaktion vor allem in den Hauptstücken und den Glomerula festzustellen. Histologisch zeigten sich hier schwere Zellschädigungen. Beim Menschen wurde die Metachromasiereaktion im Urin und Speichel geprüft. Sie trat nur nach intravenöser Applikation von SP 54 auf, nicht nach rektaler oder oraler. Nach spektrophotometrischen Messungen des Urins wird das sulfatierte Polyanion innerhalb weniger Stunden biphasisch ausgeschieden.

Ultrastructural Histochemistry in Diurnal Rhythm

MÜLLER, OTFRIED (Anatomisches Institut, Medizinische Hochschule, Hannover)

As earlier investigations have shown (Müller *et al.*, 1966; Bünning, 1967), striking changes in the ultrastructure of liver cells occur during the circadian cycle. Their appearance will be outlined. In connection with this, rhythmic ultrastructural changes of histochemically demonstrable enzymes have been studied.

In this investigation, recommended methods for alkaline and acid phosphatases, ATP-ase and succinodehydrogenase (after Seligman *et al.*, 1967) modified for electron microscopical purposes, have been used. It can be shown that there are significant differences between the inner and the outer zone of the liver lobuli in respect to the content and distribution of the enzymes and their relations to the inner structure of the cell. It can be shown further that these substances do change during the diurnal cycle. The significance of this change in respect to bile production and glycogen accumulation and mobilization will be discussed.

Enzyme-labeled Antibodies for Electron-immuno-cytochemistry *

NAKANE, PAUL K. (The University of Colorado Medical Center, Department of Pathology, Denver, Colorado)

Antibodies, labeled with enzyme of small molecular weight, are able to diffuse through tissues which ferritin labeled antibody cannot penetrate because of its large size (1, 2). The enzyme-labeled antibody method takes advantage of the amplifying effect of enzymatic reactions. Enzymes are not consumed in the reaction with substrate, consequently, each molecule of enzyme-labeled antibody bound to the antigenic sites deposits many molecules of reaction product. These are visible with the light and electron microscope.

Horseradish peroxidase was conjugated to antibodies employing p,p'-difluoro-m,m'-dinitrodiphenyl sulfone as a bifunctional reagent. The enzymatically and immunologically active conjugate was reacted with tissues which were then stained histochemically for the enzyme. Horseradish peroxidase was employed since it was obtainable from commercial firms in relatively pure form, and the histochemical reactions for light and electron microscopic studies had been developed (3). Moreover, the endogenous localization of peroxidase is limited to certain sites.

To demonstrate the method at the ultrastructural level, studies were made of basement membrane of the kidney of the mouse as an example of extracellular antigen and of luteinizing hormone in the pituitary gland of the rat under various physiological conditions as an example of intracellular tissue antigens.

The indirect immunohistochemical method was employed. Thin slices of kidney were fixed briefly in buffered formalin and those of pituitary gland were fixed briefly in buffered formalin with picric acid (4). Tissues were washed in phosphate buffered saline with 10% DMSO and frozen sectioned 20-30 microns in thickness. DMSO protected the tissues from damage caused by freezing and improved preservation. The sections were reacted with rabbit antisera and the controls with normal rabbit serum washed in phosphate buffered saline, reacted with peroxidase-labeled sheep anti-rabbit gammaglobulin and washed. The sections were postfixed in 5% glutaraldehyde, washed and stained for peroxidase using the method of Graham and Karnovsky. The stained sections were osmicated, dehydrated and embedded in either Epon or Araldite.

In the mouse kidney, epithelial basement membrane was found adjacent to the plasma membrane of tubular epithelium, but not in adjacent vascular basement membranes. Luteinizing hormone was found in granules and endoplasmic reticulum in cells of the pituitary gland of the rat.

References:

(1) Nakane, P. K. and G. B. Pierce: J. Histochem. Cytochem. *14,* 929 (1966).
(2) *Ibid:* J. Cell Biol. *33,* 307 (1967).
(3) Graham, G. C. and M. J. Karnovsky: J. Histochem Cytochem. *19,* 291 (1966).
(4) Zamboni, L. and C. De Martino: J. Cell Biol. *35,* 148A (1967).

* Supported in part by American Cancer Society Grant ♯ E105 and United States Public Health Service Grants AI 07758 and CA 08201.

Immunoenzymo-histochemistry: Localization of Hormones in Pituitary Gland of the Rat *

NAKANE, PAUL K. (The University of Michigan, Department of Pathology, Ann Arbor, Mich.)

Antibodies labeled with enzymes have been used to localize tissue antigens. The sites of antibody-antigen reactions are marked by localizing the enzyme histochemically. The preparations are permanent and may be observed with an ordinary light microscope. When the reaction products of the enzymes are electron opaque, the method is directly applicable to ultrastructural localization of tissue antigens (1, 2).

Sheep anti-rabbit gamma globulin labeled with horse-radish peroxidase and the indirect immunohistochemical methods were used in this study (3). Rabbit antisera against human thyrotropic hormone (TSH) (a gift from Dr. W. D. Odell), ovine luteinizing hormone (LH) (a gift from Dr. Niswender), ovine follicular stimulating hormone (FSH) (a gift from Dr. Gay), porcine adenocorticotrophic hormone (ACTH) (a gift from Dr. P. Vague), rat growth hormone (GH) and rat prolactin (gifts from Dr. J. Furth) were used to localize the respective hormones in the anterior pituitary gland of the rat.

Either single hormones were localized in sections of pituitary gland using the conventional method (1, 2), or two or three hormones were identified simultaneously using a modification of the method. This modification was based upon the idea that if the first antigen were localized by the indirect method and the antisera were removed from the section by elution, leaving the colored reaction products identifying the antigenic sites, the second and subsequent antigens might be localized using substrates that develop reaction products of different colors.

Specifically, sections of Bouin's fixed pituitary glands of rats were reacted with the first antiserum followed by peroxidase-labeled anti-rabbit gamma globulin (P-anti-RGG) and stained for peroxidase utilizing 3,3'-diaminobenzidine as substrate (4) (yellow reaction product). The sections were eluted for 1 hour in glycine-HCl buffer, pH 2.3, washed, reacted with the second antisera, then with P-anti-RGG, and stained with α-naphthol pyronin as substrate (5) (pink reaction product). The sections were eluted in acid buffer, washed, reacted with the third antisera, then with the P-anti-RGG and stained utilizing 4-Cl-1 naphthol as substrate (blue reaction product).

GH, Prolactin and ACTH were each contained in separate cells. GH was localized in small round cells, which were distributed throughout the pituitary glands. Most of them lay adjacent to sinusoids. Cells containing prolactin were also small and distributed in a manner similar to the cells containing GH. Although cells containing ACTH were distributed throughout the gland they were most numerous in the stalk. These cells were often located at the center of a cord with extensions of their cytoplasm lying between other types of cells to reach the sinusoids. The antisera against glycoprotein hormone, TSH, FSH and LH reacted with large cells. Cells reacted with anti-TSH were angulated and were found more in the center than the periphery of the gland. Cells reacting with anti-LH and anti-FSH were oval, were more numerous at the periphery of the gland and did not react with anti-TSH. Anti-LH and anti-FSH frequently reacted in the same cell.

References:

(1) Nakane, P. K. and G. B. Pierce: J. Histochem. Cytochem. *14*, 929 (1966).
(2) Ibid: J. Cell Biol. *33*, 307 (1967).
(3) Coons, A. H.: *In:* General Cytochemical Method. J. F. Danielli ed. Academic Press, N. Y. (1958).
(4) Graham, G. C. and M. J. Karnovsky: J. Histochem. Cytochem. *19*, 291 (1966).

(5) Pearse, A. G. E.: *In:* Histochemistry, Theoretical and Applied. J. A. Churchill, Ltd., London (1960).

* Supported in part by American Cancer Society grant # E105 and United States Public Health Service Grants AI 07758 and CA 08201.

Etude Cytochimique des Cellules Juxtaglomérulaires Rénales

NENCI, ITALO, GIANMARIO MARIUZZI, and GUIDALBERTO FABRIS (Istituto di Anatomia e Istologia Patologica dell'Università di Ferrara, Italy)

L'étude cytochimique systématique du matériel de secrétion (identifié avec la renine) des cellulus épithélioïdes granuleuses des appareils juxtaglomérulaires chez le rat, nous a permis de démontrer que dans sa composition chimique ont part trois differentes fractions: 1) la prémière glucidique neutre dont la pas-positivité est prevenible par l'acétilation, mais résiste à l'action de la diastase, sialidase, hyaluronidases et des enzymes protéolytiques, et pour celà pas rattachable au glycogène, aux mucopolysaccharides ou mucoprotéines neutres; cette fraction pas-positive n'est plus présente dans les tissues après extraction prolongée méthanolchloroformique; 2) une fraction lipidique soudanophilique après efficace lipophanérose hydrolytique, positive aux réactions pour les phospholipides et pour la choline, histochimiquement semblable aux lécithines;3) une troisième fraction protidique positive aux réactions de la tyrosine, du tryptophane, de l'histidine et de la cystéine.

Ces résultats donnés, on peut conclure que les granulations des cellules épithélioïdes juxtaglomérulaires du rat sont composées par une glycolipo-proteine complexe, dans laquelle les radicaux glicidiques font partie d'un glycolipide qui comprende assez de choline, tandis que dans le protide il y a beaucoup de tryptophane, de même que autres produits de cecrétion doués d'une action biologique fort spécifique.

Computer Aided Interactive Image Processing

NEURATH, PETER W., ZAY B. CURTIS, WILLIAM SELLES, and HENRI G. VETTER (New England Medical Center Hospitals, Boston, Mass.)

Backing up visual observations in cytology with quantitative measurements very quickly leads to an overabundance of data. Even a very moderate resolution picture has a million points, each having e.g., 100 possible density values. A variety of approaches are required to extract the desired information. Our hardware and software, the design of which has been discussed previously (1,2) deals with this at various levels. One program transforms density level data to emphasize features such as

192

chromosomes from noise and background. Light pen intervention programs allow an operator to select areas, curves or single points for input to the computer. Data management programs provide automatic orderly storage and recall capabilities using the computer disk storage. Each problem program extracts a certain feature from an image and converts it into a number. The operating systems program allows one to combine a suitable small or large portion of all these programs to deal with specific aspects of an image.

Finally, statistical analysis methods are needed to arrive at a conclusion based on the numbers obtained, e.g., the conclusion that a particular object is a D14 chromosome.

This approach will be illustrated with concrete examples.

References:

(1) Neurath, P. W. *et al,*: Annals of the New York Academy of Sciences, in press (1968). (1967 Conference on Data Extraction and Processing of Optical Images in the Medical and Biological Sciences).
(2) Neurath, P. W.: Digest, 7th Int. Conf. Medical & Biological Eng. (Stockholm) p. 118 (1967).

Interfacing Photomicrographic Images with the Digital Computer

NORGREN, P. E. (The Perkin-Elmer Corporation, Norwalk, Conn.)

In order to permit economical, high speed digital analysis of micrographs it is necessary to optimize the interface between the television microscope and the digital computer. Methods for optimizing this interface in a scanning microscope system intended for use in automatic leukocyte pattern analysis are discussed. Other approaches applicable to chromosome analysis and to automatic grain counting in autoradiographs are also treated.

Cytochemical Staining Reactions ("Marker" Enzymes) in the Study of Rat Hepatomas

NOVIKOFF, ALEX B. (Albert Einstein College of Medicine, Bronx, N. Y.)

The following enzymatic activities have been used to "mark" cytoplasmic organelles in rat liver and differentiated ("minimal deviation") and undifferentiated hepatomas:
Microbodies—Diaminobenzidine oxidation (1)
      —Urate oxidase * (2)
Plasma membrane (bile canaliculi, etc.)—Nucleoside phosphatases (3)
Endoplasmic reticulum—Nucleoside diphosphatase (3,4)

Golgi apparatus—Thiamine pyrophosphatase (4)
Lysosomes—Acid phosphatase (5)
Mitochondria—NADH-TNBT tetrazolium reductase* (6)
Except for the two indicated by asterisks, the methods have been employed at the electron microscope level as well as for light microscopy. Except for urate oxidase, all may be used with frozen sections of aldehyde-fixed tissue.

The "minimal deviation" Reuber H-35, Morris 5123, 7795, 9618, and 9633 hepatomas, and the undifferentiated Novikoff hepatoma have been studied. All but the latter show the presence of microbodies, atypical bile canaliculi, well-developed endoplasmic reticulum, and large Golgi apparatus. These features may be seen by light microscopic examination of appropriately incubated sections (7, 8, 1, unpublished observations).

References:

(1) Novikoff, A. B. and S. Goldfischer: 19th Annual Meeting, Histochem. Soc. (New Orleans) Abstr. J. Histochem. Cytochem. (1968).
(2) Graham, R. C. and M. J. Karnovsky: J. Histochem. Cytochem. *13*, 448 (1965).
(3) Novikoff, A. B., E. Essner, S. Goldfischer, and M. Heus: Symp. Intl. Soc. Cell Biol. *1*, 149 (1962).
(4) Novikoff, A. B. and S. Goldfischer: Proc. Natl. Acad. Sci. *47*, 802 (1961).
(5) Novikoff, A. B.: *In*: Ciba Foundation Symposium on Lysosomes, pp. 36-73 (de Reuck and Cameron, eds.) Boston: Little, Brown (1963).
(6) Novikoff, A. B., W.-Y. Shin, and J. Drucker: J. Biophys. Biochem. Cytol. *9*, 47 (1961).
(7) Essner, E. and A. B. Novikoff: J. Cell Biol. *15*, 289 (1962).
(8) Novikoff, A. B. and L. Biempica: Gann Monograph *1*, 65 (1966).

Studies of Stained Cell Suspensions with the Rapid Cell Spectrophotometer

O'BRIEN, REGINA (IBM Watson Laboratory, New York, N. Y.)

Stained suspensions of cells have been analyzed with the Rapid Cell Spectrophotometer (RCS) which was developed at this laboratory. The aim of these studies was to develop rapid automatic methods for characterizing populations of cells according to those optical properties by which the individual cells varied from one another. Two systems have been under investigation: the first, a method for viability assay of cultured cells and the second, a method for discriminating granular from agranular leukocytes. Details of the procedures for preparing the cell suspensions, the set-up of the RCS peculiar to these experiments, and the results of analyses will be presented.

Comparisons of manually obtained cytophotometric data with RCS output for cell suspensions stained for DNA by the Feulgen reaction will be discussed with the aim of evaluating the RCS as a potential general research instrument for quantitative investigations of fundamental cellular constituents.

Microbiological Specimen Transport Techniques

OHRINGER, PHILIP and VINCENT SPITALERI (Department of Medical and Biological Physics, Airborne Instruments Laboratory division of Cutler-Hammer, Inc., Farmingdale, Long Island, N. Y.)

Automatic optical measurement and image analysis of microbiological specimens has become increasingly frequent over the past several years as a result of advances in scanner technology, staining procedures, and utilization of computers for data processing. In conjunction with this, methods for rapidly and continuously transporting the microbiological specimens through the optical system have been developed.

Several specimen transport techniques will be discussed, with special emphasis given to two basic systems; (1) disposition of material on moving filter tape, and (2) forcing suspended particles through a flow cell. Considerations relative to designing tape transports are mechanical fragility, clarification of the tape in transmission microscopy, suppression of fluorescence in fluorescent microscopy, maintenance of focus and tracking, and particle concentration factors. With a flow cell design considerations are the size and depth of field, flow velocity, particle concentration, suspension media, and optical aberrations.

Practical application and performance results for both methods will be described, including a Reynolds-type flow cell used to achieve very small flow-cell dimensions.

Histochimie de l'Endosperme chez Iris pseudoacorus aux Différentes Étapes du Développement

OLSZEWSKA, M. J., B. GABARA and L. KONOPSKA (Laboratoire de Cytochimie, Université de Lódz, Lódz, Pologne)

Dans l'endosperme chez *I. pseudoacorus* se distinguent deux pôles—l'un chalazal, et l'autre—micropylaire qui entoure l'embryon. Au début du développement de ce tissu nutritif il y a des différences entre ces deux pôles: la vague mitotique est initiée dans le pôle chalazal, dont certains noyaux présentent un haut degré de polyploïdie; le pôle chalazal est plus riche en ER/1,2/.

Nous avons comparé les deux pôles aux étapes successives du développement de l'endosperme. Au fur et à mesure que l'embryon et la graine mûrissent, les différences entre le pôle chalazal et micropylaire disparaissent. La quantité de la matière sèche augmente, surtout en conséquence de la formation des parois cellulaires progressivement enrichies par les dépôts des hémicelluloses de réserve.

Dans le pôle chalazal P total diminue de 7,6 à 5,0 mg/g/matière sèche/, tandis que dans le pôle micropylaire il monte de 3,7 à 4,1. Le Pdu DNA monte dans le pôle chalazal de 0.2 à 0.4; dans le pôle micropylaire— de 0.2 à 0.3. e P du RNA augmente dans le pôle chalazal de 0,6 à 1,4, dans le pôle micropylaire—de 0.7 à 1.2. L'électroforèse des RNA indique que dans le pôle chalazal le pois moléculaire des RNA décelables aug- mente au cours du développement de l'endosperme; le pôle micropylaire, à partir des stades moyens du développement, en contient toutes les trois espèces. L'uridine ^{3}H est incorporée dans le pôle chalazal par les noyaux et par le cytoplasme, tandis que dans le pôle micropylaire les noyaux se marquent d'avantage.

Le N protéinique dans le pôle chalazal diminue de 33.9 à 23.7; dans le pôle microplylaire il reste au même niveau—de 27,3 à 27.5. Parmi les protéines ce sont seulement les globulines dont la quantité augmente dans tous les deux pôles. Les protéines de réserve sont élaborées sous forme de globules.

Au début du développement de l'endosperme les activités de la phos- phatase acide, d l' estérase non spécifique et de la β-galactosidase, locali- sées au niveau des granulations, sont 2 fois plus élevées dans le pôle chalazal. Au cours des étapes ultérieures, les dosages des extinctions des produits des réactions enzymatiques montrent une diminution de l'ac tivité de la phosphatase acide dans le pôle chalazal de 70 à 4,1 /E/g matière sèche/, dans le pôle micropylaire de 62 à 5. L'activité de la β-galactosidase baisse dans le pôle chalazal de 50,5 à 16,5; dans le pôle micropylaire—de 50,5 à 17. La diminution relativement peu accentuée de l'activité de cette enzyme pourrait être rapportée à la synthèse des hémicelluloses qui arrive dans l'endosperme aux stades envisagés.

References:

(1) Olszewska, M. J. and B. Gabara: Acta Soc. Bot. Pol. *35*, 557 (1966).
(2) Mikulska, E., B. Gabara and M. J. Olsewska: Acta Soc. Bot. Pol. *36*, 699 (1967).

Possibilities and Limits in the Differentiation of Oxidases and Dehydrogenases Based on Tetrazolium Salt Reactions

ONICESCU, DOINA and G. SZEGLY (Catedra de Histologie, Facul- tatea de Medicina, Bucharest, Romania)

Investigations were carried out on some oxidases and dehydrogenases acting upon the same biologic substrate and for the histochemical local- ization of which tetrazolium salts were used.

The histochemically obtained data (topochemical differences) are compared with the modifications (followed by way of microchemistry) undergone by the substrate during histochemical incubation.

Comparison is made also between the results obtained under histochemical and biological conditions (as action of the enzyme upon the substrate used in histochemistry).

The Autoradiographic Study of Measles-Virus RNA and Protein Penetration into the Cell and Intracellular Site of Their Synthesis

PARFANOVICH, M. I., N. N. SOKOLOV, O. N. BEREZINA, and L. L. FADEEVA (The D.I. Ivanovsky Institute of Virology, Academy of Medical Sciences, Moscow, USSR)

In recent years the wide application of radioisotopes in virology has made it possible to demonstrate the principle course of intracellular virus synthesis. In this paper, autoradiography was employed to investigate the measles virus RNA and protein penetration into the cell and the intracellular site of their synthesis. It was found by using 5-iodo-, 5-brom-, 5-fluorodeoxyuridine, selectively affecting DNA synthesis, that measles virus contains RNA (1). It was also demonstrated that cellular DNA did not take part in measles virus synthesis since actinomycin D, selectively inhibiting DNA-dependent synthesis of RNA, did not suppress measles virus multiplication (2, 3).

The investigation of measles virus RNA and protein penetration into the cell was carried out by use of virus labelled by H^3-uridine and S^{35}-methionin, respectively. The labelled virus was purified of nonspecific radioactivity and preserved its biological activity. In this part of the investigation, autoradiographs were made by using paraffin sections of roller-suspension culture cells infected with labelled measles virus. It was demonstrated that measles virus RNA penetrates the cytoplasm and extranucleolar part of nuclei. The data allow one also to think that the proteins of the virus envelope remain to a great extent out of the cell and the protein of virus nucleoprotein penetrates into the cytoplasm and extranucleolar part of nuclei.

The site and onset of measles virus RNA synthesis in infected cells was investigated by autoradiography using H^3-uridine (1 mC/ml) as precursor of RNA synthesis and actinomycin D (0.5 mg/ml) selectively inhibiting the cellular RNA synthesis. The study was made in a) uninfected cells; b) uninfected cells grown in the presence of actinomycin D; c) cultures infected with measles virus and d) cultures simultaneously inoculated with measles virus and treated with actinomycin D. In this case, the autoradiographs were made using the cell monolayer without embedding in paraffin and without cutting cells. The intensity of RNA synthesis was evaluated according to the grain number distribution above the nucleoli, the extranucleoli part of nuclei and cytoplasm in 100 cells for each group of cells at different time of investigation. It was demonstrated that within the first 4 hours after infection, measles virus strongly inhibited the host cell RNA synthesis in nucleolus and only little in the extranucleolar part of nuclei. The synthesis of measles virus RNA started 5-6 hours after inoculation with simultaneous renewal of cellular RNA synthesis. The measles virus RNA synthesis proceeds in the extranucloelar part of nuclei.

References:

(1) Lam, K. S. K. and J. G. Atherton: Nature (Lond.) *197*, 821 (1963).
(2) Parfanovich, M. I. *et al.:* Proc. Soc. Exp. Biol. & Med. (N. Y.) *120*, 604 (1965).
(3) Matumoto, M. *et al.:* Japan J. Exp. Med. *35, 319* (1965).

The Innate Rhythmic Nature of Several Processes or Events Involved in the Total Mitotic Cycle of the Dividing Corneal Epithelial Cell in the Rat *

PAULY, JOHN E. and LAWRENCE E. SCHEVING (Departments of Anatomy, University of Arkansas, Little Rock, Ark. and Louisiana State University, New Orleans, La.)

Circadian rhythmicity characterizes most mitotically labile tissues of both plants and animals. The phenomenon can be demonstrated by evaluating the rate of cell division in a tissue at frequent intervals along a 24-hour time scale.

It is not always recognized that some tissues may show a flux in division rate of as much as 1200% over this time period. Corneal epithelium is one example of a tissue that has an exceptionally high amplitude rhythm. In rats housed under a light-dark cycle with 12 hours of light (0600-1800) followed by 12 hours of darkness, the crest for the rate of cell division in corneal epithelium occurs between 0600 and 1100, the trough between 2000 and 2300.

Phasing of the rhythm in relation to local clock time can be predicted, once determined, for other animals of the same species when exposed to similar intervals of sampling and light-dark cycles identical to those initially used to establish the rhythm. Any phase, such as the crest or trough of activity, can be shifted by shifting the light-dark cycle, a feature which would enable an investigator to explore any phase of the rhythm without repeatedly "staying up" 24-hours to obtain his samples. It also has been demonstrated that the incorporation of injected ^{3}H-thymidine or ^{3}H-lysine into this same tissue will fluctuate significantly when evaluated along a 24-hour time scale. Such fluctuation is, of course, reflected in the autoradiographic pattern obtained. The crest of incorporation of ^{3}H-lysine occurs between 1000 and 1200. This crest is significantly higher when compared to any other hour within the 24-hour time period. The highest level recorded represents a 125% increase over the lowest recorded level.

Adrenalectomy, hypophysectomy and thyroidectomy do not abolish the rhythmic pattern of cell division in corneal epithelium of the rat. The effect of continuous illumination, darkness, or blinding on the rhythm will be discussed as well as the importance of considering the time structure of the animal when designing any biological experiment.

---

* Supported by U.S. Public Health Grant #4659.

Cytochemical Contributions to the Functional Cytology of Polypeptide Production (The APUD Cells)

PEARSE, A. G. E. (Royal Postgraduate Medical School, London)

Over a number of years, the application of a broad spectrum of cytochemical techniques to endocrine and other glandular tissues from various mammalian, avian, amphibian and reptilian species has suggested that a common and distinct metabolic pattern is shared by cells with apparently widely different functions. Two factors appear as possible links between the different cell types. First, in a high proportion of the cells, from different regions, the principal product appears to be a low molecular weight polypeptide. Secondly, in the majority of cases the cells are clearly derivatives of the primitive digestive tube (foregut).

The following is a list of the common cytochemical characteristics observed:

1) High α-glycerophosphate dehydrogenase
2) High esterase or cholinesterase
3) Fluorogenic amine content (usually 5-HT)
4) Amine precursor uptake
5) Amino-acid decarboxylase
6) High C-terminal or side chain carboxyl groups

For the sake of brevity, these six characteristics have been referred to comprehensively by the letters APUD, which distinguish two of the most prominent. All six characteristics are not necessarily present in all APUD cells, in all species, or at one and the same time.

Endocrine cells of the APUD series include the following types, given here with their principal or presumptive product: pituitary corticotroph (ACTH), pituitary melanotroph (MSH), pancreatic islet β-cell (insulin), pancreatic islet α_2-cell (glucagon), pancreatic islet α_1-cell (?pancreozymin), thyroid and ultimobranchial C cell (calcitonin), intestinal enterochromaffin cell (?secretin, kinins), intestinal or gastric argyrophil cell (?gastrin). Other endocrine cells, which are not of foregut origin, share many of the features of APUD cells. Chief among these are the nor-adrenaline-secreting cells of the adrenal medulla.

Ultrastructural investigations of polypeptide-secreting APUD cells have revealed a number of essential similarities: 1) low levels of rough ER, 2) high levels of smooth ER, 3) high content of free ribosomes, 4) electron dense, fixation-labile mitochondria, 5) membrane-bound secretion vesicles, best preserved by glutaraldehyde, of average size 100-200 mμ (during storage phase). Moreover, microtubules are a characteristic feature in many species. Electron cytochemical investigations have confirmed and extended some of the findings determined by optical cytochemistry.

The precise meaning of these cytochemical and ultrastructural character-istics remains to be elucidated but common metabolic processes are clearly implicated. Possibly also there may be involvement of a common ancestral cell type. Techniques derived from a number of disciplines need to be applied, and some of the work in progress on these lines will be described.

The Nature and Localization of Fluorogenic Amines in the Pituitary Gland

PEARSE, A. G. E. (Royal Postgraduate Medical School, London)

Treatment of freeze-dried pituitary glands from man, pig and rat with hot formaldehyde vapour (Falck, 1962) produces a fast fading yellow fluorescence in the cells of the pars intermedia (when present) and in a proportion of the R-type mucoid cells (Pearse and MacGregor, 1964; Pearse 1966 a and b). This fluorescence was originally attributed to 5-hydroxytryptamine although spectrophotometric confirmation was not available. Similar observations were made by Dahlström and Fuxe (1966) who found also that the mucoid cells of the mouse pars distalis would take up and decarboxylate 5-hydroxytryptophan. This was reported for dog pituitary corticotrophs and melanotrophs by Pearse (1966a) and these properties, together with a number of other cytochem-ical and E. M. characteristics, bring the R-type mucoid corticotrophs and the R-type mucoid melanotrophs into the category of polypeptide pro-ducing APUD cells.

Investigations by Falck and Owman (in press) have suggested that in the pig pars intermedia the fluorogenic amine is dopamine rather than 5-HT, whereas in the pars distalis the contrary is the case.

Spectrofluorimetric investigations have now been carried out on the pituitary glands of rat, dog and pig (in the first two species both before and after administration of 5-HTP or DOPA) and the results will be reported.

References:

Dahlström, A. and K. Fuxe: Acta Endocrinol. *51*, 301 (1966).
Falck, B. and C. Owman: Advanc. Pharmacol. (*in press*).
Falck, B.: Acta Physiol. Scand. *56*, Suppl. 197 (1962).
Pearse, A. G. E.: Nature *211*, 598 (1966a).
Pearse, A. G. E.: Proc. International Endocrinological Symposia—Rome 1964. Academic Press (1966b).
Pearse, A. G. E. and M. M. MacGregor: Annual Report (Part II) Brit. Empire Cancer Campaign 665p. (1964).

Histochemical Studies of Aminopeptidase by Means of L-N(5-Bromo-indol-3-yl) Leucinamide

PEARSON, BJARNE and WILLIAM BENNETT (Department of The Army, Fort Detrick, Frederick, Md.)

Aminopeptidase in tissue and component cells thereof can be demonstrated by a new synthetic indolyl substrate with leucine attached to a halogen-substituted indolyl moiety. The tissue enzyme hydrolyzes the peptide bond adjacent to the free amino group to form a chromogenic bisindigo at enzymic sites within the cells and tissues. High enzyme activity was seen in the kidney, parathyroid gland and connective tissue cells of the lamina propria. The enzyme activity in the kidney is most intense in the lower cortex with the activity of the upper cortex about one half. In the upper cortex, the enzyme is present mainly in the basilar region of the cell, whereas, in the lower cortex, it is more intense in the luminal portion. The medulla of the kidney shows only a slight reaction. Conspicuous and high enzyme activity was seen in certain connective tissue cells of the lamina propria. These cells tend to exhibit certain fibrillary processes. The stroma of the endometrium of the uterus reveals cells with high activity. These cells have an elongated cytoplasm with a dense nucleus and are morphologically similar to other stromal cells that occur in the proliferative stage of the endometrium. The parathyroid gland shows a strong enzyme reaction. This is uniformally distributed throughout the cells of the gland. The thyroid gland shows a medium enzyme activity. The reaction is seen as granules in the follicular epithelial cells. The liver shows a medium activity. The thymus shows activity mainly in the reticuloendothelial cells. There is a great deal of variation in enzyme activity which seems to depend upon physiological states. Thus, in the ovary, there is marked activity in the atretic follicle. Pulmonary alveolar macrophages show medium to strong activity.

β-Glucuronidase and Related Glycosidases Demonstrated by Indolyl Substrates in Lymphatic Tissue

PEARSON, BJARNE, JOHN R. ESTERLY, and ALFRED C. STANDEN (Department of The Army, Fort Detrick, Frederick, Md.)

Histochemical and cytochemical techniques for detecting enzyme activity have been difficult to apply to the lymphatic system. We have been able to demonstrate tissue glycosidases with techniques that show sharp intracellular localization and that are highly specific, although there still may be interpretive difficulties because of the heterogenicity and functional changes in component cells. We have developed halogen-substituted indolyl substrates for β-galactosidase, β-glucosidase, β-2-deoxyglucosidase and β-fucosidase. The most recently developed substrate for β-glucuronidase is 5-bromo-4-chloroindol-3-yl-β-D-glucopyruroniside. The final reaction product in all cases is identical: 5,5'-bromo-4,4'-chloro-indigo. This intensely blue-green, finely granular precipitate is substantive and insoluble so that tissues can be dehydrated and permanently mounted. Specific inhibition for each substrate has been demonstrated with analogue lactones.

Differences in the distribution and intensity of the resultant staining in the various cells and tissue of the lymphoid system for these enzymes offer a unique opportunity to study the development of functional changes in reticular cells, including enzymatic induction and antigenic response.

Turnover of DNA in the Adrenal Medullary Cells

PELC, S. R. and M. P. VIOLA-MAGNI (Dept. of Biophysics, King's College, London and Ist. Patologia Generale, Università Pisa, Italy)

A decrease of DNA content per nucleus in the adrenal medullary cells of rats (Italico strain) intermittently exposed to cold has been reported (1). An incorporation of H^3-thymidine takes place during the recovery period (2).

The present experiment was performed to study subsequent fate of label incorporated during the recovery period, both in rats kept at room temperature and exposed intermittently to cold.

33 animals were exposed intermittently to cold for a total of 300 hours (15 hours at $+ 4°C$ and the remaining 9 hours at room temperature, $+ 20°C$). All animals were injected intraperitoneally with H^3-thymidine (Radiochemical Centre, Amersham), 1 $\mu c/g$ of body weight, after 6 hours of recovery. 3 rats were killed after 3 hours; the others were divided into two groups: one group was exposed intermittently to $+ 4°C$ and the second was kept at room temperature. The animals of the two

groups were killed at different intervals after injection of H^3-thymidine (1, 3, 7, 14, 20 days). After fixation in Carnoy's fluid, the adrenal glands were embedded and sectioned. The sections (5 μ) were processed for the radioautographic technique with Kodak AR 10 stripping film. The time of exposure of the film was 14, 28, 150 days.

In the animals killed 3 hours after the injection of H^3-thymidine, the percentage of nuclei labeled in the adrenal medulla was 5.0%, 7.8%, 9.8% respectively after 14, 28, 150 days of film exposure. The majority of the nuclei were weakly labeled.

In the animals killed at different intervals after injection of H^3-thymidine, the percentage of labeled nuclei decreases with time. This decrease amounts to 25% in the animals left at room temperature, whereas in the animals exposed to cold, the decrease of labeled nuclei falls to 75% of the values observed 3 hours after injection.

This finding cannot be accounted for by cell loss and renewal, since the mitotic activity of the adrenal medulla is very low (3). The experiments suggest that a turnover of DNA takes place in the adrenal medulla nuclei in a normal environmental situation; this turnover is accelerated three times upon exposure of the animals to low temperature.

References:

(1) Viola-Magni, M. P.: J. Cell. Biol. *25*, 415 (1965).
(2) Viola-Magni, M. P.: J. Cell Biol. *28*, 9 (1966).
(3) Malvaldi, G., P. Mencacci, and M. P. Viola-Magni: Experientia, *in press* (1968).

Post Mortem Changes in the Fluorescence and Staining Properties of Enterochromaffin Cells

PENTTILÄ, ANTTI (Department of Anatomy, University of Helsinki, Helsinki, Finland)

Barter and Pearse (1) were the first to observe formaldehyde-induced fluorescence (FIF) in the enterochromaffin cells (EC), presumably due to 5-hydroxytryptamine (5-HT). Also, other staining characteristics of EC are suggested to be caused by the cytoplasmic 5-HT (2), but the correlation of argyrophilia and argentaffinity in EC to 5-HT is obscure (3,4). The present investigation was carried out to compare the gradual disappearance of five staining reactions and the 5-HT content using FIF in the EC of the pig duodenum.

Fluorescence, argyrophil, argentaffin, ferric ferricyanide, diazo coupling and indophenol reactions were studied in freeze-dried and paraffin sections from small tissue pieces which were allowed to stand exposed to air in a tightly closed glass vessel with a thin layer of water on the bottom or immersed in Krebs-Ringer solution for 1 to 24 hours at

4, 23, or 37°C. Before 'incubations' the pieces were thoroughly washed with sucrose solution (0.3 M, pH 6.9, at 0°C) containing 24,000 I.U. of Neomycine per ml.

After standing in air or in Krebs-Ringer solution at 37°C for 1 to 4 hours, most EC emitted a sharply delineated yellow fluorescence. With longer incubation times, the fluorescence became weaker and signs of diffusion increased. After 16 or 24 hours incubation, no fluorescence was seen in EC. When kept at room temperature for 24 hours, many EC showed a diffuse FIF, while treatment at 4°C did not yet cause diffusion.

The number of argentaffin EC decreased more quickly than that of cells exhibiting FIF. The diazo coupling, the ferric ferricyanide, and the indophenol reaction disappeared even more quickly, while the argyrophilia was more resistant. In tissues kept at 37°C for 16 to 24 hours, EC were seen showing argyrophilia but no 5-HT fluorescence. Argyrophilia was, indeed, manifest even in almost totally autolyzed pieces of duodenum.

It can be concluded that the argyrophilia of the EC does not exclusively depend on 5-HT but is due to staining properties of the granular matrix.

References:

(1) Barter, R. and A. G. E. Pearse: Nature (Lond.) *172*, 810 (1953).
(2) Pearse, A. G. E.: *In:* Histochemistry, Theoretical and Applied, pp. 641-652 (Pearse, edit.) London: Churchill (1960).
(3) Hellweg, G.: Z. Zellforsch. *36*, 546 (1952).
(4) Singh, I.: Z. Zellforsch. *62*, 121 (1964).

Changes in Amounts of the S-100 Protein During Wallerian Degeneration In Rabbit Tibial Nerve

PEREZ, VERNON J. and BLAKE W. MOORE (Washington University School of Medicine, Department of Psychiatry, St. Louis, Mo.)

The S-100 protein has been isolated and purified from beef brains in this laboratory (1) and antiserum to the protein has been prepared in rabbits. With the immunological assay of complement fixation, S-100 has been shown to be present only in the nervous systems of a wide variety of reptiles, fish, birds and mammals (2). The purpose of this investigation was to localize S-100 in the peripheral nervous system by correlating changes in amounts of the S-100 protein with cellular changes which characterize Wallerian degeneration in peripheral nerve, namely, the loss of axons and myelin and the proliferation of Schwann cells and macrophages.

Young adult male albino rabbits were anesthetized with pentobarbital sodium and the left sciatic nerve was cut in experimental animals. Rabbits were killed with air 0 (unoperated controls), 3, 7, 14, 21 and 28 days postoperatively. Segments of tibial nerve proximal and distal to the section were taken for assay as were the intact nerves from the right legs of experimental animals and from both legs of control rabbits. Nerve segments were extracted in isotonic NaCl-veronal buffer, centrifuged at 36,900 $\times$ G and S-100 was measured in the clear supernatant solution by complement fixation (3).

S-100 was reduced by 30% in the degenerating nerve segment distal to the section at only 3 days and by 98% at 28 days. No consistent change in amounts of S-100 was measured in non-degenerated segments proximal to the section in experimental animals or in uncut nerves in experimental and control rabbits. Whereas S-100 was progressively reduced during 28 days degeneration, total soluble proteins increased by 40%. No change in total soluble proteins was measured in non-degenerated nerves.

S-100 was not localized in Schwann cells or macrophages because the protein was reduced during the time of maximum proliferation of these cells. The reduction in S-100 paralleled the loss of axons and myelin. To ascertain if S-100 was localized mainly in myelin or axons, myelinated (abdominal) and unmyelinated (cervical) segments of vagus nerve were dissected from a rabbit and S-100 was measured as in tibial nerve. Unmyelinated vagus contained 50% more S-100 than did the myelinated segment. It is suggested that S-100 is localized predominantly in the axis cylinder rather than in myelin because: (a) greater amounts of the protein were measured in unmyelinated than myelinated parts of vagus nerve, and (b) the 30% reduction in S-100 at 3 days degeneration corresponded more closely to the rapid loss of axons than to the more delayed degradation of myelin.

References:

(1) Moore, B. W.: Biochem. Biophys. Res. Comm. *19*, 739 (1965).
(2) Levine, L. and B. W. Moore: Neurosci. Res. Prog. Bull. *3*, 18 (1965).
(3) Moore, B. W. and V. J. Perez: J. Immunol. *96*, 1000 (1966).

Immunofluorescent Studies on the Serum Albumin of Rat Liver Cells [*]

PETERS, THEODORE, JR., J. THOMAS DANZI, and CHARLES A. ASHLEY (The Mary Imogene Bassett Hospital (affiliated with Columbia University), Cooperstown, N. Y.)

Only about 10% of the hepatocytes of adult male rats show the presence of serum albumin, appearing as a generalized cytoplasmic fluorescence when either liver sections or isolated liver cells are tested with rabbit anti-rat serum albumin followed by fluorescent goat anti-rabbit

7S globulin. A similar finding was reported by Hamashima *et al.* (1) on sections of human liver. In animals with increased rates of albumin synthesis (table, line 1) the percentage of isolated liver cells showing albumin by this technique was found to rise as the albumin production rose (table, line 2).

| | | Type of Rat | |
| | | 5-week | 5-week |
	Adult	normal	nephrotic
(1) Rate of albumin synthesis by slices, mg/g liver/hour[1]	0.1	0.3	0.6
(2) Percentage of hepatocytes showing albumin by immunofluorescence	12	25	51
(3) Amount of albumin extractable from microsomes, mg/g liver[2]	0.3	0.45	0.4
(4) Transit time of albumin, minutes[3]	18	18	—

[1] Albumin synthesis was measured as the incorporation of ^{14}C-L-leucine into albumin by liver slices incubated 3 hours in dialyzed rat serum.
[2] Isolated cell fractions were extracted with deoxycholate and the albumin determined by immunoprecipitation (2).
[3] Transit time was measured as the lag in appearance of labeled albumin in the circulation after intravenous injection of ^{14}C-L-leucine.

The amount of albumin bound to microsomal membranes was assayed in the same animals in an attempt to identify the nature of the material reacting with the fluorescent antibodies. The values found for micro-somal albumin (table, line 3) changed very little, however, as the rate of albumin synthesis increased. Mitochondrial albumin was about 0.2 mg/g for each of the three groups of rats. Faster movement of newly-formed albumin through the cytoplasmic membrane system of the cell could not account for these results, as the transit time (table, line 4) was unaffected by the rate of albumin synthesis.

Thus, the percentage of liver cells showing albumin by immunofluores-cence correlates with the rate of albumin synthesis, but does not corre-late with the albumin content of the cell organelles. A possible inter-pretation, which requires further investigation, is that albumin is present in an inactive state in some liver cells, and is only available for detec-tion by antibodies *in situ* in those cells in which it is actively being synthesized.

References:

(1) Hamashima, Y., J. G. Harter, and A. H. Coons: J. Cell. Biol. *20,* 271 (1964). (1964).
(2) Peters, T., Jr.: J. Biol. Chem. *237,* 1181, 1186 (1962).

* Supported by U.S. Public Health Service Research Grants HE-02751 and and CA-07261.)

206

Cytochemical Characteristics of Bone Tumors

PETROVA, A. S. and N. A. PROBATOVA (Institute of Experimental
& Clinical Oncology, Academy of Medical Sciences of USSR, Moscow)

To extend cytological diagnostic possibilities the authors studied alkaline and acid phosphatases in bone tumors by a histochemical method.
Bone can serve as a starting point for tumors of various histogenesis.
Modern oncology demands precise diagnosis of this type of tumor for
the choice of method of treatment. Due to difficulties of approach in
many tumors of the locomotor apparatus, puncture biopsy is especially
widely used of late for diagnostic purposes. Acid and alkaline phosphatases were studied in prints from 70 different bone tumors: osteogenic
sarcoma, chondrosarcoma, osteoblastoclastoma, Ewing's sarcoma, malignant synovioma, etc.

The authors were able to detect the "enzymatic profile" of bone tumors
which enables them to recommend the method of cytochemical determination of acid and alkaline phosphatases as a test in differential
diagnosis. Alkaline phosphatase is extremely active in cells of osteogenic
sarcoma. Its activity as well as that of acid phosphatase are relatively
high in the cellular elements of Ewing's tumor. Activity of both hydrolases is low in the mononuclear elements of osteoblastoclastoma.
Chondrosarcoma has varying activity in different cellular elements, especially in regard to alkaline phosphatase. Cells of malignant bone
synovioma are rich in acid phosphatase, and as a rule contain no alkaline phosphatase.

Results of investigation are discussed also from the point of view of the
participation of the phosphatases in osteogenesis.

Twenty-four Hour Periodicity of Succinodehydrogenase in Rat Liver

PHILIPPENS, KAREL (Anatomisches Institut, Medizinische Hochschule, Hannover)

Histochemical standard assay of succinodehydrogenase with the NBT-
method on kidneys and livers of mice and rats has given us considerable
variations when the tissues have been harvested at different hours of
the day. These results have lead us to systematic histochemical examinations to investigate how far the topographic distribution of succinodehydrogenase is influenced by the circadian rhythm of animals. Furthermore, investigation by biochemical methods should be done if the
activity of succinodehydrogenase changes in the tissues during the day.
For this investigation, highly standardized animals were used. For the

histochemical assays, the NBT-method was used; for the biochemical investigations, several indicator-systems for SDH were applied on livers. The studies showed definite changes of SDH activity depending on the time in which the animals were sacrificed during a 24-hour period. Histochemical results have brought some views about the fluctuations in distribution of SDH in respect to the liver morphology and the metabolic state of the liver lobules, while biochemical results have shown that NBT is an excellent indicator for quantitative estimations. The studies also show that NBT may demonstrate SDH activity outside of intact mitochondria.

Problems of Regulation of Enzyme Activities During Organogenesis

PILGRIM, C. (Inst. of Anatomy, University of Wuerzburg, Germany)

Histochemical methods seem to be promising tools for the investigation of regulatory mechanisms which control the process of chemodifferentiation. It will be shown that estrogen treatment of infant female rats causes a precocious appearance of the mature distribution pattern of several phosphatases in the kidney. An increase of enzyme activity can be observed which affects mainly the outer layer of the outer medullary zone and medullary rays (terminal straight portions of the proximal tubule). As to alkaline phosphatase activity, the degree of activation is shown to depend upon the age of the animals, the increase being most pronounced at the 18th day of life. Another possibility for altering experimentally the normal course of development is the application of drugs interfering with the cellular mechanism of protein synthesis. Therefore, the influence of actinomycin and cycloheximide on the development of alkaline and acid phosphatase reactions in the kidney was investigated in male and female rats aged 5 days to 2 months. In 14 to 20 days old animals, actinomycin produces an increase of alkaline phosphatase activity which can again be localized mostly in the terminal portions of the proximal tubule. In older animals (25-35 days), a similar effect on acid phosphatase activity can be observed whereas alkaline phosphatase is now inhibited. Cycloheximide, however, shows a strong inhibitory action on both alkaline and acid phosphatase in all age groups investigated. It is discussed that actinomycin may act by blocking repressor formation and thus increasing *de novo* synthesis of enzyme protein while cycloheximide inhibits overall protein synthesis. Possibly, the control of enzyme protein synthesis by repressor formation and inactivation is a regulatory mechanism which applies also to the normal development of enzyme patterns in mammalian organs.

Utilization of Microchemical Methods with Excised Embryonic Axes as a System for Use in Botanical Histochemistry

POLLOCK, B. M. (U. S. Department of Agriculture, National Seed Storage Laboratory, Fort Collins, Colorado)

Growth in plant systems is normally localized in small groups of cells where it can be studied only by the application of cytochemical techniques to thin sections. Unfortunately, such techniques are of limited versatility when compared with the methods available to the biochemist. However, embryonic axes excised from seeds can, as a first approximation, be considered to consist of a homogenous population of cells during imbibition of water and the initiation of growth. Physiological evidence indicates that this is a period of extreme importance in the later development of the plant. The quantities of material available are adequate to permit the use of micromodifications of such conventional methods as column chromatography to obtain quantitative separation of cell components. With the versatility allowed by such methods, it is possible to study the details of metabolic control in the system. Some specific micromodifications will be discussed.

Histochemical Methods in Investigations of Transplantable Carcinomas and Their Changes Under Chemotherapy with Alkylating Agents

PRESNOV, M. A. (Institute of Experimental and Clinical Oncology, AMS, Moscow, USSR)

Methods of histochemistry have great value and wide application in experimental oncology to study of different questions. We have been occupied with systematic investigations of transplantable tumors and their changes under chemotherapy. 9 strains of transplantable carcinomas of mice, rats and hamsters were studied in this work. Adenocarcinoma of mouse forestomach (OG-5 strain) was investigated in detail. 16 methods of histochemistry were used to demonstrate nucleic acids, protein groups, lipides and enzyme activity.

It was found that tumor cells in zones of tumor infiltration and those in contact with connective tissue had high concentration of nucleic acids, proteins and acid mucopolysaccharides. The highest activity of oxidative enzymes were observed in the basal parts of cytoplasm that contact connective tissue.

After chemotherapy with alkylating agents, distrophic changes of tumor cells were observed. The cell metabolism was shifted to catabolic processes. Giant and polyploid cells were present in treated tumors. The earliest changes were found in DNA by Feulgen's reaction and in mitochondria by reactions revealing oxidative enzymes. Activation of some oxidative enzymes, esterases and leucine aminopeptidase was observed in damaged cells. Simultaneously, with a decrease in size of treated tumors, the changes of histochemical reactions were observed in the tumor stroma (cells, fibers and blood vessels) and also in the surrounding connective tissue.

Pattern Recognition Techniques for Automatic Leukocyte Identification

PRESTON, KENDALL, JR. (The Perkin-Elmer Corporation, Norwalk, Conn.)

Measurements of leukocyte image properties taken by the CELLSCAN System have been analyzed by a method which combines multivariate statistical analysis with linear programming. Several aspects of this methodology are discussed and illustrated by describing its application to some 10,000 measurements made on 250 leukocyte images.

Comparative Histochemical Study of Stratified Squamous Epithelia in Different Conditions of Keratinization

PRETO PARVIS, V. and F. CISOTTI (Istituto di Istologia ed Embriologia generale, Università, Milano, Italy)

The content of lipids, SH and SS groups, and alkaline and acid phosphatase activities were compared in the epidermis of different areas of the body and in different conditions of keratinization, as well as in cornified and non-cornified zones of the oral epithelium.
For the study of lipids, Sudan black B, Nile blue sulphate, the Baker's test, the PAS and PFAS reactions with relative controls, fluorochromes such as 3,4-benzopyrene, primuline and rhodamine were used on formol-calcium fixed specimens. For SH and SS groups the method of Chèvremont and Frederic, the PFAS reaction, the DDD method after Barrnett and Seligman were performed on conveniently fixed specimens. The alkaline and acid phosphatase activities were studied by techniques of azo-conjugation and naphthol AS-MX phosphate, according to Burstone. The studies were carried out in the rat on epidermis of the common skin of the back, on the skin of the sole of the paws, and on that of the tail with relative hair; in man, on the thick epidermis of the palm of the hand in normal conditions and in hyperkeratosis; in the cat, on the non-cornified and highly cornified ("corneae papilleae") of different parts of the tongue.
In all the areas and conditions studied, a direct relationship was found between the content in complex unsaturated Baker's positive lipides, particularly phospholipids, which were detectable in the precornified

layers where the cell structure is disappearing, and a retarded desquamation is occurring.

On the contrary, sudanophilic lipids could be found in the most superficial layers due to diffusion of the sebaceous secretion or to more advanced processes of degradation of cell structures.

The acid phosphatase activity was generally high in the layers where the lipophanerosis of phospholipids was evident, as well as in the immediately deeper layers, and it was higher the more thick was the zone where such lipids were present.

The alkaline phosphatase activity was more variable and did not seem to have a constant relationship with the layers above those rich in phospholipids, that is, with the layers where the histochemical reactions for such substances were disappearing.

The distribution and intensity of the reactions attributable to SH and SS groups at the different sites are also demonstrated and discussed, and a comparison is made between the data relative to these groups and to the lipids and the enzyme activities.

Histochemical Study of the Epithelia of Human Epididymis

PRETO PARVIS, V. and G. E. MAZZA (Istituto Istologia ed Embriologia generale, Università, Milano, Italy)

The study was carried out on sufficiently fresh epididymides obtained from 4 individuals of 20, 25, 40 and 77 years, deceased from severe road accidents. From each individual, one epididymis, fixed in Helly's fluid and embedded in paraffin, supplied groups of serial sections at regular intervals, which were stained by morphological and histochemical methods for polysaccharides (PAS and relative controls) and aminopolysaccharides (Hale; Alcian blue and toluidine blue at different pH). Frozen sections obtained from the contralateral epididymis, fixed in formol-calcium, were used for the study of lipids (Sudan black B, Nile blue sulphate, Baker's test, lipophilic fluorochromes).

A scant mucous secretion was demonstrated along the whole epididymal tract of the spermatic ways. Moreover, differential data on 3 cellular types were obtained: a) Cells of the epithelium of the ductuli efferentes which constituted most of the head of the epididymis; b) "principal" cells of the epididymal canal, tall, prismatic, stereociliated; c) cells "intercalated" among the last ones, probably corresponding to the so called "olocrine" cells, known for their orgyrophilia.

The *mucous secretion* was characterized by acid aminopolysaccharides containing sulfate and carboxylic groups. The *cells of the epithelium lining the ductuli efferentes* showed, in variable proportion, granules with a complex lipid content, which were partially positive to Sudan black B, Nile blue, Baker's test, PAS (amylase-resistant), and were also partially pigmented. The *principal cells of the epididymal canal* showed signs of protidic-lipidic secretion in the basal zone and in the Golgi area. The product consisted of fine sudanophilic and Baker's positive granules present in the distal cytoplasm, whose Baker's positive component seemed to dissolve along the stereocilia.

These data are in agreement with those obtained by the electron microscopic (Horstmann, 1962) and support the hypothesis proposed for the corresponding cells in the rat (Preto Parvis, 1967) that these cells produce enzymes which form secretion granules with a lipidic component which later dissolves in the zone granule emission.

The *intercalated cells*, as the argyrophilic olocrine cells, are PAS positive and show different pictures to Sudan black B and to the Baker's test. They appear under aspects which can be interpreted as different phases of cycle of lipid stuffing and elimination into the lumen.

A Morphological Interpretation of Machine-Oriented Parameters for Leukocyte Determination *

PREWITT, JUDITH M. S., BRIAN H. MAYALL, and MORTIMER L. MENDELSOHN (Dept. of Radiology, University of Pennsylvania, Philadelphia, Pa.)

Machine perception of stained leukocytes is being studied by a system which extracts 50 numerical parameters from digital cell images produced by a scanning cytophotometer. These parameters may be organized into five families on the basis of logical and statistical independence, and each family may be related to criteria of visual discrimination used by man. In conventional morphological terms, the families can be designated: 1) contrast and texture, 2) concentration, 3) size and content, 4) shape, and 5) nuclear-cytoplasmic relationships. In each family except the last, the parameters may be organized further into three interrelated subsets which describe the entire cell, the cytoplasm, and the nucleus. Many subsets containing relatively few parameters effectively discriminate the five major types of leukocytes (neutrophils, eosinophils, basophils, lymphocytes, and monocytes).

The role of the parameter families in constructing such discriminatory sets will be discussed, as well as the relationships of the families to conventional morphological criteria.

References:

(1) Prewitt, J. M. S. and M. L. Mendelsohn: Annals of the New York Academy of Sciences *128*, 1035 (1966).

(2) Mendelsohn, M. L., B. H. Mayall, J. M. S. Prewitt, R. C. Bostrom, and W. G. Holcomb. *In*: Advances in Optical and Electron Microscopy, pp. 77-150 (Cosslett and Barer, *edits*.) New York: Academic Press (1968). *in press*.

* Supported by Grant 4 KO6 CA18540 and Contract PH 43-62-432 from the National Cancer Institute, National Institutes of Health, Bethesda, Maryland.

A Histochemical and Biochemical Assessment of Endometrial Monoamine Oxidase in the Normal Cycle and Following Oestrogen/Progestagen Administration

PRYSE-DAVIS, JOHN and MERTON SANDLER (Bernhard Baron Memorial Research Laboratories of Queen Charlotte's Maternity Hospital, London, W.6)

Endometrial monoamine oxidase (MAO) has been shown histochemically to change from a scanty particulate stain to an intense stain with the onset of the later secretory phase of the menstrual cycle: it has been suggested that the diffuse reaction may represent an inactive form of the enzyme allowing monoamines to accumulate leading to vascular spasm and menstruation (1). In the rat variations of MAO activity in the brain and liver have been reported to occur during the oestrus cycle (2). Oestrogen/progestagen preparations used in fertility control are sometimes associated with depressive mood changes and this may be related to increased MAO (3). This paper reports parallel histochemical and biochemical studies of 3 normal endometrial biopsies, 21 rat uteri (4) and 120 biopsies from women taking oral contraceptives (3) to investigate further these three aspects of MAO activity.

MAO was assessed histochemically by the tetrazolium method and biochemically by a radioactive technique (5): both methods employed tryptamine as substrate. Alkaline and acid phosphatase were also assessed on all biopsies as marker enzymes known to vary during the menstrual cycle (6).

Three degrees of endometrial staining were recognised in human biopsy material: weak (scanty granules), intermediate (granules + weak diffuse colour), strong (intense diffuse colour — granules variable). Non-secretory and early secretory biopsies showed mostly a weak with an occasional intermediate stain while all later secretory endometrium developed a strong reaction. Biochemical assays on the normal endometrium were low during the non-secretory and early secretory phases but increased up to tenfold with the onset of the later secretory phase in agreement with the histochemical grading. Weak MAO was accompanied by strong alkaline and weak acid phosphatase activity, while strong

MAO was associated with weak alkaline phosphatase and strong acid phosphatase staining. This *in vitro* evidence of increased MAO in the latter part of the cycle does not support the hypothesis of low MAO activity allowing monoamine concentrations to increase and initiate menstruation (1) but the granular and diffuse stain may have a compartmental basis affecting *in vivo* availability. Acetone extraction of sections, however, suggests that the granular and diffuse type of staining is related to lipid formation.

The rat uteri showed no biochemical correlation of MAO activity with phases of oestrus and MAO staining was uniformly weak unlike the findings of cyclic variations in rat brain and other tissues (2).

Following the administration of oestrogen/progestagen compounds in fertility control there was an alteration of endometrial MAO and phosphatase activity according to the relative dosages of each component. The strongly progestagenic combined preparations with the highest incidence of mood changes, caused an early prolonged rise in MAO while the less progestagenic sequential regimes resulted in weak MAO throughout the cycle.

References:

(1) Cohen, S., L. Bitensky, J. Chayen, and G. J. Cunningham: Lancett *ii,* 56 (1964).
(2) Kobayashi, T., J. Kato, and H. Minaguchi: Endocr. Japon. *11,* 283 (1964)
(3) Grant, Ellen and J. Pryse-Davies: *Submitted for publication* (1968).
(4) Southgate, Jennifer, Ellen Grant, W. Pollard, J. Pryse-Davies, and M. Sandler: Biochem. Pharmac. *accepted for publication* (1968).
(5) Wurtman, R. J. and J. Axelrod: Biochem. Pharmac. *12,* 1439 (1963).
(6) McKay, D. G., A. T. Hertig, W. A. Bardawil, and J. T. Velardo: Obst. & Gynec. *8,* 22 (1956).

Specificity and Sensitivity of Techniques for Biogenic Amines Based on Fluorescent Reaction Products with Carbonyl Compounds *

QUAY, W. B. (Department of Zoology, University of California, Berkeley, Calif.)

Methods for measurement and tissue localization of particular catechol-, indole- and imidazoleamines have advanced greatly in recent years by means of the fluorescent products of reactions with specific carbonyl compounds. Certain of these, especially those with formaldehyde, have already been studied in detail. Systematic study of others is needed. Improvements in sensitivity, specificity and stability of the methods and their reaction products are possible and additional carbonyl reagents can be employed.

A most sensitive method for 5-hydroxytryptamine (5-HT), based on

214

reaction with 1,2,3-indantrione hydrate (= Ninhydrin) at pH 7.0 was found to differ in specificity from what had been reported in other laboratories. Thus, 6-hydroxytryptamine and certain harmala alkaloids as well as 5-hydroxytryptophan and 5-HT produce notable fluorescence (max. at 490 mu; activation max. 380-385 mu), and bufotenine does not. Over 30 related compounds are negative. Constitution of the reaction medium is shown to be critical for specificity. With appropriate controls and extraction procedures the method is nearly specific for 5-HT in most vertebrate tissues, but remains unsatisfactory for those of insects. The latter contain other, as yet unidentified, fluorigenic reactants.

Indandiones (1,3-indandione, 2-nitro-1,3-indandione) in contrast with 1,2,3-indantrione do not produce fluorescent products with 5-HT or related amines at pH 7.0. Other carbonyl compounds are shown to be effective, but only in certain carefully controlled conditions. Summary and analysis of these will be presented.

---

* Supported in part by U.S. Public Health Service research grant NB-06296.

Histochemical Characteristcs of Benign and Malignant Growths in the Breast and Thyroid Tumors in Human Glands

RAIKHLIN, N. T. (Institute of Experimental & Clinical Oncology, Academy of Medical Sciences of USSR, Moscow)

The author studied histochemically the localisation and activity of the redox enzymes, non-specific phosphatases, adenosinetriphosphatases, 5'-nucleotidase, thiaminpyrophosphatase, glucoso-6-phosphatase, esterase, aminopeptidase and phosphorylase in approximately 100 cases of new growths in the breast and thyroid glands.

Analysis shows that there can exist histoenzymatic differences between the epithelial growths in fibrous-cystic mastopathy and breast cancer. As a test permitting one to judge with a definite degree of probability the malignisation of epithelium with excessive proliferation, one can recommend titration of phosphorylases, adenosinetriphosphatase, esterase, aminopeptidase, succinatedehydrogenase whose activity in cancer cells, as compared to epithelium in hyperplastic processes decreases as a rule or disappears entirely. Thiaminpyrophosphatase and lactic acid dehydrogenase activity increase in cancer cells.

In thyroid cancer, in contradistinction to adenomas, there is usually a much higher activity of adenosinetriphophatase, 5'-nucleotidase, glucose-6-phosphate dehydrogenase, 6-phosphogluconate, NAD- and NADP-diaphorases.

The activity of succinic acid dehydrogenase and esterase in adenomas varies, but is usually lower than in cancers.

Activity of the above-mentioned enzymes in benign as well as malignant tumors of the breast and thyroid can vary. Hence, the decision on possible presence of malignant growth must be made on the strength of the sum total of histoenzymatic signs with obligatory consideration of morphological data.

The aggregate of histochemical signs which help to distinguish between benign growths in a given organ and malignant ones varies from one organ to another. Due to this peculiarity of tumors, it is essential to have specific histochemical criteria for each organ or tissue (or groups of organs and tissues) which can aid in making the differential diagnosis.

Nucleoprotein Metabolism During Larval Development in Sciara **

RASCH, ELLEN M.* (Department of Biology, Marquette Univ., Milwaukee, Wisc.)

Detailed analyses of schedules of larval maturation in the highly inbred, monogenetic, dipteran species *Sciara coprophila* (Lintner) were combined with microchemical determinations of nucleoprotein levels in whole larvae and with cytophotometric estimates of relative amounts of DNA in Feulgen-stained nuclei from whole mounts of salivary glands, Malpighian tubules, or gastric caeca and smears of hemocytes, neural ganglia, or developing limb anlage. Differences in DNA content of individual polytene chromosomes and specific chromosome regions before, during, and after formation of the dense, DNA-rich, heterochromatic puffs which arise prior to pupal molt in *Sciara* were assessed using the two-wavelength method of Patau (1) and Ornstein (2), the two-area method of Garcia (3) or the Deely scanning procedure (4). Finally, differences in schedules of larval growth and puffing patterns of polytene salivary chromosomes were evaluated after various dietary supplements, including mushroom powder, brewer's yeast, cortisone acetate (5) or gibberellic acid.

Characteristic differences were found between female and male larvae in (a) over-all rates of growth, (b) durations of last larval instar, (c) amounts of total protein (Lowry), (d) amounts of total RNA (orcinol), and (e) amounts of total DNA (diphenylamine) when animals of similar stages were compared. The average DNA-Fuelgen content of salivary gland nuclei from female larvae was about twice that found for similar nuclei from male larvae during late fourth instar period of growth.

Estimated DNA content of the paired, polytene X homologues from salivary gland nuclei of females was twice that found for the single polytene X chromosome of male salivary nuclei, in agreement with a 8-10% difference in DNA content of presumptive diploid hemocyte nuclei from males and females. A value of 0.4 $\mu\mu$g/cell was estimated for diploid hemocytes of *Sciara,* using comparably fixed and stained sections of rat liver or smears of chick blood as standards of known DNA content.

Elevated DNA levels were found in each of the regions of chromosomes II and III forming dense, heterochromatic puffs during the prepupal period. The net increase in amount of DNA-Feulgen was characteristic of specific puff loci, was maintained through the time of pupal molt, and was closely correlated with stage of larval maturation, whether chromosomes analyzed were sampled from male or female salivary glands.

Although highly significant differences in schedules of larval maturation were found, depending upon the nutritional history of sibling cultures, overall patterns of larval growth and patterns of salivary gland chromosome puffing were remarkably similar when experimental and control animals were sampled at similar stages of maturation assessed by morphological criteria of larval development, rather than by absolute time in days since hatching of eggs.

After more than 40 generations of inbreeding by single pair matings, the differences observed in *Sciara* between female and male larvae in growth rates and total nucleoprotein levels seem clearly a reflection of selective elimination of X chromosomes of paternal origin during early cleavage divisions of male soma (6). The expressed potential in female larvae for twice the net growth and nucleoprotein synthesis over that shown by males of identical autosomal constitution would seem due in large part to capacity for genetic activity by a paternally-derived X chromosome, since factors responsible for the selective retention of one of these by nuclei of the female soma during early embryogeny provide the only known major genetic variable as yet defined in this system (6).

References:

(1) Patau, K.: Chromosoma *5*, 341 (1952).
(2) Ornstein, L.: Lab. Invest. *1*, 250 (1952).
(3) Garcia, A. M.: J. Histochem. Cytochem. *13*, 161 (1965).
(4) Deely, E. M.: J. Sci. Instrum. *32*, 263 (1955).
(5) Goodman, R. M., J. Goidl, and R. M. Richart: Proc. Nat. Acad. Sci. *58*, 553 (1967).
(6) Crouse, H.: Genetics *45*, 1429 (1960).

* U.S. Public Health Service N.I.H. Development Career Awardee (K3-GM 3455).
** Supported in part by U.S. Public Health Service grants (GM 10503 and GM 14644).

Submicroscopic Structure of Developing Sperm in the Teleost Fish Poecilia **

RASCH, ELLEN M.* (Department of Biology, Marquette Univ., Milwaukee, Wis.)

Fine structural changes during spermatocyte maturation and development of functionally competent sperm were studied in two species of viviparous fish, *Poecilia mexicana* and *Poecilia velifera*. Testicular crypts with immature germ cells and freshly shed spermatophores containing mature sperm were fixed in 4% glutaraldehyde buffered at pH 6.8, postosmicated, and embedded in Vestopal-W. Images obtained with an RCA 3G electron microscope from ultrathin sections stained with uranyl acetate or lead hydroxide were compared with those seen in adjacent 1μ sections by light microscopy after staining with (a) the Feulgen reaction for DNA, (b) the periodic acid-Schiff reaction for polysaccharides, (c) the Morel-Sisley reaction for tyrosine, (d) the Sakaguchi reaction for arginine, or (e) toluidine blue at pH 9 for total nucleoprotein.

In primary spermatocytes at pachytene and diplotene, synapsed homologues display prominent, DNA-containing, chromatin complexes like those described for other animal species (1). During both meiotic metaphases, attachment of spindle fibers occurs only at or near the ends of small, rod-shaped chromosomes. Throughout spermatid differentiation there is a constant and specific orientation of the central fibril pair of the developing axial filament complex with relation to position of the distal centriole and eventual planes of bilateral symmetry of the mature sperm head. These relationships allow designation of dorsal, ventral, dextral, and sinstral surfaces as well as definition of anterior and posterior nuclear poles in very young spermatids. Selective mobilizations of all major cytoplasmic organelle systems during spermatid maturation are accompanied by dramatic changes in nuclear fine structure. An extensive region of rarified karyoplasm occurs along the median ventral surface of young spermatid nuclei, apparently associated with formation of a large, lucent vesicle by the outer membrane of the nuclear envelope and extrusion of some segregated nuclear component. Aggregation of the remaining 100 Å chromatin filaments into coarse, twisting 500 Å fibrils is followed by their progressive coalescence to form a dense shell some 100 mμ thick of DNA-containing material near the nuclear surface. Concurrently, progressive invagination of the nuclear membrane on the dorsal surface occurs to extend a central channel from the base of the flagellum at the posterior nuclear pole to the extreme anterior tip *Lebistes* (3), both the centriolar complex and the flagellar base thus come to lie within the sperm head in a bulbous region of the central of the nucleus. In *Poecilia*, as described for sperm of *Jenynsia* (2) and channel below the mid-dorsal surface of the nucleus. Further nuclear

218

elongation and chromatin compaction occur as the nucleus folds about the flagellar base to enclose the central channel, producing the homogeneously dense, but caniculate, lance-shaped head of mature sperm in poeciliid fishes (4). Lack of PAS staining in these sperm is consistent with apparent absence of a definitive acrosomal vesicle or other acrosomal materials at the anterior nuclear pole. A marked preferential orientation of developing sperm and their eventual aggregation in groups of 80 to 100 heads within the highly modified apical cytoplasm of individual Sertoli cells of the crypt wall occur prior to formation of spermatophores, the specialized sperm packet which will be transferred to the female genital tract by a modified anal fin, the male gonopodium. The unusual transformations found during maturation and packaging of these sperm probably reflect an evolution from the more primitive type of sperm development shown by oviparous teleosts, since viviparity in *Poecilia* involves a prolonged, separate existence of male gametes within a female environment prior to their use in fertilization.

References:

(1) Moses, M. J.: J. Biophys. Biochem. Cytol. *4*, 633 (1958).
(2) Dadone, L. and R. Narbaitz: Z. Zellforsch. *80*, 214 (1967).
(3) Porte, A. and E. Follenius: Bull. Soc. Zool. Frances *85*, 82 (1960).
(4) Rasch, E. M., R. M. Darnell, K. D. Kallman, and P. Abramoff: J. Exp. Zool. *160*, 155 (1965).

* U.S. Public Health Service N.I.H. Development Career Awardee (1-K3-GM 3455).

** Supported in part by grants from the National Science Foundation (GB 4712) and the U.S. Public Health Service (GM 10503) and (ST1-HD27).

The Influence of Low Temperature and Storage Time on Some Histochemically Demonstrable Enzyme Activities in Liver, Kidney, and Jejunum of the Rat

RIECKEN, E. O., H. GOEBELL, and C. BODE (Medizinische Klinik, Marburg, Germany)

For comparative histochemical studies, it is necessary to preserve biological material in a way which maintains its enzyme activity for long periods. Several methods are used before biochemical assays (i.e., drying in high vacuum at low temperature, freezing and storage in a deep-freeze, stabilization by high concentration of salts as precipitated suspensions in saturated ammonium sulphate). Not all of these are suitable for histochemical purposes. A study was therefore carried out to investigate the influence of low temperature and storage time on histochemical enzyme preparations of comparable tissues specimens after varying conditions of storage.

Female Wister rats, 160 to 180 g in weight, were used. Three animals were killed at different time intervals, constant conditions being strictly maintained. Liver, kidneys and comparable segments of upper jejunum were quickly removed and processed for histochemical investigation in the following ways: a) freezing of small pieces of the tissues selected (4 to 6 mm³) on cork discs in liquid nitrogen and airtight sealing in small plastic bags; b) fixation in formalin-calcium (4%) at 4° C for 18 hours with consecutive washing for 4 days in gum-sucrose solution at 4° C, followed by freezing and sealing as in a. For biochemical purposes, pieces of liver from all animals were processed as in a. The different plastic bags with the three tissues of each animal enclosed were stored at —18° C to —20° C, —75° C to —80° C and in a liquid nitrogen storage chamber (—160° C). After storage periods of 236, 137, 46 and 17 days, all plastic bags were stored together with tissues of another set of three animals at —75° C to —80° C and histochemical as well as biochemical enzyme studies were undertaken. A number of lysosomal, mitochondrial, microsomal and membrane enzyme activities were demonstrated histochemically using 5 simultaneously incubated cryostat sections of each storage period and storage temperature. The preparations were evaluated microscopically by two independent investigators. Succinate dehydrogenase (SDH) and non-specific acid phosphatase, as well as DNA, extractable protein, and mg protein/g liver tissue were measured biochemically. Histochemically we observed that storage for up to 236 days at —75° C to —80° C, and also at below —160° C, resulted in complete preservation of enzyme activities; at —18° C to —20° C considerable loss of enzyme activity occurred in a number of enzymes after 137 and 236 days of storage, especially with respect to SDH. Biochemical assay showed a similar reduction in SHD activity but this was considerable even after 17 days, when no reduction was detected histochemically. Non-specific acid phosphatase showed no change, biochemically or histochemically, under any of the experimental conditions.

Quantitative Study of Glucose-6-phosphatase Inactivation by Fixatives

RIGATUSO, JOSEPH L. (Department of Anatomy, University of Minn.)

In the present study, liver glucose-6-phosphatase (G-6-Pase) activity was determined in paired biochemical and histochemical studies following treatment with various routine fixatives.

G-6-Pase activity was determined biochemically in preweighed blocks of normal rat liver treated in the following ways: 1) immediately homogenized in 0.25 M sucrose at 2° C, (control); 2) maintained 24 hours in 10% neutral formalin at 2° C; 3) maintained 24 hours in formol-calcium at 2° C; 4) quenched in liquid nitrogen for 15 minutes and kept at —25° C for 24 hours; 5) slowly frozen at —25° C for 24 hours; 6) immersed in 5% glutaraldehyde for 2 and 4 hours, then washed in several changes of isotonic buffer-sucrose for the remainder of 24 hours, all at 2° C. Fixed tissues were washed, then homogenized in 0.25 M sucrose and analysed for G-6-Pase activity by a modified Fiske-Subbarow method.

Both quenching and slow-freezing destroyed less than 10% of the G-6-Pase activity. Formalin and formol-calcium fixations inactivated 85% and 86% respectively, of the G-6-Pase activity when compared to control liver. Glutaraldehyde fixation for 2 hours inactivated 41% of the activity, 4 hours fixation inactivated 57%.

Qualitatively similar results were obtained in the histochemical studies, except in the case of glutaraldehyde fixation. In contrast to the uniformly diminished enzyme activity seen in sections of formalin fixed tissues, the periphery of the glutaraldehyde fixed sections were negative, but significant G-6-Pase activity was present in the interior of the sections.

Since inactivation by glutaraldehyde appeared to depend upon the degree of penetration of the fixative into the tissues, studies were carried out using isolated liver microsomes. These were incubated with varying concentration of glutaraldehyde for 2 hours at 2° C. Glutaraldehyde in concentrations as low as 1% inactivated over 90% of the microsomal G-6-Pase activity.

The specificity of the G-6-Pase reaction was determined by comparing the rate of hydrolysis of beta-glycerophosphate with that of glucose-6-phosphate. At pH 6.5 (near optimal for G-6-Pase) the phosphate liberated from the nonspecific substrate was only 4.5% of that with glucose-6-phosphate. When beta-glycerophosphate was substituted for glucose-6-phosphate as a substrate in the histochemical reaction no cytoplasmic staining was evident.

Structural Changes of the DNP-complex of PHA Stimulated Human Leukocytes and Their Relation to RNA Synthesis

RIGLER, R. and D. KILLANDER (Institute for Medical Cell Research and Genetics, Karolinska Institutet, Stockholm 60, Sweden)

Microfluorimetric analysis of nucleic acids by acridine orange (AO) (1,2) indicate that stimulation of human leukocytes by PHA results in a structural alteration of the DNP complex. Within a few minutes after addition of PHA, a marked increase in the AO-binding sites of the DNA in the DNP-complex is observed (3,4). This increase in dye binding sites is accompanied by a decreased thermal stability and lengthening as well as by an increased structural order of the DNA-molecule. These changes are best explained by an alteration in the binding between

DNA and nuclear proteins resulting in a liberation (unscreening) and mutual repulsion of negatively charged dye binding sites on the DNA molecule. The observed alterations of the DNP-structure seem to be a prerequisite for the initiation of RNA-synthesis in human leucocytes. Significant amounts of RNase sensitive single stranded nucleic acids are not observed before a certain amount of dye binding sites in the DNP-complex has been liberated. If this critical level is exceeded, a correlation between the liberation of dye binding sites in DNP' and the increase of single stranded nucleic acids in 1 hour cultures is observed. The implications of the changed DNP structure for an increased template activity of the DNA-molecule are discussed.

References:

(1) Rigler, R.: Acta Physiol. Scand. *67*, Suppl. 267, 1, (1966).
(2) Rigler, R.: Ann. N.Y. Acad. Sci., *in press* (1968).
(3) Killander, D. and R. Rigler: Exptl Cell Res. *39*, 701 (1965).
(4) Killander, D. and R. Rigler: *These abstracts.*

Nucleoprotein Changes in Hen Erythrocyte Nuclei Undergoing Reactivation

RINGERTZ, N. R. and L. BOLUND (Institute for Cell Research and Genetics, Karolinska Institutet, Stockholm 60, Sweden)

In birds, the mature erythrocytes retain their cell nuclei in an inactive form. The chromatin of these nuclei is condensed into dense heterochromatin-like clumps. The synthesis of protein is very much reduced and RNA and DNA formation has been halted. Recently Harris (1) demonstrated that hen erythrocyte nuclei can be reactivated by fusing the erythrocytes with HeLa cells to form heterokaryons. In collaboration with Dr. Harris we have analyzed the cytochemical properties of erythrocyte nuclei during this reactivation process (2). Scanning microinterferometry on erythrocyte nuclei in heterokaryons representing different time intervals after cell fusion demonstrated a 5-6 fold increase in nuclear dry mass during the first 48 hours after cell fusion. Part of this dry mass increase took place before DNA synthesis was initiated. DNA synthesis started approximately 18 hours after cell fusion and at 48 hours, about 40% of the erythrocyte nuclei contained G 2 amounts of DNA. RNA synthesis is initiated shortly after cell fusion and accelerates gradually as the chromatin clumps are dispersed and the nuclei increase in dry mass and volume (1, 2). The cytochemical properties of the nuclear DNP component change at an early stage of the reactivation process. The Feulgen reactivity is altered, suggesting changes in the susceptibility of DNP to acid hydrolysis. By microfluorimetry a more than 4-fold increase in the binding of intercalating dyes (acridine orange (AO) and ethidium) can be demonstrated already before DNA replication has been initiated. Using heat denaturation in combination with AO-microfluorimetry (3) it is demonstrated that the DNP of activated erythrocyte nuclei is more labile to heat denaturation than is the DNP of control nuclei.

DNP-changes similar to those found when hen erythrocyte nuclei are reactivated in heterokaryons can also be induced in the nuclei of normal erythrocytes and erythrocyte ghosts by washing in salt solutions and tissue culture media which do not contain serum proteins (4). The induction of the DNP-changes can take place at low temperatures and is not blocked by iodoacetic acid or p-dinitrophenol. The DNP-changes can, however, be blocked by 0.1 M Mg^{++} or 0.6 mM Mn^{++}. If serum proteins are present at a low concentration, the erythrocyte cell membrane is stabilized towards lysis and the DNP-changes are prevented. The DNP-changes can, however, be induced by EDTA also in the presence of the serum factors. The results obtained suggest that the removal of divalent metal ions from the DNP complex may be an essential step in the mechanism by which the dormant erythrocyte nucleus is reactivated.

References:

(1) Harris, H.: J. Cell Sci. 2, 23 (1967).
(2) Bolund, L., N. R. Ringertz, and H. Harris: J. Cell. Sci., *in press*.
(3) Rigler, R., D. Killander, L. Bolund, and N. R. Ringertz: Exptl. Cell Res., *in press*.
(4) Ringertz, N. R. and L. Bolund: *to be published*.

UV Microspectrophotometry of the Crabtree Effect *

RITTER, CARL and BO THORELL (School of Vet. Medicine, Univ. of Pennsylvania and Patol. Inst. Karolinska Inst., Stockholm)

Changing 270 μ absorption of intracellular mitochondria was measured in single ascites tumor cells with a sensitive microspectrophotometer. Spectra (240-260 μ) were run before and after each metabolite-stimulated response. In ascites cells, glucose produced a 0.02 to 0.05 OD decrease in 270 μ absorption in 4 to 5 minutes ($T_{1/2}$ 1.5 to 2 minutes). The response was inhibited by pretreatment with either deoxyglucose or dicumarol. Deoxyglucose causes a 0.01 OD decrease in 270 μ absorption, and dicumarol a 0.02 to 0.05 OD decrease. In Chang liver cells glucose produces a 0.02 OD decrease in 270 μ absorption which is partially reversed when the glucose is removed. In the ascites cells, removal of glucose did not stimulate such a reversal. The microspectrophotometric method is sufficiently sensitive to measure pyridine nucleotide reduction at both 270 and 340 μ if it occurs. Such reduction did not occur in ascites cell or Chang cell mitochondria, but was seen in Chang cell cytoplasm in response to glucose. Glucose seems to decrease the adenine nucleotide content of ascites cell mitochondria. The lost adenine nucleotide is then apparently diluted into the nonmitochondrial area of the cell and tightly sequestered to a molecular 'compartment' possibly phosphofructokinase.

* Supported by grants from USPHS and Swedish Medical Research Council.

Histochemical and Biochemical Evidence of Sulfated Sialomucins

RIZZOTTI, M., G. AURELI, C. BALDUINI and A. A. CASTELLANI (Istituto di Anatomia degli Animali Domestici-Istologia ed Embriologia. Università di Milano; Cattedra di Biochimica. Facoltà di Scienze, Università di Pavia, Italia)

Histochemical investigations were undertaken with metachromasia with toluidine blue at pH values between 0.5-3.0 Alcian Blue, Alcian Blue Alcian Yellow and a colloidal iron method. Prior to staining, sections were exposed to testicular hyaluronidase (1 N HCl at 80°), neuraminidase, hydrolysis with dilute HCl at pH 2.5, methylation, methylation-demethylation, methylation-demethylation-neuraminidase in various combinations carried out on mucus cells of dog submaxillary and of dove crop glands which showed the presence of a carbohydrate material containing either sialic acid or sulfate ions.

By biochemical methods, sulfated mucins have been extracted from both the glands. Galactose, glucosamine, galactosamine, fucose and sialic acid are present, in different ratios, in the carbohydrate moieties of both the mucins. The protein moieties, the protein-polysaccharide linkages, and the sequence of monosaccharides have been also studied in these macromoloecules.

The results obtained either by histochemical or biochemical methods agree in demonstrating the presence of sulfate ions and sialic acid; the biochemical evidence suggests the real existence of sulfated sialomucins.

Visualization and in vitro Correlates for Nucleoside Phosphatase Activities in Kidneys from Winter Hibernating, Summer Active, and Cold-stored Bats (Myotis lucifugus)

ROSENBAUM, ROBERT M. (Albert Einstein College of Medicine, Bronx, N. Y.)

Previous publications from this laboratory have described a seasonal variation in activity and localization of several nucleosidases and ATPase within proximal and distal renal tubular epithelium of the bat, Myotis *lucifugus.* Electron microscopic visualization of reaction product following use of metal-salt substrate mixtures for ATPase activity at near neutral pH's showed localization within a complex vacuolar system which, at the light microscope level, appeared as a stained Golgi region in renal tubular epithelium from hibernating and cold-stored bats. This localization could not be demonstrated in tubular epithelium from summer active bats.

The Golgi-like localization of ATPase activity can be correlated to an increase in pooled urine phosphate from hibernating animals as compared to active animals. Pooled serum ATPase activity could also be demonstrated as increased in kidney homogenates from winter animals. Methods for assay of nucleoside phosphates were applied to serum and to kidney homogenates using anion exchange and treatment of the eluate with formate-formic acid systems for separation of the total acid soluble extract. Assurance of the presence of nucleotide bases was obtained by paper chromatography or, in some cases, by enzymatic assay employing firefly luminescence on pooled serum or tissue samples. Levels of other nucleosides including inosine and cytosine phosphates were also determined although these were very much lower than were the phosphates of adenosine.

The increase in levels of phosphate in urine and of adenosine phosphates in serum of hibernating bats can be correlated with the intracellular localization of ATPase activity and activity of other phosphatases in vacuoles in renal proximal epithelium. Since these cells also show increased accumulation of exogenous peroxidase during hibernation, it is suggested that increased substrate is acquired by the proximal epithelial cell by a similar mechanism at the expense of a phosphate pool known to exist in tissues such as muscle during hibernation.

References:

(1) Rosenbaum, R. M. and A. Melman: J. Cell Biol. *21*, 325 (1964).
(2) Rosenbaum, R. M. et al.: "Mammalian Hibernation," Oliver and Boyd, London, p. 295 (1967).

Diurnal Variation of Mouse Hepatic Parenchymal Cell Nucleus-to-Cytoplasm Ratio *

ROSENE, GORDON L. (Dept. of Physiology and Health Science, Ball State University, Muncie, Ind.)

Mouse hepatic parenchymal cell nucleus-to-cytoplasm ratio (NCR) was measured at two-hour intervals during the 24-hour period. This ratio is generally accepted to be an index of relative metabolic activity. Increases in the parameter indicate increased metabolic rates (1). Jerusalem

(1964) has mentioned rhythmic changes in the nuclear diameter of hepatic parenchymal cells during the 24-hour period.

Animals were maintained on a 12 hour light, 12 hour dark illumination regimen. Livers were removed at two-hour intervals from 96 Strain A mice. Eight animals were sacrificed at each two-hour interval. Livers were fixed for 12 hours in 10% buffered formalin, paraffin embedded, sectioned at 5 micra and stained with hematoxylin and eosin.

The NCR was measured using a Zeiss integrating eyepiece equipped with a Hennig reticule. This technique is a modification of the method of Chalkley (3). Nucleus and cytoplasm "hits" were recorded and expressed as the NCR.

The hepatocyte NCR was significantly elevated during the forenoon hours. A second, less pronounced elevation was seen between the hours of 6:00 p.m. and midnight. Low points appeared on the curve between noon and 4:00 p.m., and again between midnight and 6:00 a.m. Relations between this study and previous work will be discussed, as well as the significance of the findings.

References:

(1) Wilson, M. E. *et al.*: Cancer Res. *13*, 86 (1953).
(2) Jerusalem, C.: *In:* II Internat. Kongr. Histo. U. Cytochem. Frankfurt/M, 1964. Berlin-Gottingen-Heidelberg: Springer (1964).
(3) Chalkey, H. W.: J. Natl. Cancer Inst. *4*, 47 (1943).

---

* Supported by grant 2362 from the Delaware County Cancer Society.

Histochemical Investigations on Enzyme-activities of Human Mucous Membranes in Different Biological Conditions

ROSSI, FERDINANDO (Istituto Anatomia Umana Normale, Università Genova, Italia)

Biopsy specimens of human mucous membranes often show structural patterns that do not correspond to those considered as normal. Cytological and histochemical modifications occur in the wall of various organs and, in particular, the digestive and respiratory tracts where a direct relation to the external environment must be considered. In some instances, these modifications can be regarded as reactions to abnormal stimuli and, therefore, as the expression of defensive or adaptative mechanisms.

The human being has wide possibilities for specific and unspecific defense. Enzymatic reactions provide, in some instances, useful tools to detect organismic reactions; they can also be employed for diagnosis. Histochemical equivalents are still incomplete and need further improvement; however, they should express dynamic features of the living protoplasm.

We are trying to establish in various organs those histochemical patterns which can be regarded as the expression of normal situations. In comparison, we are investigating specimens which are taken in the proximity of inflammatory or neoplastic zones. Some modifications can be detected by this comparison; they could be defined as "alarm-signs." Variations in the histochemical patterns are shown, not only in epithelia but also in various components of the tunica propria. In particular, histiocytes in the proximity of pathologic areas show significant enzymatic activities. Enzyme-histochemical localizations in epithelial zones are often found, in a spotty appearance, in the larynx, joined with malignant neoplasms.

A number of histochemical patterns will be considered in the mucous membranes of the mouth, pharynx, stomach, intestine, nose, larynx, bronchi. The application of histochemical reactions for the demonstration of hydrolytic and oxidative enzymes will be discussed.

Quantitative Studies of Fluorescence Fading in Tissue Sections

ROST, F. W. D., and A. G. E. PEARSE (Dept. of Histochemistry, Royal Postgraduate Medical School, London)

The fading of fluorescence in tissue sections excited with ultraviolet or blue-violet light has been recorded by several authors (e.g., Popper, 1944; Falck, 1962; Yamada et al., 1966). The rate of fading of fluorescence under standardised conditions has been found to be characteristic of the fluorophore (e.g., Caspersson et al., 1966). However, few quantitative studies of fluorescence fading have been made although Thaer (1966) demonstrated that the fluorescence emission spectrum of Acridine Orange changes during excitation, and Ritzén (1966, 1966a) has published curves for the fading of the formaldehyde-induced fluorescence of noradrenalin and 5-hydroxytryptamine.

The present work is concerned with the quantitative study of fluorescence fading in tissue sections and models by microspectrophotometry of fluorescence excitation and emission and also of absorption, at intervals during prolonged excitation of fluorescence. Absorption, excitation and emission measurements are carried out with a Leitz microspectrograph (Ruch, 1960) modified for use as a fluorescence microspectrophotometer (Rost and Pearse, 1968). With this instrument, continuous recording is possible of absorption or of emission spectra during excitation with ultraviolet or blue-violet light. Applications of this technique to the fading of formaldehyde-induced fluorescence of monoamines and of fluorochromes will be described, and the photochemical implications discussed.

References:

Caspersson, T., N.-Å. Hillarp, and M. Ritzén: Exp. Cell Res. *42*, 415 (1966).
Falck, B.: Acta Physiol. Scand. *56*, suppl. 197 (1962).
Popper, H.: Physiol. Rev. *24*, 205 (1944).
Ritzén, M.: Exp. Cell Res. *44*, 505 (1966).
Ritzén, M.: Exp. Cell Res. *45*, 178 (1966a).
Rost, F. W. D. and A. G. E. Pearse: Proc. Roy. Microscop. Soc. *3*, 22 (1968).
Ruch, F.: Z. wiss. Mikrosk. *64*, 543 (1960).
Thaer, A.: Haus der Technik-Vortragsveröffentlichungen, (E. Leitz G.m.b.H.), *69*, 49 (1966).
Yamada, M., A. Takakusu, K. Yamamoto, and S. Iwata: Arch. Histol. Jap. *27*, 387 (1966).

Coordinated Spectrofluorometry of the DNA of Intact Cells

ROTH, DANIEL (N.Y.U. Medical Center, Dept. of Pathology, New York, N. Y.)

Conditions were established for the specific binding of acriflavine to the DNA of intact cells. Of primary importance was maintenance of a high ratio between polymer and dye concentration. To fall within the sensi- tivity limits of a conventional filter fluorometer, samples of 4×10^5 cells were stained with 1.1×10^{-7}g of dye. It was estimated that approxi- mately one molecule of dye was bound for each 100 nucleotide sequence on the native DNA complement in each cell. Other biopolymers capable of binding acriflavine under conditions offering more abundant dye, were shown to remain unstained when competing against DNA for sharply restricted amounts of the dye. With sterile technique and suit- able pH (phosphate buffer pH range 6.0-7.4) the experimental cells remained viable.

Two frames of reference were used in the determinations. One, di- rected toward the supernatant (free) dye which remained after cells had been stained and then separated by centrifugation, gave extremely pre- cise yet rapid measurements of the DNA content of the whole cell population. The other, involving microspectrofluorometry of individual cells from the sediment, provided quantitative data related to the dye- binding range among the cells and the distribution of the dye as a fluorescent label within individual cells. For measurement of the light flux emitted by the estimated 10^8 molecules of dye within a single cell, modification of conventional instrumentation was required in order to gain a satisfactory signal to noise ratio.

Quantitative Fluorescence Microscopy

RUCH, FRITZ (Institute for General Botany, Technische Hochschule, Zurich, Switzerland)

Quantitative determination of substances in the cell can be advantageously done with the fluorescence microscope (1). This method is very sensitive and its application is often simpler than light absorption techniques. Under proper conditions, the fluorescent light emitted by an object is proportional to the amount of substance and independent of its distribution. Scanning methods as used in measurements by absorption are therefore not necessary. Furthermore the influence of light scattering is negligible. Fluorometry gives relative values of the amount of substance present. Therefore, it can be used to compare the contents in various cells or cellular structures. Absolute values may be obtained if a standard object of known contents is also measured.

The instrument for cytofluorometry consists of a fluorescence microscope with an illuminating system for excitation in incident light and a photometer attachment with variable diaphragm, observation tube, photomultiplier and galvanometer. The primary fluorescence of cellular compounds such as pigments and cyclic amino acids may be assayed. Of course, staining reactions using fluorochrome dyes have a much wider applicability. Examples for the determination of DNA, proteins and amino acids (1, 2, 3) will be discussed.

References:

(1) Ruch, F.: *In:* Introduction to Quantitative Cytochemistry, pp 281-294 (Wied, edit.) New York: Academic Press (1966).
(2) Bosshard, U.: Z. wiss. Mikroskopie *65*, 391 (1963).
(3) Rosselet, A.: Z. wiss. Mikroskopie *68*, 22 (1967).

Replication-Precisions of Length and Areal Measurements on Normal Human Chromosomes by Means of Manual, Semi-Automated, and Automated Techniques *

RUDDLE, FRANK H., HERBERT A. LUBS, and ROBERT S. LEDLEY (Departments of Biology and Medicine, Yale University, New Haven, Conn. and National Biomedical Research Foundation, Silver Spring, Md.)

Normal human chromosome preparations were obtained by phytohemagglutinin mitotic stimulation of whole peripheral blood cultivated in medium 199 plus 15 percent fetal calf serum. Metaphase cells were accumulated by means of colcimide, and microscopic preparations made by the air-dry technique coupled with weak orcein staining. Photographs of metaphase spreads were photographed on high contrast microfile film under phase contrast microscopy. Two different hand replication studies were performed, both on identical cells. In the first, 30 enlarged prints (8 X 10) were made from a single negative phototransparency. This was termed a replication of prints. In the second, a single cell was rephotographed 30 times and care was taken to move the object and refocus the microscope before each exposure. An enlarged print was made from each negative. Each of 30 prints was then measured. This was termed a replication on frames. Hand measurements were recorded by means of dividers and alternatively by means of a low power magnifier with a built-in scale. Semi-automated analysis will be performed by means of a digital X-Y plotter, the PF-10 Pencil Follower produced by Edwin Industries. Replication precision measurements will be made on the basis of prints and frames. Areal measurements will also be performed. The same data will also be recorded completely automatically by the Fidac-Fidacsys system. The replication-precision values obtained by the three methods will be compared.

---

* Supported by Contract PH-43-67-1463 from the Division of Biologics Standards, NIH and Public Health Service Grant GM 9966-06.

Electrophoretic Variant of the Most Anodally Migrating Serum Esterase in Mus musculus *

RUDDLE, FRANK H., THOMAS B. SHOWS, and THOMAS H. RODERICK (Department of Biology, Yale University, New Haven, Conn. and The Jackson Laboratory, Bar Harbor, Maine)

An electrophoretic variant form of the most anodally migrating serum esterase was detected in wild mice collected in the Duke Forest, North Carolina. This variant form migrated more slowly than the typical esterase phenotype represented in wild mice and inbred stains. We have designated the fast form b and the slow form c. Genetic analysis of F_1, F_2 and BC progeny have indicated that the two allelic genes control the expressions of the b and c esterases. The alleles are codominant. A hybrid enzyme could not be detected in the heterozygote. Preliminary linkage tests have demonstrated close linkage with the serum esterase locus Es-1. This suggests that the variant electrophoretic forms are possibly segregating at the Es-2 locus. A close linkage between Es-2 and Es-1 has recently been established by Popp (1) employing a silent gene variant which segregates at Es-2 (2). Genetic crosses are now being performed in order to determine whether the b and c alleles do or do not segregate at the Es-2 locus. These data will be reported. A general survey of the murine electrophoretic variants will be presented.

230

References:

(1) Hered, J.: *58*, 186 (1967).
(2) Petras: Proc. Natl. Acad. Sci. *50*, 113 (1963).

* Supported by Public Health Service Grant GM-9966-06.

The Histochemical Demonstration of γ-Glutamyl Transpeptidase

RUTENBURG, A. M., H. KIM, E. BRODIE, J. FISCHBEIN, and C. L. ROSALES (Yamins Research Laboratory, Beth Israel Hospital, and University Hospital, Department of Surgery, Boston University Medical Center, Boston, Mass.)

A simultaneous coupling azo dye method for the histochemical demonstration of γ-glutamyl transpeptidase activity using the new substrate γ-glutamyl-4-methoxy-2-naphthylamide was developed. In comparison to previous reported methods for this enzyme, the present technique yielded an azo dye of more intense color density and could easily be modified for the electron microscopic localization of the enzyme. The optimum conditions for the histochemical reaction were developed and the distribution of enzymatic activity in tissues of the rat described.

Glycogen in Migrating Cells in Brucellosis and Leukemia

SADAUSKAS, P., L. LUKSYS and V. DABKEVCIUS (Dept. Cytol. and Cytochem. Institute of Biochemistry, Acad. Sci. of Lithuania, Vilnius, USSR)

Glycogen in migrating cells has been one of the indicators for the study of immunogenesis of brucellosis and reactivity of an organism (cows) ill with leukemia. Cytological preparations for cytochemical analysis of glycogen (PAS) were obtained from cattle horn vital tissue erosion according to the method worked out by the authors ("horn opening") which was established at the basis of the Rebuck and Crowley method. Carbohydrate metabolism in the case of immunogenesis of brucellosis varies in all migrating cells (neutrophils, macrophages and lymphocytic cells). In the immune organism, synthesis of glycogen increases sharply which, in spite of active phagocytosis of brucelli by micro- and macrophages, is preserved in quantity in present cells. In phagocytes of unimmune cattle, a sharp decrease of glycogen was found up to its complete disappearance. A considerable suppression of carbohydrate metabolism within the cells and well defined glycolysis with evidence of structureless lyoglycogen was observed in cattle ill with brucellosis.

Micro- and macrophages lose glycogen and become colorless during the process of phagocytosis of brucelli.

In migrating cells of cows ill with leukemia in the lymphatic form, a great amount of glycogen accumulates which is 1.5-2 times greater than in well cattle. In cells, glycogen is mainly disposed along the edge of cytoplasm and more often than not does not have a small granulated structure; it is a structureless lyoglycogen. In connection with the accumulation of glycogen in quantity in migrating cells, increased activity of phagocytes is observed related to the specific irritation. While immunizing cattle ill with leukemia, a sharp decrease of the amount of glycogen in migrating cells and the weakening of their phagocytic function was detected. This shows that carbohydrate metabolism in migrating cells of horned cattle ill with leukemia is changed and, as a result, their functional properties are distorted.

The "horn opening" method presents an opportunity to investigate cytochemical transformation of phagocytes with respect to their dynamics related to reactivity of an organism ill with different diseases.

Visualization of Intraneuronal Monoamines by Treatment with a Formaldehyde Solution. The Role of Reaction Conditions

SAKHAROVA, A. V. and D. A. SAKHAROV (Institute of Neurology, Acad. Med. Sci., and Institute of Developmental Biology, Acad. Sci., Moscow, USSR)

A short fixation in an aqueous solution of formaldehyde (F) was reported to be capable of converting intraneuronal noradrenaline into a fluorescent compound (1). Different attempts were made to apply this simple technique to tissue sections (2-4). To develop an effective procedure, it is important to know factors affecting the histochemical reaction. These were studied with the use of the nervous tissue of pulmonate snails, *Lymnaea* and *Helix*. The preparation does not require oxygen and metabolic substrates and thus is more suitable for experimentation than pieces of mammalian brain. Monoaminergic central neurones of snails are known to contain a primary catecholamine, dopamine, and 5-hydroxytryptamine.

Invariably, an isolated ganglionic ring was immersed for a desired period in a F solution made in snail Ringer, then sections were cut in a cryostat at $-15°$ and mounted on slides for examining in a fluorescent microscope. Experiments were carried out to discover the importance of the following conditions: the concentration of F, the duration of the fixation period, the way of drying the sections, the mounting medium (if any), the temperature regime at different stages of the procedure.

The development of fluorescence was shown to be dependent upon both conditions during the fixation and those after drying the section. It was concluded that these two main stages of the procedure correspond to the different steps of chemical reaction.

In the first step, low temperature is essential as a factor preventing the dislocation of amines. The concentration of F by itself is not of great importance provided its decrease is compensated for by prolongation of the fixation.

In the second step, the absence of water is essential though not sufficient for development of fluorescence, the way of drying being of little importance. The reaction is easily controlled by the temperature lasting for many hours under 0° and for 3-4 minutes at 80°. The development of fluorescence can be controlled by mounting medium.

Taking this into account, one can achieve a high intensity and sharp localization of the specific fluorescence when using fairly variable conditions.

Some general conclusions can be drawn. It is known that the F condensation of monoamines is a two-step reaction (5). The advantage of the technique with the use of F solution, as compared with that using F gas (the Falck-Hillarp method), is that the two steps are separated. Hence, each of them can be carried on under conditions far from being critical. This permits the demonstration of enzymes in the same sections. We succeeded in the consecutive demonstration of dopamine and glucose-6-phosphate dehydrogenase.

References:

(1) Eränkö, O. and L. Räisänen: J. Histochem. Cytochem. *14*, 690 (1966)
(2) Laties, A. M., R. Lund, and D. Jacobwitz: Ibid. *15*, 535 (1967).
(3) El-Badawi, A. and E. A. Schenk: Ibid. *15*, 580 (1967).
(4) Sakharova, A. V. and D. A. Sakharov: Citologia *10*, 389 (1968).
(5) Corrodi, H. and G. Jonsson: J. Histochem. Cytochem. *15*, 65 (1967).

Fine Structural Localization of Acetylcholinesterase in Neuromuscular Junctions

SALPETER, M. M. (Cornell University, Ithaca, N. Y.)

Recent studies in our laboratory have been aimed at improving the yield of quantitative information from electron microscope autoradiography. This applies both to the localization of radioactivity (resolution) and to determination of the amount of radioactivity present (sensitivity). A study of acetylcholinesterase (ACh'ase) and other esterases at motor endplates will be used to demonstrate such quantitative procedures.

The active sites of ACh'ase were labeled with H^3-diisopropylfluorophosphate (H^3-DFP) according to the procedure of Rogers *et al.* (Nature *210:*1003, 1966). Glutaraldehyde fixed muscle blocks were first incubated with non-radioactive DFP to posphorylate all active esterase sites. Pyridine-2-aldoxime methiodide (2-PAM) was then used to selectively reactivate active sites of ACh'ase, making them available for a subsequent phosphorylation by radioactive DFP. Esterases other than ACh'ase were labeled by incubating the muscle, first in H^3-DFP and then removing the label from the ACh'ase sites with 2-PAM.

The amount of label found at the endplates was related to molecules of bound DFP, using values for sensitivity previously established (Bachmann and Salpeter, J. Cell Biol. *33*:299, 1967) and the distribution of the label was related to expected grain distribution around radioactive sources of known geometry (Salpeter *et al.*, in preparation).

The Study of Localization of Myosin in Striated Muscle by the Ferritin-Conjugated Antibody Method

SAMOSUDOVA, N. V., M. M. OGIEVECKAJA, and M. B. KALAM-KAROVA (Inst. of Biophysics, Academy of Sciences, Moscow, USSR)

The ferritin-conjugated antibody technique has been used to study localization of myosin in fresh chicken breast muscle. Myosin was prepared by the method of Szent-Georgyi, then purified by repeated precipitation (3 times) at low ionic strength and by additional removal of actomyosin. Ferritin was conjugated to antibodies by a modified Singer method (1). Equimolar ratios of globulin and ferritin were used. To make the method more sensitive, a purified antibody fraction obtained on cellulose absorbent was used instead of the whole gamma globulin fraction. Immunologic analysis of the conjugates was carried out. To improve penetration of the large ferritin-antibody-conjugates, the muscle was excised in strips, tied to rods, treated with digitonin (4×10^{-5}M for 15 minutes) fixed in 1% glutaraldehyde (5 minutes), washed (12 hours) and homogenized. A suspension of myofibrils was incubated with ferritin-conjugate for 1 hour at 37° C. For examination under the electron microscope, the specimens were prepared with the usual techniques. The antimyosin "staining" pattern was studied as a function of sarcomere length. Four "stained" bands are observed in the sarcomere of the relaxed myofibril. The two "stained" regions in the centre of the anisotropic band are seen on both sides of the M-line. In this part of the sarcomere, the distribution of antimyosin coincides with the H-zone borders, where there is no overlap of thin and thick filaments and where, probably the antigenic sites of myosin are not blocked by the interaction with actin. The two other bands are localized in the isotropic band symmetrically to the Z-line, as suggested at the place of N-lines. As the sarcomere contracts, the width of the central "stained" bands decreases corresponding to that of the H-zone. "Stained" lines of the I-band move to the border of the A/I bands. At the complete overlap of H-zone, the antimyosin disappears in the center of the A-band, while the "stained"

234

lines of the I-band are clearly seen. The latter appear to contact with the anisotropic band.

The specificity of the "antigen-antibody" reactions is confirmed by negative control experiments. They were carried out with labelled normal γ-globulin, pure ferritin and also with the myofibrils extracted with Weber-Edsall solution and treated with ferritin-antibody conjugate. The structural organization of the sarcomere will be discussed.

References:

(1) Singer, S. Y.: Nature *183*, 1523 (1959).

Histochemistry of Human and Primate Vascular Tree in Relation to Atherosclerosis

SANDLER, MAURICE and GEOFFREY H. BOURNE (Dept. of Urological Surgery, Upstate Medical Center and Yerkes Regional Primate Center of Emory University)

Histochemical studies in this and other laboratories have outlined the development of the atherosclerotic lesion in man. The problem of experimental studies in man are obvious yet most of the animal species studies have proven to be quite different histochemically from man. This report will compare the histochemical localization seen with the enzymes nucleoside triphosphatase, nucleoside diphosphatase, 5' nucleotidase, esterase, lactic dehydrogenase, DPN diaphorase, succinic dehydrogenase and aminopeptidase in man as compared to sub-human primate species (Chimpanzee, Wooly Monkey, Squirrel Monkey, Marmoset, *Rhesus* and Baboon). Evidence will be presented to show the striking similarity of the chimpanzee to man and the differences between these two species when compared to the others studied. Specific differences noted were the absence of 5'-nucleotidase in certain species, decreased ATPase in others and varying levels of activity noted in the adventitial vessels which, in most studies prior to this have been ignored. Naturally occurring atherosclerotic lesions in the chimpanzee have been studied and the similarities of these lesions to those in the human will be described.

Hydrolysis of Halogen Substituted Carboxylic Acid Naphthylamides by Kynurenine Formamidase

SANTTI, R. S. and V. K. HOPSU-HAVU (Department of Anatomy. University of Turku, Finland)

Halogen substituted carboxylic acid naphthylamides were introduced as histochemical substrates by Gomori (1954) and they have been considered to demonstrate nonspecific amidases. On the basis of rough tissue fractionation studies it was suggested that kynurenine formamidase, an enzyme active in the catabolism of tryptophan may be involved in hydrolysis of this type of substrate, e.g., chloroacethyl-naphthylamide (1). This enzyme has now been purified from guinea pig and rat liver and its capability of hydrolyzing histochemical substrates has been analyzed (2).

Kynurenine formamidase was purified over 200 fold by means of ammonium sulphate fractionation, gel filtration on Sephadex G 100, chromatography on calcium phosphate and DEAE-cellulose columns. The final preparation was shown to contain one main protein fraction and to be free of amino acid acylases and naphthylamidases as well as of several other hydrolytic enzymes. The preparation hydrolyzed readily formylkynurenine, acetyl and chloroacetyl derivatives of aniline and of naphthylamines. Corresponding derivatives substituted with a trihalogen group were hydrolyzed very slowly. The hydrolytic reactions were shown to be activated by the presence of aliphatic alcohols. The formation of esters in the presence of alcohols, hydroxylamine and acyl naphthylamides in the presence of free naphthylamine were also demonstrated. This proves that the enzyme carries also transacylation reactions. The enzyme was optimally active at neutral pH and was inhibited by E 600. The same basic reactions were carried out by the enzyme purified from the rat liver even while the transacylation was carried out to a markedly less degree.

These findings prove that kynurenine formamidase, among others, is capable of readily hydrolyzing halogen substituted acyl-naphthylamides. Since this enzyme is present in the soluble fraction of the liver and thus readily lost from the section, appropriate fixation of the enzymic protein is a prerequisite for a histochemical demonstration of this enzyme.

References:

(1) Hopsu, V. K., R. Santti, and G. G. Glenner: J. Histochem. Cytochem. *14*, 653 (1966).
(2) Santti, R. and V. K. Hopsu-Havu: Hoppe Seyler's Z. physiol. Chem. *in press* (1968).

Electron-histochemical Observation of Glycogen Phosphorylase Localization

SASAKI, MITSUO and TADAO TAKEUCHI (Department of Pathology, Kumamoto University School of Medicine, Kumamoto, Japan)

The histochemical electron-micrography of phosphorylase activity was reported separately by two methods in which the polysaccharide synthesizing technique was used by Sasaki and Takeuchi (1963) and the metal precipitation procedure was applied by Deimling *et al.* (1960),

Hori (1964, 1966), and Dorn (1965). The intracellular localization shown by these techniques, however, did not agree respectively. Even localization of metal precipitates in the latter method was different. Further study using these two methods simultaneously is presented in this paper. Rat skeletal muscles were used in the experiments. Fresh, small pieces of muscle tissue were incubated in the substrate mixture of the polyglucose synthesizing method consisting of 50 mg glucose-1-phosphate, 10 mg muscle adenylic acid, 10 ml 0.2 acetate buffer (pH 5.7), 10 ml distilled water, 2 mg glycogen, 2.1 gm sucrose and 5 ml ethanol. Pieces from the same material were incubated in the other substrate of the lead precipitation method consisting of 100 mg glucose-1-phosphate, 20 mg muscle adenylic acid, 10 ml 0.2 M acetate buffer (pH 5.7), 10 ml distilled water, 20 mg lead nitrate, 10 mg sodium fluoride, 4 mg glycogen, 2.1 gm sucrose and 5 ml ethanol. After incubation at 20° C for 15-30 minutes, the muscle tissues were fixed with 6.5% glutaraldehyde and refixed with 2% osmium tetroxide. Following dehydration with a graded alcohol, they were embedded in Epon 812. Control procedures without substrate were repeated respectively. Before electron microscopic use, the embedding blocks were cut about 1 micron thick on an ultramicrotome for recognizing reaction products by the phosphorylase activity at the light microscopic level.

The newly synthesized polyglucose was confirmed histochemically both with the iodine reaction and by periodic acid Schiff reactions in the synthesizing method. The lead precipitates were observed by ammonium sulfide substitution in the metal precipitation method. After trimming, the same blocks confirmed excellently; ultrathin sections cut by the Porter-Blum ultramicrotome were examined with Hitachi HU-11 A electron microscope.

Polyglucose formed histochemically only by phosphorylase and stained blue with iodine appeared to be fine granules showing about 100-150 A in diameter, different from native glycogen particles of about 300 A in diameter. These granules stained less with lead in electron micrography than native glycogen particles which stained densely with it. They were less resistant to the electron beam than native glycogen. The polyglucose granules were mainly formed in the matrix of intermyofibrillar spaces and in subsarcolemmic areas. Sometimes they appeared in a small numbers among intramyofibrillar myofilaments. No polyglucose granules were synthesized in mitochondria and sarcoplasmic reticulum, and on the myofilaments. They were observed in the same areas containing native glycogen particles.

The finer precipitates of lead were also found in the matrix of intermyofibrillar spaces and the subsarcolemmic areas. The precipitation did not occur in mitochondria and in myofibrils both at the A band level as well as A-I junction level; different from Hori's and Dorn's results. No reaction products were found to be associated with the membraneous structures. The nonspecific, erratic precipitates of lead could appear very often in uncertain locations in the fine structure under the unsuitable conditions.

In our experiments, the reaction products appeared to be formed in the same sarcoplasmic matrix, both in the polyglucose synthesizing method and in the metal precipitation technique.

It could be interpretated by the results that phosphorylase is localized in the sarcoplasmic ground substance which may contain native glycogen particles.

References:

(1) Sasaki, M. and T. Takeuchi: J. Histochem. Cytochem. *11*, 342 (1963); 6th Intern. Congr. Election Micros. (Kyoto) p. 77 (1966).
(2) Hori, S. H.: Stain Tech. *39*, 275 (1964); Acta Histochem. *25*, 111 (1966).
(3) Dorn, A.: Acta Histochem. *21*, 406 (1965).

Localization of Pyroantimonate Precipitate in Mussel Gill Ciliated Epithelium *

SATIR, P. (Dept. of Physiology-Anatomy, University of California, Berkeley, Calif.)

Electron microscopy of gill epithelium of the freshwater mussel (e.g., *Elliptio complanatus*) reveals localized deposits of electron-opaque granules, usually <100Å diameter, along cellular membranes, after fixation with 1% unbuffered OsO_4 containing 1% potassium permanganate. Relatively dense deposition occurs on the cytoplasmic side of the limiting unit membrane and on microvilli of frontal, laterofrontal, intermediate and lateral cells. Deposition may be periodic through the septate junction. The membrane surfaces of certain vesicles of smooth endoplasmic reticulum of the intermediate cell and the matrix sides of the cristae walls in mitochondria also contain precipitate. The ciliary membrane, however, contains fewer deposits than do microvillar membranes or cell junctions. The nuclear membrane, the ground cytoplasm, the nucleus, and the remainder of the preparation are also relatively free of precipitate. Sometimes larger particulate deposits are found in the extracellular ground substance below the epithelium. The amount of overall deposition is variable and depends on treatment prior to fixation; usually, deposition is enhanced by soaking the living epithelium in 40mM NaCl for half an hour. Komnick (1962) and others (e.g. Kaye *et al.*, 1965) have proposed that the fixation used here localizes bound sodium. If there is a common cation binding site on the cell membrane (cf. Carvalho *et al.*, 1963), ciliary membranes may have sites with altered affinities for certain cations, e.g. sodium. This may be a consequence of

238

factors affecting general cation equilibria in the ciliary membrane, as opposed to neighboring microvillar or junctional membranes, and may reflect membrane function in ciliary motility (cf. Naitoh and Yasumasu, 1967).

References:

(1) Carvalho, A. P., H. Sanui, and N. Pace: J. Cell Physiol. *62*, 311 (1963).
(2) Kaye, G., J. D. Cole, and A. Donn: Science *150*, 1167 (1965).
(3) Kommick, H.: Protoplasma *55*, 414 (1962).
(4) Naitoh, Y. and I. Yasumasu: J. Gen. Physiol. *50*, 1303 (1967).

* Supported by grants from the USPHS (GM15859) and American Medical Association Education and Research Foundation.

Enzyme-Histochemical, Biochemical and Autoradiographic Studies in the Course of Cancerization of Rat Liver with Diethylnitrosamine

SCHAUER, ALFRED (Dept. of Pathology, University of Munich, Western-Germany)

At about the 70th day following cancerization of the rat (Sprague-Dawley) with diethylnitrosamine in a dosage of less than 4 mg/kg bodyweight, distinct areas with loss of the histochemical reaction for nucleoside-5'-triphosphatase and glucose-6-phosphatase developed in the liver. On cryostat serial sections it was found that in these areas of enzyme defects the uptake of H_3-cytidine was increased by 2 to 2,5 fold as compared with the surrounding cells. Kinetic autoradiographic studies showed that 45 minutes after administration of radioactive cytidine, the H^3-cytidine labelling was found almost exclusively above the nuclei, while after 140 minutes labelling it was increased considerably in the cytoplasm. In some areas of enzyme defects, the uptake of H^3-thymidine was many times greater than that in surrounding cells. In areas of enzyme defects, storage of glycogen was found; this was lost completely after the cancer developed.

Corresponding areas with loss of enzyme reactions were found in the liver of Wistar-rats after tube feeding with diethylnitrosamine (2 mg/kg bodyweight/die) at about the same time.

In further experiments after application of diethylnitrosamine in the drinking water in the already mentioned dosage, the activity of the nucleoside-5'-triphosphatase was measured biochemically at the isolated and simultaneously electronoptically controlled nuclei of the whole organ.

In these experiments, a decrease of the enzyme activity of the nuclear nucleoside-5'-triphosphatases was found when tumor growth developed.

Euglena Threonine Deaminase: Distribution and Properties

SCHER, STANLEY and PATRICIA L. HALEY (University of California, Berkeley, Calif.)

Threonine deaminase (TD) catalyzes the conversion of L-threonine to α-ketobutyrate. The rate of this conversion is typically under feedback control by the end product isoleucine. However, feedback properties of this enzyme may differ in their sensitivity to isoleucine. We describe here some observations on the distribution of TD in *Euglena* and a mutant strain lacking the ability to form chloroplasts. In addition, we report on some properties of the enzyme from experiments with crude cell extracts.

Log-phase light adapted *Euglena gracilis* z were grown in a synthetic medium at pH 3.5, broken in a French pressure cell, and chloroplasts were isolated by flotation on sucrose. To assay for TD activity, α-ketobutyrate was determined colorimetrically with 2,4-dinitrophenylhydrazine. When various fractions were compared, virtually all of the TD activity was confined to the cell extract (lacking chloroplasts); little or no TD activity was detected in the chloroplast fraction. The absence of TD activity in chloroplasts is consistent with the distribution of this enzyme in mutant strains. Comparable amounts of TD activity were found in wild type and mutant cells unable to develop a normal photosynthetic organelle.

O-methylthreonine (OMT) has been reported to induce permanent loss of chloroplast forming ability in *Euglena*. Since OMT may be considered an analog of both the allosteric inhibitor isoleucine, and the substrate threonine, the effect of this compound was examined on TD. Comparison of threonine and OMT as substrates for the enzyme provide evidence that the catalytic site cannot deaminate OMT to the corresponding α-keto acid; hence TD cannot use OMT as a substrate. Experiments with isoleucine and OMT as feedback inhibitors of TD show that the allosteric site is relatively insensitive to either the end product or the analog. Even at high concentrations, TD activity was only slightly depressed.

The results presented here can be interpreted as evidence that TD activity in *Euglena* is under inefficient feedback control. The absence of TD activity in chloroplasts may reflect the lack of end product control for isoleucine biosynthesis at the organelle level. According to this interpretation, the chloroplast may be dependent upon an extraplastidic control mechanism to regulate the isoleucine pool size. The failure of OMT to strongly influence TD activity suggests that the analog may exert its inhibitory effect on chloroplasts by competing for isoleucine incorporation into plastid proteins.

Circadian and Ultradian Rhythms in Several Physiological Parameters of the Rat *

SCHEVING, LAWRENCE E. and JOHN E. PAULY (Departments of Anatomy, Louisiana State University, New Orleans, Louisiana and the University of Arkansas, Little Rock, Arkansas)

Examples of biorhythms with two different frequencies will be presented. One, the circadian, with a frequency of about 24-hours, the other, a higher frequency or ultradian, with several cycles per day. When biochemical determinations are made of total protein in the plasma of adult male rats sampled at bi-hourly intervals over a 24-hour time scale, an overt rhythmic pattern can be seen. The crest for total protein occurs between 0100 and 0800, the trough between 0500 and 0800. The difference between the highest and lowest hourly mean values obtained were statistically significant. Electrophoresis done on this same plasma demonstrated similar rhythms for albumin, α 1-globulin, α 2-globulin, β globulin and γ-globulin. When the highest mean value for γ-globulin during the 24-hour period (at 0200) was compared with the lowest mean (at 2000) there was a 28% increase. The fluctuation in mucoprotein over a 24-hour period was as great as 40%.

Fluorometric measurements were made of serotonin, dopamine, and norepinephrine extracted from brains which were obtained from different subgroups of adult male rats killed at hourly intervals over three separate 24-hour periods. A synchronized circadian rhythm in serotonin was seen with a crest extending from 1130 to 1730 followed by a trough between 2030 to 2130. Although a circadian rhythm was not detected for dopamine and norepinephrine in the whole brain, the data did suggest that these two catecholamines are characterized by reproducible higher frequency, ultradian rhythms.

In all the above studies rats were maintained under a rigidly standardized routine including a light-dark cycle with 12 hours of light (0600-1800) alternating with 12 hours of darkness. The effect of continuous light and blinding on these rhythms will be discussed.

The time structure of the animal is not only important to bioassay but also one can demonstrate that the ability of an organism to survive a potentially noxious stimulus may depend on time of day the agent is administered. For example, identical doses of strychnine or amphetamine administered to separate subgroups of rats at frequent intervals along the 24-hour scale may kill as many as 80% of the animals at one time of day and as few as 5% at other times.

The importance of recognizing and considering the organism's time structure will be discussed.

---

* Supported by U.S. Public Health Grant 4659.

Problems of Chemodifferentiation

SCHIEBLER, T. H. (Anatomisches Institut der Universität, 87 Würzburg, Germany)

The application of histo- and cytochemical methods in embryology greatly enlarges our understanding of differentiation processes. By histochemical means it is, for instance, possible to detect a presumptive organ anlage at an earlier developmental stage than by morphological methods. Frequently it may be observed that enzymatic differentiation is preceded by morphological changes. This could be due to the fact that the borderline of demonstrability of enzyme activities is reached later than the structural differentiation can be visualized. It is one of the main objectives of histochemistry to investigate the quantitative and qualitative behavior of various histochemically demonstrable substances from their appearance on up to their definite state. As a result of this it can be shown, for instance, that the timing of enzymes is organ-dependent and that the chemical nature of a substance may alter during development (e.g., the acidity of mucous produced in the gall bladder and the intestine) increases in the course of development. Furthermore, histochemical methods provide better facilities to visualize the termination of differentiation than do morphological methods. In some cases, structural differentiation closely parallels chemodifferentiation; in others, the morphological development of an organ is terminated prior to its histochemical differentiation. Another major problem concerns the relation between chemodifferentiation and functional development.

Interference Microscopic Studies of Human Plasmacytoma [*]

SCHIEMER, HANS-GEORG (Institut für Experimentelle und Vergleichende Pathologie der Universität, (69) Heidelberg, Germany)

The fact that patients with plasmacytoma (multiple myeloma) react differently to therapy stimulated our study of six cases of plasmacytoma from the autopsy material of the Heidelberg Institute of Pathology. Besides clinical data, the results of interference microscopy are reported, for which the technique as described by Schiemer (Klin. Wschr. *45*, 393; 1967) was used. There are always patients with multiple myeloma who during chemotherapy with radiomimetic agents die in the first six months from a hemorrhagic diathesis (SN 19/67). The cell analyses with the interference microscope revealed values of nuclear dry-weight and volume in the tetraploid range and average results with the allometric growth studies (analyses of regression). In two patients with longer survival (SN 474/66 a. SN 1130/66), the values were hyperdiploid. The analyses of regression strongly suggested slowly proliferating and synthesizing tumor cells. In two other plasmacytoma patients there were allometric data (see above) which pointed to a high capacity for cytoplasmic synthesis; in addition, there was evidence of an excessive production of pathologic proteins (paramyloid, Bence-Jones protein). The cells of a 65 year old woman (SN 1104/66) with paraproteinosis (BSG 142/145), a carpal-tunnel syndrome, and a plasmacytoma kidney were also studied. She survived her illness only six months. Results of the studies with the interference microscope indicated that the altered cellular function was localized primarily in the cytoplasm. Another patient, a 70 year old woman (SN 652/66) likewise had a short course of her illness; she died because of plasmacytoma kidneys and pneumonia. The interference microscopic studies in this case indicated the cytoplasm of the myeloma cells synthesized abundant amounts of markedly abnormal paraprotein. On the basis of differences in cell morphology and in variations of clinical course of disease, attempts have been made with therapeutic purposes in mind to classify plasmacytomas: 1) those with the principal alteration in the nucleus; 2) those with the primary alteration in the cytoplasm; 3) those with the alterations both in the nucleus and cytoplasm. From such considerations, it seems justifiable to postulate that treatment with radiomimetics may be effective in tumor cells clearly synthesizing DNA. If, however, the biophysical relationships in the nuclei of myeloma cells are constituted like those of normal bone marrow cells, a chemotherapeutic agent which acts only on DNA synthesis will effect lethally normal hemopoietic cells just as it does myeloma cells.

---

* Mit Unterstützung der Deutschen Forschungsgemeinschaft.

Ueber Saure Hydrolasen in den Schwiegger-Seidelschen Kapillarhulsen und einigen Pulpamakrophagen der Milz des Huhnes

SCHLÜNS, JÜRGEN (Institut für Histologie und Embryologie der Freien Universität, Berlin 1 Berlin 33)

Zwei C-N-Bindung spaltende Hydrolasen konnten in den Kapillarhülsen von Huhn und Schwein nachgewiesen werden. Die vorliegende Arbeit prüft, ob bei 8-10 Wochen alten Hähnchen mit diesen Enzymen saure Phosphatase und β-Glucuronidase vergesellschaftet vorkommen. Die Aktivität der asuren Phosphatase erscheint in dens Hülsenzellen bei Anwendung der Methode nach Burstone (Naphthol AS-BI-Phosphat und Echtrotviolettsalz LB) feingranulär. In einigen Hülsenzellen ebenso wie regelmässig im Grenzgebiet zwischen Hülse und Pulpa beobachtet man grosse, stark gefärbte Granula. Individuell unterschiedlich enthält der die Hülse umgebende Pulpamantel ebenfalls derartige Granula. Das pH-Optimum liegt zwischen pH 5-6. Durch 10^{-3}M Na$_2$MoO$_4$ wird das Enzym total gehemmt. $10^{-2}M$ NaF oder $10^{-3}M$ L(+) Weinsäure hemmen stark. Ausser Naphthol-As-BI-Phosphat werden auch $\alpha-$ und $\beta-$ Glycerophosphat in diesen Strukturen umgesetzt. Damit ähnelt das Enzym den bekannten lysosomalen, sauren Phosphatasen und zeigt Parallelen zu dem kürzlich von Chersi et al. (1966) aus der Schweinmilz isolierten Enzym. Für die lysosomale Natur dieses Enzyms spricht ebenfalls die verteilung der β-Glucuronidase (Methode von Hayashi et al. . . . Naphthol AS-BI-β-D-Glucuronid und Hexazonium-pararosanilin), die völlig mit der Lokalisation der sauren Phosphatase übereinstimmt. Es wird vermutet, dass es sich bei den grossen Hydrolasepositiven Granules um Phagolysosomen handelt, die möglicherweise zum Abbau von Erythrozyten in den Hülsen in Beziehung stehen. Aus den Inhibitions-Versuchen ergibt sich, dass die Pulpa der Hühnermilz Zellen mit NaF und Tartrat-resistenter saurer Phosphatase-Aktivität enthält. Es handelt sich um Gruppen von Zellen in der Nähe grosser Arterien oder an der Peripherie von Follikeln. Sie enthalten gelbe Schollen unterschiedlicher Grosse, die hellgelbe Primärfluoroszenz zeigen, eine positive Sudanschwarz B-Reaktion geben und Berlinerblau sowie Turnbullblau negativ sind. Beim Nachweis der sauren Phosphatase nach Burstone färben sich die Zellen homogen rot an. Das pH-Optimum liegt um pH 5,0-5,2 bei Zusatz von $10^{-3}M$ L(+) Weinsäure zum Inkubationsmedium. Das Verhalten des Enzyms bzw. der Enzyme gegenübei anderen Substraten ist uneinheitlich. Während in manchen Zellen neben Naphthol AS-BI-Phosphat auch $\alpha-$ und $\beta-$Glycerophosphat umsetzt werden, findet man andere Zellen, die vorwiegend Naphthol AS-BI Phosphat umsetzen. Das in diesen Chromolipid-Makrophagen vorkommende Enzym verhält sich ähnlich wie die kürzlich von Maggi et al (1967) beschriebene saure Phosphatase.

References:

(1) Chersi, A., A. Bernardi, and G. Bernardi: Biochem. Biophys. Acta *129*, 12 (1966).
(2) Maggi, V., L. M. Franks, and A. W. Carbonell: Histochemie *6*, 305 (1966).

Ultrastructural Localization of Acetylcholinesterase Activity in Tetrahymena pyriformis (W) *

SCHUSTER, F. L. and B. HERSHENOV (Department of Biology, Brooklyn College, Brooklyn, N. Y.)

Axenic stocks of the ciliated protozoan *Tetrahymena pyriformis* (W) were examined for acetylcholinesterase (AChase) activity. Cells were fixed in glutaraldehyde and treated following Barrnett's (1) method for AChase localization. Preparations treated with eserine ($10^{-4}M$) prior to

244

incubation in the reaction mixture served as controls. In experimental cells, reaction product was found localized in the ciliary shafts, in the area surrounding the axonemal microtubules. No consistent pattern of staining was observed at the level of kinetodesmal fibers and/or kinetosome-associated microtubules in the pellicle. Sub-pellicular deposition of reaction product of an apparently non-specific nature occurred in mitochondria in both experimental and control cells.

As an alternate method for demonstration of AChase, H³-diisopropylfluorophosphate (10^{-4}M) was employed following the technique of Salpeter (2). In experimental preparations, label uptake appeared confined to the sub-pellicular area, though not associated with either mitochondria or fibrillar elements. The significance of these results in terms of a possible role of AChase in ciliary coordination will be discussed.

References:

(1) Barrnett, R. J.: J. Cell Biol. *12*, 247 (1962).
(2) Salpeter, M. M.: J. Cell Biol. *32*, 379 (1967).

* Supported by NIH research grant GR-14882.

Dye-polyanion Interactions in Histochemistry

SCOTT, J. E. (Medical Research Council Unit, Canadian Hospital, Taplow, Bucks., England)

The interaction of cationic dyes with polyanions is histochemically more valuable if use is made of (a) dye metachromasia in the presence of some substrates, or (b) selective competitive displacement of the dye from the substrate. Displacement by salts at "critical electrolyte concentrations" (CECs) (1) is the preferred method of (a) distinguishing between types of acid groups, (b) increasing the specificity of a staining procedure, and (c) selecting the best dye to demonstrate a particular substrate.

Cationic dyes are of two kinds (2), (I) with non-polarising charges (e.g., Alcian blue, methyl green, etc.) which show no particular preference for a type of acid group, and (II) with polarising charges (e.g., chelated Al^{+++} attached to haemotoxylin) which bind most strongly to more easily polarisable negative groups (e.g., phosphate esters or carboxylates). The shape and size of the dye is important in non-electrostatic strong short-range binding, which can impart extra specificity (3). Alcian blue is unique among dyes so far tested in that binding is primarily electrostatic, and it can therefore be used to distinguish between types of charged group without excessive ambiguity due to non-electrostatic binding. Results on cartilage using the CEC approach (4) have been correlated with enzymic and microchemical analyses, with good agreement. New analogues of Alcian blue have been synthesized pure and with known chemical structure (Scott and Silveyra, unpublished).

References:

(1) Scott, J. E. and J. Dorling: Histochemie 5, 221 (1965).
(2) Scott, J. E.: Biochem. J. 99, 3P (1966).
(3) Scott, J. E.: Histochemie 9, 30 (1967).
(4) Scott, J. E. and R. A. Stockwell: J. Histochem. Cytochem. 15, 111 (1967)

Feasability of Quantitating 5′-Nucleotidase Histochemically

SCOTT, T. GILBERT (Institute of Neurology, Queen Square, London, W.C.1.)

There are many theoretical considerations to be taken into account in using a slide photometric technique for the assay of an enzyme by histochemical methods. Important among these are variations in section thickness, the stoichiometry of the reactions, the solid and particulate nature of the reaction products, the non-uniform distribution of the enzyme across the incident light path and the non-specific deposition of reaction products over and above that deposited as a result of specific enzyme activity.

A quantitative method for the assay of 5′-nucleotidase in brain by slide photometry is described. Control biochemical assays of the enzyme have been made on homogenates of sections adjacent to those used for the histochemical assay. Statistical regression analysis of the results showed that the histochemical method of assay correlated highly significantly with the biochemical method ($r = + 0.890$, $p < 0.001$). In a series of 100 consecutive measurements of 5′-nucleotidase activity in mouse brain, the variability between the readings was found to be sufficiently low (mean $= 0.456$, standard deviation $= 0.0474$, coefficient of variation $= 10.51\%$) as to warrant the entire technique being investigated further. In addition, it was found as a result of these preliminary studies that there was a measurable amount of non-specific deposition of the final reaction product. It is suggested that this non-specific deposition might act as a significant source of error in the interpretation of both semi-quantitative and qualitative studies of the enzyme. This non-specific effect, however, will not act as a significant source of error in the present quantitative method since there is a simple way of obviating it. In applying this correction for the non-specific effect to the previous results, there was found to be an increase in the correlation coefficient (and statistical significance) of the histochemical method.

These preliminary results are therefore taken to imply that, in spite of all the theoretical considerations, the histochemical method for the assay of brain 5′-nucleotidase is, in practice, a valid one.

The Ultrastructural Localization of Parts of the Respiratory Chain on the Outer Surface of the Inner Mitrochondrial Membrane: Implications for Mitchell's Chemiosmotic Theory

SELIGMAN, ARNOLD M. (Departments of Surgery, Sinai Hospital of Baltimore and The Johns Hopkins University School of Medicine, Baltimore, Md.)

New methods for the ultrastructural chemical demonstration of succinic dehydrogenase activity (SDH) and cytochrome oxidase activity (CO) have been developed in our laboratory. Study of the mitochondrial membranes of rat heart and liver reveals that both SDH and CO are located on the outer surface of the inner mitochondrial membrane. Since the histochemical reagents for these enzymes serve different functions, i.e., in the case of SDH the reagent is a hydrogen acceptor and in the case of CO the reagent is a hydrogen donor, the substrates for SDH and SO (cytochrome c) received hydrogen from opposite surfaces of the cristal membrane and discharge their electrons on opposite sides of the cristal membrane. These results are consistent with Mitchell's chemiosmotic hypothesis. Final histochemical proof must await development of independent methods for cytochromes a and a_3.

Mechanism of Phago —and Pinocytosis

SENO, SATIMARU, EI-ICHI YOKOMURA, NOBUTAKA ITOH, and MICHIO YAMAMOTO (Dept. of Pathology, Okayama Univ. Med. School, Okayama, Japan)

The phenomenon of phagocytosis or pinocytosis of macromolecules is of special interest in relation to ion transport through membranes, be-caused it may give an enlarged picture of the membrane challenged by a single ion since macromolecules have a number of ionized groups on their surfaces and act as a mass of ions.

Observations on macrophages from mouse tumor ascites, after being exposed to metal colloid particles of negative charge, revealed that engulfing of the cell surface is induced by adsorption of the colloid particles being accompanied with or without pseudopod formation and being supplied with the energy from anaerobic glycolysis. If the phago-cytic vesicle was once established, the colloid particles adhered to the cell surface and are transferred and accumulated in the vesicle through the canals or tubules connecting the vesicle and the cell surface. Transportation of the particles may be accomplished by membrane flow, the sliding of the outer half of the bimolecular lipid layer.

In the vesicles, the outer layer of the membrane is dislodged, by which the molecular structure of the membrane should be changed. This may facilitate the permeation of ions and molecules. By using iron and platinum colloid particles and basic proteins including peroxidase, the processes of phago- and pinocytosis was analyzed. Cytochemical techniques serve for the demonstration of the cell membrane and the protein molecules.

Study of the Metabolic Level of Cultivated Nerve Cells at Rest and Under Functional Load

SHUNGSKAYA, V. E., S. O. ENENKO, L. D. LUKYANOVA, and E. N. SARCH (Institute of Biophysics, Academy of Sciences, Moscow, USSR)

Explants from cerebellum, spinal cord and sympathetic ganglia of adult rabbit were cultivated by roller tube method in a medium consisting of serum of human placental blood (60%) and Tyrode solution (40%) with the addition of cortisone acetate. The medium was saturated with a mixture of oxygen (95%) and carbonic acid (5%).

Some aspects of nerve cell metabolism of adult mammals were studied in tissue culture in the normal state and under functional load. The activity of the oxidative enzymes succinate dehydrogenase and cytochrome oxidase was investigated by histochemical reactions. Polarographic study of oxygen absorption by explants, detailed investigations of acetylcholinesterase of cultivated cells and their nucleoprotein metabolism were carried out for two weeks. To determine the succinate dehydrogenase activity, the reaction of Nitro BT according to Nachlas was used. In cerebellum Purkinje cells and in nerve cells of spinal ganglia in particular, high activity of the enzyme was observed at the times studied. Cytochrome oxidase was studied according to Nachlas, high enzymatic activity observed for a long time being revealed as granules of indophenol violet. The polarographic technique was used for determining the quantity of oxygen absorbed by the cultivated nerve tissue. It was possible to calculate relative velocity of O_2 consumption of 1 mg of tissue per minute, and to study the respiratory rhythm of nerve cells at various days of cultivation (intensification, weakening). Different respiration rhythm was shown to be in different nerve cell types.

The cholinesterase activity of cultivated nerve cells was studied according to the Koelle technique in a Gomori modification on frozen sec tions. It was found in all cultures as granules, located in the cells. In the cerebellum, the enzymatic activity is localized mainly in the granular layer, while it is absent in the Purkinje cells of the ganglion layer. In the spinal ganglion AChE-activity is localized in nerve cells, in some cases peripheral localization of granules is observed. High enzymatic activity is found in the sympathetic node. Nucleoprotein metabolism was studied by quantitative ultra-violet cytophotometry on spinal ganglion cells. RNA content in cultivated nerve cells was shown to decrease, compared with that in intact animal, and gets lower at cultivation days.

The effect of serotonin and strychnin (10/ml medium) on the respiration of cultivated cells was studied. The increase in oxygen absorption was shown to proceed at the decreased level respiration of the explants. The reaction of the explants from different types of nerve cells is different.

Central Nervous System Glucosaminidase and Galactosaminidase

SHUTER, ELI, FIROZE JUNGALWALA, and ELI ROBINS (Washington University Medical School, Department of Psychiatry, St. Louis, Mo.)

Glucosaminidase and galactosaminidase of the central nervous system have been studied as a continuation of previous investigations of β-glycosidases and because of their possible relationship to the metabolism of glycolipids.

Glucosaminidase and galactosaminidase catalyze the hydrolysis of glycosidic bonds containing the respective amino sugars. Their activity was assayed spectrophotometrically using para-nitrophenyl-N-acetylglucosaminide and para-nitrophenyl-N-acetylgalactosaminide, and fluorometrically with the corresponding 4-methylumbelliferone glycosides. The latter method was adapted for quantitative histochemistry.

Purification of these enzymes has disclosed that upon DEAE-cellulose chromatography of the 25-50 percent ammonium sulfate precipitate of rabbit brain homogenate, 14 per cent of the glucosaminidase and galactosaminidase activity is adsorbed and 86 per cent unadsorbed when the column is washed with 5 mM tris-PO_4 buffer pH 7.2. Further purification of the DEAE-cellulose unadsorbed enzyme on a Sephadex G-200 column, and of the DEAE-cellulose unadsorbed enzyme on a carboxymethyl cellulose column has yielded fractions in which enzymatic activity is enhanced five hundred fold. However, within these purified fractions, activity with glucosaminide and galactosaminide substrates is not separated and remains proportional to that of the original brain homogenate. Therefore, it appears that there may be only two enzymes with glucosaminidase and galactosaminidase activity in brain, one of which is adsorbed upon a DEAE-cellulose column, and one of which is not. Further purification of these adsorbed and unadsorbed fractions has not resulted in the differentiation of glucosaminidase and galactosaminidase activity into separate and distinct enzymes. Work is in progress utilizing additional chromatography and electrophoresis to separate these activities in both the partially purified adsorbed and unadsorbed DEAE-cellulose fractions.

Quantitative histochemical studies of glucosaminidase and galactosaminidase activity of cerebellar layers have been completed. In the rat, glucosaminidase activity using 4-methylumbelliferone-N-acetylglucosaminide as substrate was 493 mM/K dry weight/hr in the molecular layer, 678 in the granular layer, and 244 in white matter. With para-nitrophenyl-N acetylglucosaminide, the activity was 358, 512, and 230 in the respective layers. Activity of galactosaminidase, on the other hand, using para nitrophenyl-galactosaminide as substrate, was 43.8 mM/K dry weight/hr in the molecular layer, 44.5 in granular layer, and 37.5 in white matter. We believe that the relatively greater activity of glucosaminidase than galactosaminidase in the granular layer, when compared with activity in the other layers, gives support to the argument that glucosaminidase and galactosaminidase are distinct enzymes, despite the failure to sepa rate them in either the Sephadex G-200 fractions of the DEAE-cellulose adsorbed enzyme or the carboxymethyl cellulose fractions of the DEAE cellulose unadsorbed enzyme. Accordingly, it should be possible to separate glucosaminidase and galactosaminidase by methods of protein purification discussed above. In the rabbit, distribution of glucosaminidase and galactosaminidase in cerebellar layers is similar to that of the rat, except that galactosaminidase activity is greater in the granular layer than molecular layer or white matter.

The significance of these results in relation to previous studies of β-glycosidases, the functional lysosomal hypothesis, and glycolipid metabolism will be discussed.

Cytochemical Localization of Alkaline Ribonuclease and Phosphodiesterase I [*]

SIERAKOWSKA, HALINA (Institute of Biochemistry and Biophysics, Polish Academy of Sciences, Warsaw, Poland)

The cytochemical localization of phosphodiesterase I and alkaline RNase by the azo-dye coupling technique has been extended by further

250

detailed studies on normal rat tissues and human breast carcinomas and fibroadenomas. Activities of both enzymes in cryostat unfixed tissue sections were abolished by formalin-calcium fixation but survived similar fixation of tissue blocks. Localization patterns for both activities were supplemented by studies on fractionated rat liver cells, in particular the distribution within mitochondria. Phosphodiesterase I has been localized in the external, but not internal, mitochondrial membrane (in collaboration with M. Erecinska).

An improved procedure has been developed for the synthesis of α-naphthyl uridine-3'-phosphate, a substrate for RNase localization. The specificity of RNase localization with this substrate has been independently confirmed by the preparation of α-naphthyl inosine-3'-phosphate, which is susceptible to RNases T_1 and T_2. No activity against the purine substrate could be detected in rat tissues. The cytochemical localization obtained with α-naphthyl uridine-3'-phosphate must there-fore be due to alkaline RNase similar in specificity to the bovine pancreatic enzyme.

The specificity of phosphodiesterase I localization with α-naphthyl thy-midine 5'-phosphate was further established by the inhibitory effect of EDTA at pH 5-7 on the residual activities of rat liver, kidney and duodenum.

---

* Supported by the Wellcome Trust, the World Health Organization and the International Atomic Energy Agency.

Measurement of Protein in Sections by Quantitative Electron Microscopy

SILVERMAN, L. and D. GLICK (Stanford University School of Medi-cine, Palo Alto, Calif.)

The method of quantitative electron microscopy established by Bahr and Zeitler and others for mass measurement of whole biological parti-cles has been developed for use on thin sections. Normal human red blood cells were selected as test objects because of their relatively uniform and well defined concentration of hemoglobin. They were fixed in a solution of 1% paraformaldehyde in 0.1 M potassium phos-phate buffer, pH 7.4, washed in buffered 7.5% sucrose and stained in 5% phosphotungstic acid (PTA) in 6.25% Na_2SO_4. The red cells were then washed in 2% ammonium acetate with formic acid added to bring pH to 2.0, dried *in vacuo* over silica gel and embedded in Epon-Araldite. Shape of the cells was well maintained and their diameters were measured. Values for the hemoglobin, packed volume and diameter of the cells in buffered (pH 7.4) physiological saline were measured.

As a standard for section thickness, spheres of Dowex anion exchange resin (15-50 μ diameter), stained with 5% PTA at pH 6.4, were embedded with the red cells. The stoichiometry of the reaction between PTA and both Dowex resin and red cell protein was determined. Section thickness in the range of 500Å-700Å, and mass of protein per unit area were measured by quantitative electron microscopy, and the protein concentration was calculated to be 30 grams per 100 ml of red cell volume. The concentration derived from clinical laboratory methods, after correction for volume change, was 32.6 grams per 100 ml of packed red blood cells.

A Reduction in Ploidy Associated with the Neoplastic Transformation in Rat Liver

SIMARD, A. and R. DAOUST (Laboratoires de Recherche, Institut du Cancer de Montréal, Hôpital Notre-Dame et Département d'Anatomie, Université de Montréal, Montréal, Canada)

Radioautographic studies have indicated that a *prolongation* of the G_2 phase of the cell cycle occurs in rat hepatocytes following the chronic administration of the carcinogenic agent 4-dimethylaminoazobenzene (DAB). Later, however, a *reduction* of the G_2 phase was observed in sites of neoplastic transformation and in the induced hepatomas. These results suggested that removal of both the G_1 and the G_2 blocks contribute to the acceleration of cell proliferation in sites of tumor formation.

To verify the conclusions drawn from radioautographic studies, cytophotometric determinations of nuclear DNA contents were carried out in the livers of DAB-fed animals. Higher levels of ploidy appeared in liver parenchyma during DAB feeding and, at later stages, a reduction to lower levels occurred in hyperbasophilic foci considered as the sites of neoplastic transformation. Several hepatomas showed a low level of ploidy comparable with that observed in hyperbasophilic foci while some were highly polyploid and aneuploid. These results thus agree with the radioautographic data: a) an arrest or delay in the G_2 phase results in the accumulation of hepatocytes with higher DNA contents and b) following removal of the G_2 block in hyperbasophilic foci, the hepatocytes pass directly into mitosis and give rise to daughter cells of a lower degree of ploidy. On the other hand, the high degree of polyploidy and aneuploidy commonly observed in tumors does not seem to be directly associated with the neoplastic transformation but apparently results from tumor growth.

Serum Aminopeptidase in Pregnancy Hydatiform Mole and Choriocarcinoma

SMITH, EDGAR E. and ALEXANDER M. RUTENBURG (University Hospital, Department of Surgery, Boston University Medical Center, Boston, Mass.)

The activity of the L-leucyl-beta-naphthylamide (LNA) hydrolyzing enzymes in serum increases markedly during pregnancy (1,3). Electrophoretic studies have shown this increase in activity to be due to the appearance in serum of a new enzyme (2,4). The present report deals with the differential effect of L-methionine on the serum LNA hydrolyzing enzymes in pregnancy, hydatiform mole, and choriocarcinoma.

The hydrolysis of LNA by the enzymes in normal early pregnancy, hydatiform mole (7 cases) and choriocarcinoma (4 cases) serum was inhibited in excess of 90% by 0.1 M L-methionine. The magnitude of this inhibition dropped progressively to about 50% from the nineteenth week of pregnancy to term.

Electrophoresis on starch gel showed that the normal serum enzyme was markedly inhibited whereas the new slower moving pregnancy enzyme was only slightly inhibited by 0.1 M L-methionine. The pregnancy enzyme was absent in the serum of all patients with choriocarcinoma and in 72% of those with hydatiform mole. Therefore, its demonstration on starch gel in early pregnancy (9 weeks) by means of differential methionine inhibition may prove useful in the early differentiation of the latter disorders from pregnancy. More data are required to confirm these findings.

References:

(1) Goldbarg, J. A. and A. M. Rutenburg: Cancer *11*, 283 (1958).
(2) Page, E. W., M. A. Titus, G. Mohun and M. B. Glendening: Am. J. Obst *82*, 1090 (1961).
(3) Siegel, I. A.: Obst. Gyn. *14*, 488 (1959).
(4) Smith, E. E., E. P. Pineda, and A. M. Rutenburg: Proc. Soc. Exp. Biol. *110*, 683 (1962).

Development of a Technique for Light and Electron Microscopic Localization of Hydrolytic Enzymes Using Substituted Naphthol Substrates and p-(acetoxy-mercuric) Aniline Diazotate as the Capturing Agent **

SMITH *, R. E. and WILLIAM H. FISHMAN (Dept. of Pathology, Stanford Univ.—VA Hospital, Palo Alto, Calif. and Dept. of Pathology (Oncology), Tufts Univ. School of Medicine, Boston, Mass.)

One major objective of enzyme cytochemistry is to be able to precisely localize a variety of acid hydrolytic enzymes within cell organelles in order to identify differences and interrelationships in their subcellular localization. Within a single organelle, the lysosome, there are at least eleven acid hydrolases other than acid phosphatase. However, much of our present concept of lysosomes and their function is based on electron microscopic studies of the distribution of this "single enzyme," processed by the lead-salt (β-glycerophosphate) technique of Gomori. Clearly, there is a need to develop alternative approaches to the visualization by electron microscopy of organelle sites of hydrolase activity. We have been attracted to the possibility of utilizing a substituted mercurial diazotate to react with the substituted naphthol product of cleavage by specific hydrolases of the corresponding substrates, e.g., the glucosiduronic acid, phosphate, acetate, or sulfate conjugate. Among the desirable advantages of this approach are the highly substantive properties of the naphthol AS-BI, the opportunity to eliminate the variable of the nature of radicals released in the study of various hydrolases, and the necessity to develop only one technique for visualizing the electron density of mercury-colloid which would then apply to all hydrolases.

The technique presently is: To fix thin slices of tissue in 1.5% singly distilled glutaraldehyde containing 1% sucrose, buffered at pH 7.4 with .067M cacodylate, or in cold formaldehyde-calcium prepared from para-formaldehyde. Frozen and nonfrozen sections not exceeding 15 and 20μ are used.

A post-coupling procedure has given the best results. Sections are incubated for one to several hours in media containing substrate—no dye or metal. After a brief wash, the released naphthol is rapidly captured (two minutes) using a diazotate of p-(acetoxymercuric) aniline, forming an insoluble, intensely-red, pigment product at enzyme sites visible in the light microscope. Sections for electron microscopy are subsequently reacted with slightly alkaline (pH 7.4) 1% thiocarbohydrazide (T) at 25° C. An organomercuric sulfide is formed which decomposes on standing or heating, giving mercuric sulfide. If the heating is carried out in the presence of a second more stable metal (2% $Pb(NO_3)_2$ or 2% OsO_4) 45° C, a HgTMe reaction product of increased electron density and stability is formed. Sections are then dehydrated through alcohols, embedded in an epoxy resin, and thin sectioned.

With various experimental conditions, reaction product of acid phosphatase has been found in the endoplasmic reticulum, Golgi, and lysosomes of mouse liver. Injections of T1339 into rats causes a marked increase of acid phosphatase in liver lysosomes. In the livers of C_{57} mice, β-glucuronidase has been observed, but not in the livers of C_3H mice, a low tissue β-glucuronidase strain. Particular attention has been given to the rat preputial gland and to the mouse renal β-glucuronidase response to testosterone.

---

* Dernham Fellow, California Division, American Cancer Society.
** Supported by grants from American Cancer Society (E-415) and National Aeronautics and Space Administration (MGR 05-202-177).

The Effects of Chlormadinone Acetate Application and Withdrawal on Acid Phosphatase Activity in the Endometrium of the Estrogen Primed Ovariectomized Rabbit *

SMITH, R. E. and MILAN HENZL (Department of Pathology, Stanford University—VA Hospital, Palo Alto, Calif.)

Chlormadinone acetate is a derivative of 17-α-hydroxyprogesterone with a chlorine atom attached at the C6 position of the steroid nucleus. The compound is a highly effective progestational agent without estrogenic or androgenic properties. It is presently used in physiological fertility control. Clinical data are in agreement with experimental studies that as an oral contraceptive it is effective, while unpleasant side effects are notably less common.

Previous investigations of fine structure and AcPase localization have demonstrated that physiological levels of estrogen perpetuate a developed Golgi complex with increased enzyme activity and organization of lysosomes in the endometrium of the ovariectomized rabbit. In this report we present our findings as to changes in subcellular structure and AcPase localization during chlormadinone-induced secretory transformation, and regression following hormone withdrawal. The aim of the investigation was to establish enzymorphological patterns of AcPase activity to aid in a more complete understanding not only of steroidal contraception, but to the basic principles in endometrial physiology.

Young, mature New Zealand rabbits were used throughout the study. After 20 days castration, each animal was appointed to one of four groups: (1) controls, castrates with no medication; (2) ovariectomized animals treated with estradiol only for nine days; (3) ovariectomized animals treated with estradiol for nine days and the last three days simultaneously with chlormadinone; and (4) ovariectomized animals administered estradiol for 15 days with chlormadinone given simultaneously on the seventh, eighth, and ninth days. Estradiol was administered by injection subcutaneously as a microsuspension in methylcellulose —1 μg/Kg body weight. Chlormadinone acetate in methylcellulose was administered by lavage in a dosage of 25 μg/Kg body weight. In groups 3 and 4, animals from each group were randomly selected and sacrificed at 12, 24, 48, 96 hours, and six days after last medication. At sacrifice, uteri were removed immediately and thin segments of uterine tissue were fixed by immersion for five hours in cold 1.5% singly distilled glutaraldehyde. Nonfrozen sections (~30μ) were prepared, using a Sorvall TC-2 tissue sectioner. Acid phosphatase was demonstrated by the Barka and Anderson modification of the Gomori lead salt technique. Cytochemical controls included the incubating of sections in media without substrate with lead present, and heat inactivated sections in complete media. Sections were subsequently postfixed in 1% OsO₄ phosphate buffered, dehydrated, and embedded in Araldite. From Araldite blocks 1μ sections were cut, fixed to glass slides, treated with dilute ammonium sulfide, stained with crystal violet, and studied by phase contrast microscopy. Such preparations permitted semiquantitative evaluations of enzyme reactive material and aided in selecting tissues for thin sectioning.

(1) Castrates without medication have small Golgi systems in regressive epithelial cells of the endometrium. AcPase reaction product was found deposited as dense patches intermittently along the first to second innermost Golgi cisternae. Lysosomes containing coarse precipitates of reaction product and well-defined limiting membranes were found randomly distributed throughout the cells.

(2) In the estradiol-only stimulated endometrium, enlarged epithelial cells generally have a Golgi complex with numerous short and compact cisternae, of which one to three show a continuous deposition of reaction product. AcPase reactive Golgi vesicles and multivesicular bodies were frequently observed. Lysosomes were more numerous than in the castrate, somewhat larger, but more important they contained reaction product to a point of homogeneous lead-salt deposition.

(3) Chlormadinone acetate increased the length of Golgi cisternae, the number of Golgi vesicles, and intensity of enzyme localization within the total complex. From 24 hours to six days withdrawal, a progression of regressive enzymorphological events were distinctly observed: (a) a general decrease in size and localization of AcPase in the Golgi complex; (b) an increase first in number and then in size of lysosomes; and (c) the appearance of large AcPase-reactive autophagic bodies, and lysosomes with apparent membrane rupture and diffusion of enzyme.

(4) In animals where estrogen treatment was continued for six days beyond chlormadinone, the Golgi complex in size and enzyme localization more closely paralleled that seen in epithelial cells of estrogen-only treated animals. Lysosomes were generally more numerous than in estrogen-only treated animals, yet they were well delimited and densely reactive. Autophagic bodies were significantly less frequent and ruptured lysosomes absent.

Our observations presently suggest that chlormadinone acetate has a labilizing effect on endometrial cell lysosomes and that estrogen, if not a stabilizing effect, reduces the effect of chlormadinone. These considerations are being approached through biochemical investigations.

* Supported by grants from Syntex Research, and American Cancer Society (E-415).

Endoplasmic Reticulum Localization of Cytochemical Markers of Various Organelles: a Light and Electron Microscopic Study [*]

SOBEL, HAROLD J. and ERNA AVRIN (Beth Israel Hospital and the Max Wachstein Research Laboratory, Passaic, N. J.)

The thiamine pyrophosphatase (TPPase), inosine diphosphatase (IDPase), acid phosphatase (AcPase), α-hydroxy acid oxidase and DPNH reductase procedures were applied to frozen sections of formol-calcium fixed rat kidney for study with the light microscope. Both nitro-BT and tetranitro-BT were used as electron acceptors with the latter two methods, and AcPase activity was demonstrated by both the Gomori and Barka-Anderson procedures. The TPPase, IDPase and Gomori AcPase methods were applied to glutaraldehyde fixed tissue for study with the electron microscope. Acid phosphatase preparations were also studied following the intravenous administration of horseradish peroxidase.

In addition to the organelles expected to stain (Golgi apparatus and plasma membranes, TPPase and IDPase; lysosomes, AcPase; peroxisomes, α-hydroxy acid oxidase; mitochondrial, DPNH reductase), the endoplasmic reticulum (ER) was visualized with each of the procedures used. The light microscopic staining pattern was interpreted as indicating ER activity when the nuclear membrane and cytoplasmic strands and/or clumps continuous with the nuclear membrane were visualized. AcPase activity in the ER was most distinct and widespread following administration of horseradish peroxidase, paralleling increased AcPase activity of renal homogenates. ER staining was not observed with any of the procedures used when enzymatic activity was abolished with heat or specific inhibitors and in the absence of substrate.

These observations may be explained by the synthesis of these enzymes in the ER enabling visualization of activity in fortuitous areas with sensitive cytochemical procedures, and reinforces the thought that these organelles are interrelated. Although the concept of organelle markers is useful, it is once more apparent that they cannot be relied upon to any great extent. The importance of careful light microscopic observations and correlation of these findings with electron cytochemistry is stressed.

* This work has supported by research grant HE-05950 of the National Heart Institute.

Simultaneous Coupling Azo-dye Method in Submicroscopic Localization of Esterases at Myoneural Junction [*]

SONG, SUN K. and PAUL J. ANDERSON (Mount Sinai School of Medicine, New York, N. Y.)

Attempt was made to obtain a precise submicroscopic localization of esterases at the myoneural junction with alpha naphthyl acetate and indoxyl acetate as substrate and the hexazonium pararosanilin as capturing agent. Rat gastrocnemius, soleus and diaphragm were studied, using (1) cryo-protected cryostat sections, (2) unfrozen tissue-chopper sections, and (3) block incubation technic. All sections showed consistent and reproducible results with sufficient contrast to evaluate localization of the reaction product at a submicroscopic level. The final reaction product was visualized as an amorphous non-crystal electron-dense material distributed over the muscle plasma membrane covering the junctional folds. A portion of the plasma membrane of the axon terminal over the junctional folds also contained reaction product but none was observed on synaptic vesicles or subsarcolemmal cytoplasmic organelles. In addition, two other technics using metallic salts as a final product, (1) acetylthiocholine with copper thiocholine sulfate or copper ferrocyanide, and (2) thiolacetic acid with lead nitrate as final products, were tested and compared with the azo-dye method.

---

* Supported by USPHS Grant No. NB 05041.

Acid Phosphatase Activity of Spherosomes in Guard Cells of Campanula persicifolia

SOROKIN, HELEN P. and SERGEI SOROKIN (The Radcliffe Institute, Cambridge, Mass., and Department of Anatomy, Harvard Medical School, Boston, Mass.)

Acid phosphatase activity fluctuates in mature epidermal guard cells of *Campanula persicifolia* when plants are exposed to varying light conditions. As demonstrated both by Gomori's method and by Burstone's simultaneous coupling azo dye technique in both unfixed and aldehyde fixed epidermal strips, the enzyme is active in the spherosomes when the guard cells are flaccid and the stomata are either partially or fully closed. It is inactive when the cells are turgid and the stomata are fully open, regardless of whether opening had resulted from a photoactive or scotoactive process. In living preparations, such inactive spherosomes will gradually become active for acid phosphatase if turgid guard cells are subjected to partial plasmolysis. Marked activity for acid phosphatase becomes manifest in the vacuoles as well as in the spherosomes of the guard cells if plants are kept in the dark for many days. In senescent leaves, the vacuoles of the guard cells are reactive even though the plants are grown in the light. The walls of the epidermal cells exhibit sporadic activity for the enzyme, but the walls of guard cells have not been observed to react. Nuclear staining sometimes is present after tissues are incubated in the Gomori medium, but it is considered to result from non-enzymic binding of lead on the nuclear surface. The spherosomes evidently are one of the principal sites for acid phosphatase activity in plant cells. In guard cells, the spherosomes are considered to be enzymatically active when the cells are in a relatively catabolic phase of metabolism and to become inactive when the cells enter a relatively anabolic phase.

Histochemistry of Organogenesis in the Mammalian Lung

SOROKIN, SERGEI P. (Department of Anatomy, Harvard Medical School, Boston, Mass.)

The lung is organized around its airway, which becomes developed from a midline ventral pharyngeal bud into an arboreous system. In this system, two functionally distinct regions are recognized, the terminal buds and the airway previously laid down. Expansion and branching of the airway occur in the terminal buds principally because its epithelium divides more rapidly than the surrounding mesenchyme. Histochemically this epithelium is characterized by glycogen storage, cytoplasmic basiphilia and usually alkaline phosphatase activity. Cell division declines and branching normally ceases to occur in the epithelium previously laid down. Such cells lose their glycogen but increase in enzymic activity associated with aerobic oxidative pathways. These and other changes at first appear in the tracheal epithelium. They gradually spread distally so that a gradient of maturity extends between the trachea and the terminal buds. The surrounding mesenchymal investment also becomes differentiated centrifugally. Consequently, the operation of one continous morphogenetic process serves to establish the basic structure of the trachea, bronchi and bronchioles. As branching proceeds, however, certain details gradually become modified. The walls of the bronchi progressively become thinner. Eventually cells that are unique to the lung (Clara cells, great and small alveolar cells) arise in the epithelium. Moreover, once alveoli begin to develop, some of the previously formed bronchioles undergo remodeling to permit alveolarization of their walls. Histochemical studies have been valuable in elucidating the basic plan of development in the bronchial tree inasmuch as they have provided tissue and cellular localizations required to perceive it. Using methods that permit greater resolution than was possible a decade ago, future investigations should clarify the problem of alveolar formation and increase our comprehension of events occurring in terminal buds, in the more differentiated airway, and in associated structures.

The Ultrastructural Localisation of Hydrolytic Enzymes in a Keratinising Epithelium

SQUIER, C. A. and J. P. WATERHOUSE (Department of Oral Pathology, The London Hospital Medical College, Dental Institute, London, E. 1. and Department of Oral Pathology, University of Illinois at the Medical Center, Chicago, Ill.)

The presence of lysosomes in oral epithelium has been suggested by the histochemical distribution of acid phosphatase and non-specific esterase as discrete granules in the tissue (Ten Cate, 1963). There has been little work on the localisation of these enzymes at the ultrastructural level although Hashimoto et al. (1966) described a lysosomal distribution of acid phosphatase in non-keratinised buccal epithelium. The work reported here was concerned with the existence and function of these organelles in keratinising oral epithelium.

Two acid hydrolytic enzymes, acid phosphatase and non-specific esterase, were localised at the ultrastructural level using two independent techniques for each enzyme. Cytochemical tests were carried out on 50 micron cryostat sections of keratinised rat cheek epithelium that had been fixed with glutaraldehyde. Conventional lead precipitation techniques were used for the demonstration of both acid phosphatase (Barka and Anderson, 1963) and non-specific esterase (Miller and Palade, 1964) but the localisation of these enzymes was also determined with the alternative substrates synthesised by Seligman and co-workers (Hanker et al., 1964); these produce an electron-dense osmium reaction product in the tissue.

Both enzymes were localised in membrane-bound structures that could be considered as lysosomes from a morphological as well as a cytochemical viewpoint; acid phosphate also appeared in vesicles of the Golgi complex. The enzymes appear to participate in the breakdown and removal of cell organelles in the granular layer which precedes keratinisation; keratohyalin granules and the cell nuclei did not, however, show enzyme reaction deposits. Superficially, in the keratinised cell layer, acid phosphatase appeared extracellularly, where it may facilitate the breakdown of cell adhesion and the subsequent desquamation of cells.

Slight differences were apparent in the localisation of each enzyme by alternative methods. These differences probably reflect the varying specificities of members of a group of enzymes in the presence of different substrates.

References:

(1) Ten Cate, A. R.: Arch. Oral Biol. *8*, 747 (1963).
(2) Hashimoto, K., R. J. Di Bella, and G. J. Shklar: J. Invest. Derm. *47*, 512 (1966).

260

(3) Barka, T. and P. J. Anderson: Histochemistry theory, practice and bibliography, pp. 239-242, London: Harper & Row, Inc. (1963).
(4) Miller, F. and G. E. Palade: J. Cell Biol. *23*, 519 (1964).
(5) Hanker, J. S., Arlene R. Seaman, L. P. Weiss, H. Ueno, R. A. Bergman, and A. M. Seligman: Science, N. Y. *146*, 1039 (1964).

Chromatographically Pure Fluorescent Tracers

STEINBACH, GÜNTER (Anatomisches Institut, Medizinische Hochschule, Hannover)

All commercial available fluorescin isothiocyanates show, after chromatographical analyses, that their substance contains several fluorescent components. Therefore, methods to gain a chromatographical pure fluorescin-isothiocyanate, given in the literature, have been repeated and some modifications in dye-preparation have been added. The binding-rates of these substances have been determined and the importance of the use of pure substances in respect to non-specific staining and the simplification of the method will be outlined and demonstrated with some practical examples.

Detection of Rescued Oncogenic Virus from Transformed Cells After Cell Hybridization

STEPLEWSKI, ZENON (The Wistar Institute, Phila., Pa. and Institute of Oncology, Gliwice, Poland)

Human and primate cells transformed by the oncogenic virus SV_{40} do not yield infectious virus after the transformation process is completed. However, when these cells are fused by Sendai virus with SV_{40} susceptible green monkey kidney cells, the viral genome may become sufficiently activated to express its late functions. As a result, synthesis of viral coat protein can be detected by immunofluorescence in nuclei of heterokaryocytes originally derived from transformed cells which remained "virus free" for more than 1000 generations. Thus, the susceptible cells seem to supply a factor which either blocks the "repressor" present in the transformed cells or provides a metabolite necessary for synthesis of the complete virus. There are also SV_{40}-transformed cell systems in which infectious SV_{40} cannot be activated after fusion with susceptible cells, but which yield infectious virus after fusion with other transformed cells from which SV_{40} could not be isolated.
Through simultaneous immunofluorescent staining and autoradiography

of the same cell, it was possible to determine the derivation of nuclei in heterokaryocytes capable of synthesis of viral coat protein.

Enzyme Histochemical Study of Primary Hepatic Carcinoma in Thailand *

STITNIMANKARN, TINRAT (Department of Pathology, Faculty of Medicine and Siriraj Hospital, University of Medical Sciences, Bangkok, Thailand)

Primary carcinoma of the liver is a common disease in Thailand. From August 1965 to October 1967, a histochemical study of some of the hydrolytic and oxidative enzymes was made on 24 surgical biopsy specimens of primary hepatic carcinoma obtained immediately after surgery and on 19 autopsy specimens from postmortem examinations performed within 48 hours after death. The purpose of this communication is to present some of the findings.

Alkaline phosphatase (calcium-cobalt method) was shown to be absent from the tumor tissues but present in necrotic areas within tumor nodules and in the sinusoids, bile canaliculi and in the non-neoplastic tissues surrounding the tumors. No definite correlation between the staining reaction of the tumor and liver tissues and the level of serum alkaline phosphatase could be made.

Staining intensity was greater in the liver tissue than in the tumor nodules for glutamic and glucose-6-phosphate dehydrogenases. However, the staining intensity of glutamic dehydrogenase was slightly greater in the tumor tissue in some postmortem specimens.

Lactic and beta-hydroxybutyric dehydrogenases, DPNH and TPNH diaphorases exhibited equal intensity in both liver and tumor tissues in most cases, but with slightly greater intensity in the tumor nodules from some patients.

Among the total of 43 cases studied in this series, there were 12 instances associated with cirrhosis and 1 tumor arising in conjunction with a liver fluke (*Opisthorchis viverrini*) infestation of the liver. The tumors were classified histologically as hepatocellular carcinoma in 29 instances, cholangiocarcinoma in 8, mixed hepatocellular and cholangiocarcinoma in 1, and unclassified in 5. No differences in the staining intensity of the tumor tissues were observed whether cirrhosis was present or not.

---

* Supported by United States Public Health Service Grant No. TW-00100-01, 02, and 03; and National Research Council of Thailand Grant.

The Diverse Uses of Salicylhydrazide in Fluorescence Histochemistry

STOWARD, PETER J. (Nuffield Department of Orthopaedic Surgery, University of Oxford)

The many and unique properties of salicylhydrazide as a fluorescence histochemical reagent will be reviewed. In particular, its use for demonstrating ketosteroids, proteins possessing C-terminal carboxyl groups, and periodate-reactive mucosaccharides and polysaccharides will be described. Salicylhydrazide forms fluorescent derivatives only with compounds containing active carbonyl groups. Sometimes the derivatives (salicylhydrazones) fluoresence only after post-treatment with certain metal salts or alkalis. For instance, the derivatives formed from 3-ketosteroids emit a comparatively weak, greenish-yellow fluoresence that can be completely destroyed with dilute sodium hydroxide. Salicylhydrazones of 17-keto-steroids, on the other hand, do not fluoresce unless they have been treated with alkali; and those of simple aliphatic ketones unless they have been treated with a neutral solution of zinc acetate.

The uniqueness of salicylhydrazide will be illustrated by two properties of the salicylhydrazones of periodate-oxidized mucosubstances. Such hydrazones are the only acid arylhydrazones which can be converted into formazans by treatment with diazonium salts in the presence of aqueous pyridine; and the fluoresence of their aluminium complexes fades exponentially, usually in one or more discrete stages. The duration and half-life of each stage appears to be a characteristic of a particular mucin type.

Comparative Observations on the Location of Horseradish Peroxidase and Antibody to Horseradish Peroxidase in Lymphoid Tissue of Rabbits

STRAUS, W. (The Chicago Medical School, Chicago, Ill.)

After immunization of rabbits with horseradish peroxidase, sites of antibodies to horseradish peroxidase were detected by treating fixed tissue sections of lymphoid tissue with the antigen *in vitro* and subsequent staining of the sections for peroxidase. An intense antibody reaction was observed in plasma cells and in the reticulum of spleen and popliteal lymph nodes. Since control sections, not treated with horseradish peroxidase *in vitro,* showed the location of the antigen alone (in addition to endogenous peroxidase in leucocyte granules), the location of antigen and antibody in the same tissue could be compared. The proximity of antigen and antibody in the reticular cells of lymphoid tissue, or sites very close to them, may point to a function of these cells in antibody formation.

Low Temperature Tissue Preparation Without Solvents and Without Embedding Media for Phase Contrast, Fluorescence and Autoradiographic Studies

STUMPF, WALTER E. and LLOYD J. ROTH (Department of Pharmacology, The University of Chicago)

In order to avoid diffusion and extraction of compounds and tissue constituents during histologic preparation, it is paramount that all technical steps are eliminated which have become known sources of such artifacts, i.e., liquid fixation, solvent dehydration, clearing, embedding, de-embedding, rehydration, floating in or on liquid, wet section mounting and thawing. If translocation is to be avoided, the use of one or more of the listed technical steps is justified only if adequate controls can be provided.

The method of non-solvent low temperature tissue preparation (1,2) excludes all of the listed procedures. It is based on (a) simultaneous quenching and mounting in liquefied propane at about $-180°$ C, (b) cryostatic sectioning at $-30°$ to $-70°$ C, depending on the desired section thickness, (c) vacuum freeze-drying by cryosorption pumping with dry ice slush as tissue coolant, and (d) dry mounting of the freeze-dried section by pressure with a teflon support on a slide coated with a gelatin mixture which had been dried prior to the mounting. Handling of the freeze-dried sections requires low humidity conditions ($<40\%$ r.h.). The freeze-dried sections may be stored in a desiccator and exposed to vapor treatment for fixation or other histochemical reactions. The superiority of dry-mounted freeze-dried sections in autoradiography of diffusible compounds has been demonstrated. Not only diffusion is avoided but also high histological and autoradiographic resolution with 0.5 to 2.0 μ sections is provided (3,4,5). Phase contrast photomicrographs of various tissues and autoradiograms with ^3H azetazolamide in kidney, ^3H digoxin in heart, ^3H urobilinogen in liver and ^3H estradiol in hypothalamic neurons are presented. The use of unfixed and unembedded dry-mount, freeze-dried sections for catecholamine fluorescence in nervous tissue and tetracyclin fluorescence in undecalcified bone, cut at $-55°$ C, is also demonstrated.

References:

(1) Stumpf, W. E. and L. J. Roth: Nature *205*, 712 (1965).
(2) Stumpf. W. E. and L. J. Roth: J. Histochem. Cytochem. *15*, 243 (1967).
(3) Stumpf, W. E. and L. J Roth: J Histochem Cytochem *14*, 274 (1966).
(4) Roth, L. J. and W. E. Stumpf: *In:* International Conference on the Use of Radioactive Isotopes in Pharmacology (Geneva, 1967) John Wiley and Sons, New York, *in press.*
(5) Stumpf, W. E.: *In:* Radioisotopes in Medicine: In Vitro Studies, AEC Symposium Series No. 13 (CONF-671111) Oak Ridge, Tenn. (1968).

On the Nature of the Different LNA-splitting Enzymes in Mammalian Cells

SYLVÉN, B. (The Cancer Research Division of Radiumhemmet, Karolinska Institute, Stockholm 60, Sweden)

A review is presented on the nature and characteristics of the different enzymes hydrolysing the carbamide bond of the chromogenic substrate leucyl-β-naphthylamide (LNA). The influence by pH and added activators and inhibitors will be considered both from a chemical and histochemical level. Ways are discussed of obtaining a more specified discrimination between the main groups of enzymes involved. It would, for instance, seem possible to distinguish in sections between cathepsin B and the general group of aminopeptidases.

"Nadi" Reaction in Myeloid Leucocytes and Vitamin K

TAKAMATSU, HIDEO and KEI-ICHI HIRAI (Dept. of Cytochemistry, Chest Disease Res. Inst., Kyoto Univ., Kyoto, Japan)

The stable Nadi reaction of myeloid leucocyte granules has been reported to be a non-enzymatic reaction because of the stability against fixation and heating for 1 minute at 100° C (1). Dimethyl-p-phenylene diamine and α-naphthol solutions are oxidized spontaneously, and a combination of these solutions gives a fine Nadi reaction. Freshly prepared reagents are not so effective. Therefore, the chemical mechanism of the Nadi reaction seems to be based on a dehydrative condensation of the two reagents.

The present study demonstrated that naphthoquinone may be a Nadi reaction factor in granules of leucocytes obtained from the peritoneal cavity of rats by peptone stimulation. It was proven that the reaction in these granules was affected by acetone. Although about 10% of leucocytes showed a positive reaction to treatment with pure acetone for 2 hours (probably eosinophiles), all of the reactivity was lost when the treatment was continued 24 hours longer. The reaction was restored by vitamin K. However, vitamin K had no effect when leucocytes were treated with 96% acetone. This experiment suggests the presence of a naphthoquinone-like substance, which may be bound to phospholipids (2,3).

A clear liquid substance of high viscosity was extracted from leucocytes with 100% acetone. This was not soluble in alkali or water but was soluble in ethanol, chloroform and acid, and slightly soluble in methanol. This liquid became faint yellowish brown in color when exposed to light. In ultraviolet spectrophotometry, this acetone-extracted substance showed fine absorption peaks at 256 and 325 mμ. This absorption spectrum corresponds with that of naphthoquinones, i.e., vitamin K (4).

References:

(1) Takamatsu, H.: The Saishin-Igaku (Japan) *13*, 477 (1965).
(2) Lester, R. L. and S. Fleisher: Biochim. Biophys. Acta *47*, 358 (1961).
(3) Okui, S., Y. Suzuki, and K. Momose: J. Biochem. *54*, 471 (1963).
(4) Morton, R. A.: *In:* Biochemistry of Quinones, pp 23-89 (Morton edit.) London: Academic (1965).

Histochemistry of Fatty Acid Synthetase

TAKAMATSU, HIDEO and SHINSUKE KANAMURA (Dept. of Cyto-chemistry, Chest Disease Research Institute, Kyoto Univ., Kyoto, Japan)

The calcium salts of short-chain saturated fatty acids with less than ten carbon atoms (*n*-capric acid) are soluble in water while the other fatty acids, with more carbon atoms, are insoluble. This fact led us to develop a method for demonstrating fatty acid synthetase. Unfixed 12μ cryostat sections of mouse liver were incubated at 37°C for 4 hours in 0.2 ml of a medium containing 3 μmoles of capryl-Co-A, 6 μmoles of malonyl-CoO, 25 μmoles of NADPH, 15 μmoles of glucose 6-phosphate, about 0.5 mg of a mixture of equal amounts of calcium palmitate, myristate and laurate, 60 μmoles glyclyglycine and 55 μmoles of CaCl$_2$. The pH was adjusted to 6.7-7.0. Sections were subsequently treated with 0.1 M acetate buffer at pH 5.4 for 15 minutes, immersed in 1% lead nitrate for 13 minutes, treated with dilute ammonium sulfide for 2 minutes, and mounted in glycerin-jelly. Capryl-CoA, malonyl-CoA and NADPH underwent an enzymatic reaction and became longer chain fatty acids which were precipitated as insoluble calcium salts. Under these experimental conditions, calcium phosphate was also formed by the decomposition of CoA, NADPH and glucose 6-phosphate. However, we found that post-treatment of the incubated sections with acetate buffer (pH 5.4) eliminated calcium phosphate from the sections, while the Ca salts of fatty acids remained stable. In mouse liver, the brown reaction products could be observed in coarse granular forms in hepatic cells. They were more intense in the centrolobular areas. When capryl-CoA, malonyl-CoA, or NADPH were eliminated from the incubation mixture, the reaction was depressed markedly. The addition of iodoacetamide to the incubation mixture inhibited the reaction. The reactions in the livers of fasting animals were the same, but they were suppressed by CCl$_4$ intoxication. Fixatives such as neutral formalin, ethanol or acetone destroyed the activity completely.

Enzymohistochemical and Autoradiographic Observations on Cyclic Glycogen Metabolism

TAKEUCHI, TADAO (Department of Pathology, Kumamoto University School of Medicine, Kumamoto, Japan)

Intracellular glycogen metabolism was observed histochemically in proliferating cells. Ascites hepatoma AH 13 tumor cells which consist of free cells and contain usually rich glycogen were used. The mitotic cell cycle of AH tumor cells was investigated with ^{3}H-thymidine autoradiography. Synthesis of ribonucleic acid was observed by an application of ^{3}H-cytidine. The histochemical demonstration of glycogen phosphorylase and uridine diphosphoglucose-glycogen glucosyltransferase (UDPGGT) was performed by using the improved techniques of Takeuchi-Kuriaki's and Takeuchi-Glenner's methods respectively. For visualization of the reaction products by the enzyme activities, both iodine reaction and periodic acid Schiff's reaction procedures were used. It was noteworthy that the double demonstration technique for each enzyme and synthesized nucleic acid was newly tried.

In the mitotic cycle, it took eight hours in DNA synthetic phase (S). Post-DNA synthetic phase (G_2) lasted two hours. Mitotic phase (M) needed two hours and pre-DNA synthetic phase (G_1) was six hours.

Phosphorylase was activated during the DNA synthetic period. The reaction products appeared diffusely or granularly in whole areas of cytoplasm of these cells. This activity increased gradually and was activated maximally in the mitotic phase. The enzyme activity was intensely demonstrated in mitotic cells of various stages, particularly in the metaphase and anaphase. It was reduced just a little in the telophase and reduced suddenly in daughter cells. However, the daughter cells still contained active phosphorylase. On the other hand, the intracellular glycogen deposits began to disappear from the perinuclear areas of cytoplasm to the peripheral in the S period and gradually decreased later. Therefore, silver grains of ^{3}H-thymidine began to appear in the glycogen rich cells. Glycogen often disappeared completely in cytoplasm in the M phase. It disappeared very often during post-anaphase and telophase. No glycogen granules were observed in daughter cells in many cases. In these phases, UDPGGT was not yet activated. It was activated in the beginning of the G_1 phase and the activity reached the maximum during this later period. At the same time, glycogen granules began to deposit in the peripheral zone of cytoplasm and gradually increased in the whole areas of cytoplasm. During the G_1 period, the phosphorylase activity was remarkably reduced, but it often was still active. Sometimes it was completely inactivated. Silver grains of ^{3}H-cytidine were concentrated into the nucleolus during the later half in the G_1 phase.

It was found from the results that there is a very close relationship between cell proliferation of AH 13 tumor and glycogen metabolism, and the enzyme system related to glycogen metabolism in this hepatoma changes into the system of intracellular energy metabolism in the proliferating mechanism. A cyclic pathway of glycogen metabolism was also cytochemically proved. (This study is owed to assistance of Dr. Ieri, Y.).

References:

(1) Takeuchi, T.: Ann. Histochim. *7*, 61 (1962).
(2) Takeuchi, T.: *In:* Enzyme Histochemistry in Japanese Edition, pp 187-214 (Takeuchi *et al.* edit.) Tokyo: Asakura (1967).
(3) Takeuchi, T.: Tr. Soc. Path. Jap. *55*, 35 (1966).
(4) Takeuchi, T. *et al.*: J. Kumamoto Med. Soc. *41*, 690 (1967).

Erythrocyte Esterase Interaction with Heparin Demonstrated with Starch Gel Electrophoresis

TEMPLETON, McCORMICK (Department of Anatomy, School of Dentistry, University of Southern California, Los Angeles, Calif.)

Esterases from erythrocytes of human and of C_{57} brown mice were subjected to electrophoretic separation in starch gels and their activities demonstrated by histochemical methods. Lysed erythrocytes, plasma and serum were compared with use of alpha-naphthyl acetate, naphthyl-AS acetate, alpha-naphthyl butyrate and acetylthiocholine as substrates. Erythrocyte esterases are distinct from plasma esterases and, on the basis of substrate affinities, three different esterases were present. One band in the erythrocyte zymogram hydrolyzed alpha-naphthyl butyrate but was only slightly active to acetate esters. The fastest migrating bands in mouse erythrocyte zymograms hydrolysed acetate faster than butyrate esters but would hydrolyze both substrates. In samples from human blood, these leading bands do not appear to hydrolyze alpha-naphthyl butyrate. A third esterase, distinct in both mouse and human erythrocyte samples, was represented in the zymogram by a pair of bands which would hydrolyze only acetate esters, hereafter designated as "acetate" bands. When heparin, or plasma containing heparin, was inserted in a gel so as to overlap an erythrocyte sample, only the "acetate" bands were altered. When heparin is inserted cathodal to an erythrocyte sample, the "acetate" bands are selectively affected by the heparin in such a way as to cause the "acetate" band to migrate more rapidly toward the anode; other esterase bands were not similarly affected. A combination of heparin and the "acetate" bands would be more electronegative than the untreated esterase and would favor more rapid migration toward the anode. A decreased intensity of the esterase stain was noted where the "acetate" bands interact with heparin. When added to staining solutions, heparin had no inhibitory effect on plasma or erythrocyte esterases. The decreased staining of the "acetate" bands depends upon the presence of heparin during electrophoresis. "Acetate" bands were not inhibited by 1 mM eserine.

On Some Interesting Aspects of the Distribution of Acid Phosphatase Activity Amongst the Constituents of the Eye and the Optic Lobe of a Dragon Fly and its Physiological Significance in the Processes of Image Formation

TEWARI, H. B. and H. R. TYAGI (Department of Zoology, University of Udaipur, Udaipur, Rajasthan, India)

This communication incorporates details of the distribution of acid phosphatase activity amongst the cellular elements of the eye and optic lobe of the dragon fly at the intracellular level. The study is based on frozen sections (fixed as well as unfixed) processed through Gomori's and Burstone's techniques for acid phosphatase (see Pearse, 1961). Besides discussing the physiological significance of the distribution of the enzyme amongst the various layers of the eye, the investigation focuses attention on two areas considered to be of considerable physiological significance in the visual processes of the insect namely, the ommatidia in eye and the lamina ganglionaris in optic lobe. In eccentric cells of ommatidia, acid phosphatase positive bodies, identified to be lysosomes, are arranged in linear fashions on both sides of the rhabdoms. Such locales have been considered to be important functional units in the processes of visual formation in insects (Burtt and Catton, 1966).
Further, in the lamina ganglionaris of optic lobe, certain sac-like structures, placed from side to side, have been demonstrated. Such sites reveal enormous accumulation of acid phosphatase positive lysosomes.
The metabolic significance of the presence of a large number of lysosomes, equipped with intense phosphatase activity, in ommatidia as well as in lamina ganglionaris of the optic lobe of the insect will be discussed with reference to visual processes associated with the image formation in this insect.

References:

(1) Burtt, E. T. and W. T. Catton: *In:* Advances in Insect Physiology, Vol. 3, pp. 1-52 (Beament *et al.,* edit), New York: Academic Press (1966).
(2) Pearse, A. G. E.: *In:* Histochemistry Theoretical and Applied. 2nd edition. pp. 431-455. London: J. & A. Churchill, Ltd. (1961).

Activation of Glycogen Phosphorylase by Ischemia in Cochlear Structures [*]

THALMANN, R., L. GLISMANN, and F. M. MATSCHINSKY (Dept. of Pharmacology and Otolaryngology, Washington Univ. Sch. of Medicine, St. Louis, Mo.)

Total phosphorylase and phosphorylase *a* were measured in the organ of Corti (OC), stria vascularis (SV) and Reissner's membrane (RM) of each turn of the cochlea from guinea pigs. Specimens weighing 0.3-0.6 μg, were dissected from frozen-dried temporal bones as published (1). The enzyme was measured fluorometrically by the rate of conversion of glycogen to 6-P-gluconate with the necessary auxiliary enzymes in the presence and absence of 5'-AMP. Total phosphorylase was 685 ± 108 (OC), 152 ± 8 (SV) and 110 ± 17 (RM) mmoles per kg dry tissue per hr, irrespective of the cochlear turn. Initially, phosphorylase *a* amounted to 23 (OC), 30 (SV) and 11 (RM) % of the total activity and rose to 37, 64 and 65% when the temporal bones were frozen after 3 min of ischemia. The rate of phosphorylase activation was fastest in SV. After 30 sec of ischemia in the stria, glycogen phosphorylase *a* had increased more than 2-fold, whereas in the organ of Corti there was no significant change of activity. These enzyme analyses supplement previous studies of glycogen which was found to be distributed among the cochlear structures in the same proportions as total phosphorylase (2). Therefore OC, the structure furthest from the blood supply, appears to be well adapted to withstand lowered oxygen tension that may occur in the endolymph.

References:

(1) Matschinsky, F. M. and R. Thalmann: Laryngoscope 77, 292 (1967).
(2) Matschinsky, F. M. and R. Thalmann: Ann. Otol. Rhinol. Laryngol. 76, 638 (1967).

---

* Supported by USPHS grants AM 10591 and NB 06575.

The Histochemistry of Early Changes of Wallerian Degeneration in Peripheral Nerves

THOMAS, E. (Max-Planck-Institut f. Hirnforschung, Neuropath. Abt., u. Neurologisches Institut d. Universität, Frankfurt a.M.)

The chemistry of Wallerian degeneration is divided into three stages, i.e., the stage of axon collapse and physical destruction of myelin, the stage of cellular proliferation with chemical degradation and removal

270

of myelin lipids and the stage of fibrosis (Rossiter, 1961). Our results demonstrate the early loss of plasmalogen which is a myelin lipid. This begins already 48 hours after a nerve lesion. In the normal myelin layer alkaline phosphatase is localized within the Schmidt-Lanterman clefts. This enzyme is found also in the inner lamellae of the perineurium. Barrier functions in connection with these localizations will be discussed. During the myelin break-down, alkaline phosphatase disappears from the 2nd day onwards (Pinner and Campbell, 1965). Other hydrolases which are distributed evenly within the myelin lamellae (e.g., aryl sulphatase) become diminished together with myelin destruction. Early axonal enzyme accumulations proximal and distal from a local nerve injury are also seen when a nerve continuity is left intact. This is achieved by freezing or by mild coagulation with heat In this case, the rapid enzyme increase in a nerve fiber around a local lesion can hardly be due to axoplasma movements as a result of mechanical alterations in the axon. Chemical changes, therefore, may locally induce a potency of the axoplasma to produce enzymes rapidly, e.g., oxydoreductases and acid phosphatase (Thomas, 1965). This would be in correspondence with the accumulation (proliferation) of mitochondria and dense bodies seen by electron microscopy (Webster, 1962). It is only then that 'la mort du cylindraxe' (Nageotte, 1910) in the nerve separated from the perikaryon begins.

References:

Nageotte, J.: C. R. de la soc. de Biol., *68*, 463 (1910).
Pinner, B. and J. B. Campbell: *In:* Proc. 5th Intern. Congress Neuropathol., Excerpta Medica Found., Amsterdam 1966, pp. 871-880.
Rossiter, R. J.: *In:* Chemical Pathology of the Nervous System. J. Folch Pi *edit.*, Pergamon Press 1961, pp. 207-227.
Thomas, E.: *In:* Proc. 5th Intern. Congress Neuropathol., Excerpta Medica Found., Amsterdam 1966, pp. 369-404.
Webster, H. de F.: J. Cell Biol. *12*, 361 (1962).

Studies on Lead Inhibition of Adenosine Triphosphatases

TICE, LOIS W. (Laboratory of Experimental Pathology, National Institute of Arthritis and Metabolic Disease, National Institutes of Health, Bethesda, Md.)

The inhibitory effects of lead were investigated on ATPase activity associated with rat liver microsomes and a 'membrane' fraction of rat kidney. In both cases, a mixed type of inhibition was observed, which could be partly overcome by suitable increases in substrate concentration. When liver microsomes were incubated with 1 mM ATP and 10 mM Mg (conditions used in the Wachstein-Meisel medium) the rate of inorganic phosphate release in the absence of Pb was 7 μM Pi/mg prot/hr. This fell to 1.5 μM Pi/mg/hr when 4 mM Pb was added to the system. If the ATP concentration was increased to 10 mM, net phosphate release increased to 4.0 μM Pi/mg prot/hr. The Na-K dependent ATPase of kidney membranes was more sensitive to Pb inhibition, and little Na-K activation could be observed in the presence of 1 mM Pb and up to 10 mM MgATP. With 0.5 mM Pb, again little activation could be observed at low ATP concentrations, but Pb inhibition again was partly overcome by increasing ATP to 10 mM.

In addition to direct interaction with enzyme proteins, Pb also can affect ATPase activity through chelation by ATP (k_f approx. 15,000 at pH 7.0, 24° C in 0.1 M NaCl). As a result of 'lead ATPate' formation, Pb effectively competes with Mg for ATP. This phenomenon may partly account for the disappearance, in liver microsomes incubated with 4 mM Pb, of substrate inhibition normally occurring above 4 mM ATP, and an increase, in the presence of Pb, in the Mg concentration optimal for enzyme activity.

The consequences of these findings for the design of histochemical media for demonstrating ATPase activity will be discussed.

Hypophysaires en Culture Organotypique

TIXIER-VIDAL, A. (Laboratoire de Biologie Moléculaire, College de France, Paris 5e)

Dans le but d'analyser les mécanismes réglant la biosynthèse des hormones hypophysaires, la radioautographie de haute résolution a été appliquée à des hypophyses de canard préalablement maintenues en culture organotypique pendant 10 à 12 jours. Dans ces conditions, on sait que l'adénohypophyse ne contient plus que deux types cellulaires (c. à prolactine et c. de type MSH) et élabore des quantités constantes de prolactine (1). Le précurseur marqué (D-Lou 1-Leucine 4-5T) est incorporé pendant 10 minutes. Ensuite, on suit par des expériences de chasse prolongées pendant 48 h. le déplacement des protéines néosynthétisées sur les différentes structures cytoplasmiques. Parallèlement, on mesure l'évolution de la radioactivité des protéines précipitables par le TCA dans les tissus et dans les milieux de culture.

L'étude quantitative des autohistoradiographies permet de déterminer les proportions relatives de la radioactivité cytoplasmique localisées respectivement dans chacun des compartiments (ergastoplasme-cytoplasme, appareil de Golgi, granulations protéiques) et de suivre leur évolution en fonction du temps de chasse (15 min., 30 min., 1 h., 4 h., 24 h. et 48 h.). On détermine également la proportion des granulations marquées par rapport au total des granulations de la cellule (RSg.). On recueille séparément les données relatives aux deux types cellulaires présents dans les explants. Les résultats numériques sont soumis à un contrôle statistique par analyse directe de contingence 2X2?

Le processus des synthèses obéit à un schéma commun pour les deux types cellulaires, chacun d'eux se distinguant par des modalités particulières. Les protéines, synthétisées en quelques minutes dans l'ergastoplasme (2) se concentrent en 30 minutes dans la zône golgienne. Elles migrent ensuite hors de cette zône dont la vidange totale demande 4 h. environ. Elles se localisent alors en parties égales, d'une part sur les granulations protéiques, d'autre part sur le cytoplasme. La RSg atteint son maximum à l'achèvement de la vidange golgienne, mais celui-ci est très bas (3%). La RSg décroît ensuite faiblement (de 1% en 40h.). On encconclut que le renouvellement des granulations hypophysaires est très bas (3%). La Rsg décroît ensuite faiblement (de 1% en 40h.). On continus. La cellule à prolactine se distingue de la cellule de type MSH par une migration plus lente des protéines néosynthétisées, mise en relation avec l'existence de citernes ergastoplasmiques dilatées offrant une possibilité de stockage (3).

L'étude de l'évolution de la radioactivité des protéines précipitables par le TCA dans les explants et dans les milieux de culture montre que: 1° la radioactivité des protéines dans les cellules est maximum 15 et 30 minutes après le début de la chasse 2° dans les milieux de culture, la radioactivité des protéines atteint un maximum à partir de 4 h. de chasse.

Ces faits seront discutés et confrontés avec les données biologiques relatives aux quantités de prolactine et de MSH élaborées par les explants.

References:

(1) Tixier-Vidal, A., et D. Gourdji: Compt. Rend. Acad. Sci. *261,* 805 (1965).
(2) Tixier-Vidal, A., R. Fiske, R. Picart, et F. Haguenau: Compt. Rend. Acad. Sci. *261,* 1133 (1965).
(3) Tixier-Vidal, A. et R. Picart: J. Cell. Biol., *35,* 501 (1967).

Enzyme-histochemical Studies on Rat Thymus after X-ray Irradiation

TÖRÖ, I., E. BACSY, GY. VADASZ, and GY. RAPPAY (Institute of Experimental Medicine, Budapest, Hungary)

The different radiosensitivity of the various thymic cells permitted study of the changes of the enzyme pattern in thymic reticular cells following irradiation. The strongly radiosensitive cells were damaged and partly underwent phagocytosis and digestion by cells of the thymic parenchyma. During this process, intense activity of acid hydrolases was demonstrated histochemically within the phagocytes. In order to decide

273

whether this activity was more intense than it had been before the
phagocytosis and whether any qualitative changes concerning the enzyme pattern of the cells occurred, the izozymes of the non-specific
esterases were also studied by means of starch gel electrophoresis. Three
days after irradiation we found qualitative and quantitative changes of
the izozyme pattern in the homogenates of the thymuses.

Cytochemical Localization of Deoxyribonuclease by an Indigogenic Method *

TSOU, K. C., MILDRED Y. CHANG, and S. MATSUKAWA (Univ. of
Pa. Hospital, Harrison Dept. Surgical Res., Phila., Pa.)

5-Iodoindoxyl thymidine-5'-phosphate (I), m.p. 210°C. (dec.) and 5
iodoindoxyl thymidine-3'-phosphate (II), m.p. 225°C. (dec.) have been
synthesized. These new synthetic substrates are useful for histochemical
study of phosphodiesterase hydrolysis of the 3' or 5' bonds in DNA and
are thus potentially useful also for the study of deoxyribonuclease. The
advantage of this method lies in the low diffusion of the 5-iodoindoxyl
moiety and the short half-life of the 5,5'-diiodoindigo radical to the
corresponding indigo. Thus, this method in part overcomes the diffusion
problem of earlier attempts in localizing DNAse and supplements the
well-known film method. An additional advantage in the use of these
new substrates is the electron density of the indigo dye formed so that
it could lead to an electronhistochemical method. This paper reports
synthesis, substrate specificity when treated with several phosphodiesterases and preliminary histochemical experiments. Thus, venom phosphodiesterase hydrolyzed I and not II. In contrast to expectations, spleen
phosphodiesterase hydrolyzed II only, slowly. Histochemical study showed
the striking difference in activity between these two substrates especially
in nuclei reaction. Thus, I is hydrolyzed by the nucleus in endothelial
cells and II is not at pH 9. Both kidney and liver hydrolyze I and not
II as expected. The activity in kidney is restricted to the glomerulus
region and the activity in liver is very strong in the lining cells of
sinusoid. Rat leucocytes are positive for the T-5'-substrate, and appear
to be restricted to only a small number of polymorphonuclear neutro-
philes. While spleen does hydrolyze both I and II, it does to a much
lesser extent for I. Both light and electronmicrographs will be shown
and the discrepancy between biochemical and histochemical findings will
be discussed. Preliminary study with Hela and glioma cells are negative
for both types of phosphodiesterase.

* Supported by USPHS Grant CA-07339 and AEC Contract AT (31)-1384.

274

Intracristae Localization of Succinic Dehydrogenase Activity (SDH) with a New Osmium Containing Tetra-tetrazolium Salt (Os-TNST) *

TSOU, K. C., CLEON GOODWIN, DWO LYNM, and BETTE SEAMOND (Univ. of Pa. Hospital, Harrison Dept. Surgical Res., Phila., Pa.)

In a continuous effort to improve the specificity and localization of SDH by the tetrazolium method, we have converted a new synthesized ditetrazolium salt, 2,2',5,5'-tetra-p-nitrophenyl-3,3'-stilbene ditetrazolium chloride (TNST) to its corresponding dimeric osmate, Os-TNST, by careful reaction of the ethylenic bond with an equivalent of osmium tetroxide. The localization of SDH with this new osmium-containing tetrazolium salt was then tested in comparison with TNBT and TNST in a standard procedure at both the light and electronmicroscopic level. TNST and TNBT are comparable in electrondensity and localization. TNST shows a very interesting pattern resembling the well-known Fernandez-Moran-Green particles on sonicated mitochondrial membranes, but no improvement over TNBT in intact tissue. Os-TNST, however, yields very clear intracristae localization within the inner membrane. The Os-TNST formazan deposit can be seen within the cristae, arranged often in a helical-like manner with possible separation of about 100 Å spacing. The use of such osmium-containing and enzyme-noninhibitory substrates also opens a new possibility of synthesizing other substrates based on the same principle. Improvement in ultra-structural detail without losing contrast can be done by post-staining with PTA or uranyl acetate, but not with Reynold's lead. It is suggested that this difference is due to metal-metal bonding of uranyl to osmium. Electronmicrographs will be shown.

* Supported by USPHS Grant CA-07339.

Alkaline Phosphatase Activity of Individual Neutrophilic Leukocytes Determined Cytophotometrically by use of a Biochemically Calibrated Model System

VAN DER PLOEG, M. and P. VAN DUIJN (Laboratory for Pathology, Biochemical Section, University of Leiden, The Netherlands)

In the preceding communication, investigations concerning the reliability of the azo dye coupling method for quantitative demonstration of alkaline phosphatase activity have been discussed. From the results obtained with a model system consisting of polyacrylamide films containing intestinal alkaline phosphatase, it appeared that this method could be used for cytophotometrical quantification of the enzyme in individual neutrophilic leukocytes.

For this purpose, polyacrylamide films were prepared in which a sonicate of leukocytes from a guinea pig exudate was incorporated. These films were assayed for linearity between amount of reaction product and incubation time, proportionality between amount of end product and enzyme concentration or film thickness, and optimal reaction conditions. Since the enzyme activity of the films could also be studied biochemically—using disodium phenyl phosphate as a substrate—a direct relation between cytochemical and biochemical activity could be determined. For the quantification of the enzyme activity in the cells, microscopical preparations of exudate neutrophils were stained together with films containing sonicate from the same cell suspension. The amount of dye formed in the cytoplasm of individual leukocytes was measured with a cytospectrophotometer based on the two-wavelength principle. By reference to the relation between biochemical and cytochemical activity in the model films this cytochemically determined enzyme activity could be expressed in biochemical units.

Independently, the average alkaline phosphatase activity per cell was determined by direct biochemical assay in the leukocyte suspension. The results of both assays showed good agreement.

References:

Van Duijn, P., E. Pascoe and M. van der Ploeg: J. Histochem. Cytochem. *15*, 631 (1967).

A New Microfluorometer for Rapid Measurement of Cells Stained with Fluorochromes *

VAN DILLA, M. A., J. R. COULTER, and P. F. MULLANEY (Los Alamos Scientific Laboratory, University of California, Los Alamos, New Mexico)

A new method for high-speed, quantitative measurement of fluorescence of cells stained with acridine orange or fluorescein has been developed; information on spectral distribution is obtained with interference filters. The method has potential as a dual sensor for simultaneous measurement of cellular fluorescence and size (by small angle light scattering) and as a new sensor for cell sorting (1).

Cells in suspension stained with fluorochrome are lined up in a stream of very small diameter (about 5 microns) and flow individually across an intense, narrow beam of violet light (about 400 to 500 nm) at a rate of about 50,000 per minute. The fluorescence emission pulse (about 15 microseconds long) from each cell is counted, and overall light intensity

or intensity in selected wavelength bands is also measured. Counts of rag-
weed pollen (diameter 19 microns) stained with acridine orange (AO),
compared with Coulter counter results, show that each fluorescent particle is counted. Chinese hamster ovary (CHO) cells grown in suspension
culture show fluorochromasia (2); fluorescein accumulation per cell has
been measured after exposure to fluorescein diacetate concentrations as
low as 5×10^{-7} M. When stained with AO, CHO cell nuclear fluorescence is green, while cytoplasmic fluorescence is orange-red. Nuclear
fluorescence is measured in the 520- to 560-nm region with sharp cutoff
filters; fluorescence emission above 600 nm is measured with a second
filter combination and is taken as mainly cytoplasmic. When interference from proteins is excluded by staining technique and other dye-
binding substances like mucoplysaccharides are absent, intracellular
nucleoproteins bind AO in a characteristic way; DNA-AO complex
emits green fluorescence, while RNA-AO complex emits orange-red
fluorescence (3). Experiments with CHO cells in random culture and
synchronized at the beginning of the G_1 and S phases of the life cycle
are underway to determine the feasibility of measurement of DNA and
RNA distribution over a cell population with these methods.

References:

(1) Fulwyler, M. J.: Science *150*, 910 (1965).
(2) Rotman, B. and B. W. Papermaster: Proc. Natl. Acad. Sci. *55*, 134 (1966).
(3) Rigler, R.: Acta Physiol. Scand. *67*, Suppl. 267, 3 (1966).

---

* This work was performed under the auspices of the U. S. Atomic Energy
Commission.

Enzyme Kinetics in a Cytochemical Model System of Polyacrylamide Films Containing Alkaline Phosphatase

VAN DUIJN, P., E. PASCOE, and M. VAN DER PLOEG (Laboratory
for Pathology, Biochemical Section, University of Leiden, The Netherlands)

Cytochemical quantification of enzyme activity can be complicated by
diffusion gradients of the reagents around the enzymic site. The limitations imposed by too slow a penetration of the reagents into the cell
or tissue can have a direct influence on the enzyme reaction kinetics,
since concentration of substrate and trapping agent at the enzymic site
will then be lower than measured in the incubation medium.

The problem is similar to that encountered in heterogeneous catalysts
where chemical reactants must diffuse into the pores of solid catalysts
before they can react. For such processes an effectiveness factor n has
been defined as the ratio of actual rate of reaction when the catalyst

is geometrically constrained to that which would occur if the catalyst were free in solution (Satterfield and Sherwood, 1963). Factors governing diffusion of reactants into structure-bound catalysts are formalised in the dimensionless Thiele modulus Φ. The relationship between n and Φ is such, that to avoid complications due to slow diffusion, Φ must be smaller than a given value—depending on the order of the reaction.

A theoretical treatment of the influence of slow diffusion on kinetics of an enzyme reaction in a cytochemical system is presented.

The effect of several parameters on diffusion of naphthol AS-MX phosphate and 4-aminodiphenylamine diazonium sulphate (Variamine Blue RT) was studied experimentally with a model system consisting of polyacrylamide films into which intestinal alkaline phosphatase was incorporated. The amount of azo dye formed in the films was measured quantitatively with a special colorimeter.

The experimental results confirmed the theoretical relationships. Incubation conditions can be obtained for which the amount of azo dye precipitated during a given reaction time is proportional to the enzyme activity present. Under such conditions, problems for enzyme determination are not different from those in biochemical systems, and quantitative cytophotometrical determination of enzyme activity in individual cells is possible.

References:

(1) Satterfield, C. N. and T. K. Sherwood: The role of diffusion in catalysis pp 56-63, Redding, Massachusetts: Addison-Wesley Publishing Co. Inc. (1963).
(2) Van Duijn, P., E. Pascoe, and M. van der Ploeg: J. Histochem. Cytochem. *15*, 631 (1967).

Genetics of Enzyme Localization in the Production of Polyacetylenes and Polyenes in Plants

VAN FLEET, D. S. (Department of Botany, University of Georgia, Athens, Ga.)

Wavelength maxima and extinction values proportional to the number of conjugations of unsaturated bonds were used to determine the quantity of known poly-ynes and polyenes and to localize the cells producing the oils. Diagnostic maxima between 2200Å and 3900Å, corresponding to the stretching frequency and conjugation of the bonds, provided an analysis in microgram samples held in hexane in microcuvettes. Causal factors for the number, type and position of cells responsible for acetylene oil formation were found to be linked with genes that determine the formation and localization of enzymes responsible for dehydrogenation and conjugation. Phenotypic variation was rare in the many plants under study. New genotypes were obtained in species hybrids in some genera; in many genera, in the same tribe, no genotypic differences were found. The greatest variation in acetylenic oils was obtained in octoploid perennial plants as somatic mutants. The greatest genetic drift in type of poly-ynes was obtained in weak plants as bottom recessives for several characters. In all examinations to date, poly-ynes were found to be produced by multinucleate or ploided cells with the greatest varia· tion and genetic drift in octoploids (autotetraploids). Homozygous or fixed acetylenic types were found in many annual diploids in which other genetic characters are rarely if ever homozygous. In general, the association of poly-ynes and polyenes was found to be with a tissue genome of high DNA and metabolism associated with vigor. No genetic linkage with anthocyanin or flavanoid markers could be found, but in sectored "ab whites," lacking anthocyanin, there was evidence of lack of high calories required to make the triple bond. Loss in mean un· saturation through failure of triple bond formation was found in weak recessive plants and was induced in standard vigorous plants grown under conditions of reduced light. The latter had a high frequency of somatic mutants. Somatic mutations were found to be associated with changes in chromosome number and chromosome imbalance and not with gene mutations.

Histochemical Investigations of Myelinating Tissues

VanHOUTEN, WIECHER H. (Mental Health Research Institute, University Medical Center, Ann Arbor, Mich.)

Myelination of the pyramidal tract of the rat was studied at the cervical level with the help of histochemical techniques for oxidoreductases (oxidative metabolism) and lipids (myelin). The rat was used because the central nervous system of this animal contains hardly any myelin at birth, whereby it is possible to study not only actual deposition of myelin but also events preceding myelination.

During myelination, an increase in oxidative enzyme activity was seen in cellular elements, which could be identified as oligodendroglial cells. The reaction for glucose-6-phosphate dehydrogenase showed a stronger staining intensity than the other enzyme reactions. In the pyramidal tract of the 150-day-old rat, the reaction for the several oxidoreductases was weak in glial cells and axons.

In addition, a quantitative study was made of the number of cells in sections stained with gallocyanin-chromalum and stained for $NADH_2$-tetrazolium reductase. The total number of cells stained with gallocyanin-chromalum showed a sharp increase and decreased just before the beginning of myelination. During myelination the total cell count remained high. The number of cells stained for $NADH_2$-tetrazolium reductase increased during myelination, but the total number of cells counted was much higher than the cells which stained for activity of this enzyme.

Chemical Changes of Potato Tuber Plastids by the Action of Light

VECHER, A. S., A. A. MASKO, K. I. PREDKEL, V. N. RESHETNIKOV, and M. T. TCHAIKA (Institute of Experimental Botany, Academy of Sciences of the Byelorussian SSR)

Due to the formation of chloroplasts, there takes place in the light the pigmentation of potato tubers which is more intensive in surface layers. From potato tubers of different degrees of pigmentation, it is possible to isolate plastids and to investigate them.

By modern techniques of isolation of structural components of plant cells, different fractions of plastids were obtained. It was established by dividing tuber chloroplasts into fractions in the saccharose density gradient (40-50 per cent), the main mass of plastids is represented by chloroplasts of small sizes having a rounded form, which are located in the test-tube on the same level with leucoplasts at similar conditions of division. The major part of chloroplasts is likely to be formed of leucoplasts. Since the total yield of plastids from pigmentized tubers is higher than that of tubers which were in the dark, there is an increase in plastid substance at the transformation of leucoplasts into chloroplasts or a portion of the chloroplasts are formed anew from proplastids.

At the first stage of chloroplast formation in the light there is a fast accumulation of RNA, the quantity of which increases on the 7th-9th day of greening by 2.0-2.5 times as compared with the quantity of RNA in leucoplasts.

The study of photochemical activity and the chemical changes of potato tuber plastids has shown that, at the active greening stage, the tuber chloroplasts are capable of primary photochemical reactions of the photosynthesis process. At the discontinuation of pigment accumulation, the Hill reaction and photosynthetic phosphorylation activity decreases greatly.

This suggests that the photosynthetic activity of potato tuber chloroplasts depends upon the stage of their development and that the "half-life" of a chloroplast is the time from the beginning of the exposure to light till it reaches the maximum of its photochemical activity.

Histochemical Attempts to Analyse the Structural Stability of Macromolecular Aggregates

VELICAN, C. and DOINA VELICAN (Institute of Internal Medecine, Bucharest 10, Rumania)

An attempt has been made to analyse, on microscopical sections, the structural stability of biopolymers, using a limited number of convenient histochemical techniques, adequate to determine : a) *decrease or increase of intramolecular, intermolecular and intermacromolecular aggregate* links (10% NaCl, an electrostatic bond breaker by ionic competition; urea 8 M and distilled water at 90°C to break hydrogen bonds; 1% KOH to rupture the covalent linkages between carbohydrates and proteins; 10% formalin and 5% tannic acid to introduce new cross-links between macromolecules); b) *specific elimination of some constitutive substances of the macromolecular complex* (sialic acid after sialidase digestion, hexosamine after lysozyme digestion, etc.); c) *removal of some reactive groups* (sulphate groups subsequent to methylation and saponification).

Starting from a histochemical "normal" pattern of the structural stability of ground substance, cartilage and bone matrices, epithelial mucins, collagen, elastic and reticulin fibers, basement membranes, secretory granules, etc., some alterations which lead to a conversion of the normostability into hyperstability (increase of the macromolecular complex resistance to the usual methods of degradation) or hypostability (decrease of the macromolecular complex resistance to the usual methods of degradation) were pointed out. The examples given in the present work indicate that alterations of the structural stability of macromolecular complexes could be related to pathogeny of numerous chronic diseases: hyperstability as a basical step in the morphogenesis of various types of fibrosis, hypostability as a basical step in the abnormal degradation of epithelial mucins, ground substance, matrices, fibers and cellular organelles.

A Study of Variations of Enzymatic Activities in the Course of the Different Phases of the Generation Cycle of Cultivated Cells

VENDRELY, C., A. LAGERON and P. TOURNIER (Institut de Recherches sur le cancer, Villejuif et Unité de Recherches de Physiopathologie Hepato-biliaire Hôpital St-Antoine, Paris, France)

Histoenzymological techniques applied to tissue cultured cells often give a very heterogeneous pattern of highly positive and nearly negative reactions. Since every cell of the population is in the same medium condition as the others, it is possible to imagine that these striking differences may be due to the fact that the cells are in different phases of their life cycle. For that reason we tried to relate histoenzymological studies with cytophotometric and autoradiographic investigations on the very same cells submitted first to a thymidine pulse. We perform first the histoenzymological reaction: lactate dehydrogenase, malate dehydrogenase, glucose-6-phosphate dehydrogenase and tetrazolium reductase; a suitable field in the preparation is determined and photographed, then the staining is removed by chloroform, a Feulgen reaction is performed, and the DNA content of the nuclei is determined by cytophotometry. From this DNA content, it is possible to recognize cells in G_1 and in G_2 phases; an autoradiography made afterwards on the same preparation allows the determination of the cells in S phase. The comparison of these data with the histoenzymological response which can be seen on the microphotographs makes possible for each cell the determination of its phase and of its enzymatic activity.

Our preliminary studies on BHK cells indicate that S phase and G_2 phase are particularly active for LDH and MDH. A comparison of normal cells and cancer cells with such techniques will be discussed.

Preliminary Observations on the Photodecomposition of the Fluoresence of Enterochromaffin Cells

VIALLI, MAFFO and GIOVANNI PRENNA (Istituto di Anatomia Comparata dell'Università di Pavia e Centro di Studio per l'Istochimica del C.N.R., Italia)

The characteristic fluorescence of enterochromaffin cells seems an excellent means for the quantitative histophotometric determination of their 5-HT content. In preliminary experiments we carried out quantitative studies on photodecomposition of the fluorescence of enterochromaffin cells (guinea-pig duodenum) thus treated: 1) freezing-drying followed by formaldehyde treatment a) *in toto* (Falck) for two hours, b) in sections for 30 minutes, 3, 24 and 48 hours; 2) normal fixation with 10% neutral formalin for 4 hours.

Microfluorometric determinations by means of a M.P.V. Leitz with E.M.I. photomultiplier 6094 A, with power supply Knott (type NSHM BN600), with a microvolt-ammeter (Keithley 150B) and with a recorder (Microcord 44—Photovolt). Fluorescence excitation in transmitted and incident light by means of a stabilized Xenon lamp (Osram—150 W/1). The following filters were used: excitation filters UGl/2 mm. and BG12/5 mm., barrier filters K430 and K530 and a line interference filter at peak of emission of the fluorescent product of 5-HT. Microspectrofluorometric measurements were done in a Leitz-microspectrograph equipped with television-oscilloscopic system under standardized conditions of measurement for both cells and background.

The first results show photodecomposition in both cells and background under all experimented conditions with quantitative variations. The best ratio of cells to background in terms of ffuorescence intensity is obtained with Falck's method measuring at the peak. The first microspectrofluorometric results after different irradiation periods show spectrum variations with peak displacement towards longer wavelengths. The beginning of fading and half-life time vary in relation to numerous factors.

It seems likely that, under conditions to be determined, quantitative measures of the fluorescent product of 5-HT in single cells can be made.

Blocking Mechanism of Alcian Blue Staining of Acid Mucin by Methyl Alcohol in the Presence of Hydrochloric Acid

VILTER, VOLDEMAR (Laboratoire d'Ecologie Histo-physiologique, 92, Meudon-la-Forêt, France)

The suppression of Alcian Blue staining of anionic mucin by treatment with methyl alcohol in HCl is not due to methylation. The blockade of basophilia does not imply esterification of anionic groups of mucin by CH_3OH. Hydrochloric acid, dissolved in different organic solvents, has the same effect as methyl alcohol. The blockade of basophilia is essentially due to the lactonizing action of HCl, when this acid reacts in a completely anhydrous medium. The anionic group responsible for Alcian Blue staining is reattached to a neighboring CHOH group with the elimination of a water molecule. Most spectacular abolition of basophilia is obtained by gaseous and anhydrous HCl dissolved in a totally nonpolar solvent as cyclohexane. The "demethylating" action of KOH is interpreted as a delactonization phenomenon.

The desulfatation of very acid mucopolysaccharides does not necessarily assist in the intervention of hydrochloric-methyl alcohol (Kantor and Schubert, 1957). The $-HSO_3$ group may be liberated by anhydrous HCl dissolved in a non-polar medium such as benzene or cyclohexane.

Histophotometric Study of the Architectonic of the Polyanionic Mucins of the Oviduct

VILTER, VOLDEMAR (Laboratoire d'Ecologie Histo-physiologique, 92, Meudon-la-Forêt, France)

Modifications of the basophilia of acid mucins in relation to pH of the stain appear to be more complicated than what current research may suggest. Attenuation of pH does not necessarily lead to diminution of basophilia; it may even increase the stainability of mucins. A systematic study of Alcian Blue affinity for mucin in relation to pH, shows that a slight difference in acidity of the stain (in the order of 0.1 to 0.2 pH units) may lead either to a fall or to an increase of staining intensity. Between pH 0.3 and 3.0, alcianophilia of polyanionic mucin shows nine peaks of absorption separated by deep grooves. A typical saw-teeth absorption curve is obtained by very acid mucins ($-HSO_3$) as well as by moderately acid mucins ($-COOH$). The variation of basophilia reflects the structural differences of the protein more than the difference in ionic force of acid groups responsible for histochemical basophilia. Each absorption peak of alcianophilia is probably due to the opening of a particular type of internal linkage within the protein macromolecule, resulting in "demasking" of an anion group which becomes accessible to a specific stain for acid mucin, as Alcian Blue or Pararosaniline-Paraldehyde.

Molecular Aspects of the Regeneration of Bone Tissue After Application of Bone Tissue Extract and Royal Jelly

VITTEK, JOSEF (Clinic and Research Laboratory, I. Clinic of Stomatology, Faculty of Medicine of Comensky University, Bratislava, CSSR)

The healing process of injured bones of stromal mandible and after application of materials accelerating the regenerative process (extract of bone and Royal Jelly) has been studied in rabbits (Chinchilla).

The regenerative process was evaluated by histological and histochemical methods, by quantitative cytological methods (cytograms, mitotic and differentiation index), changes in vacuolar system of osteoblasts, volume changes of nuclei and of cytoplasm and by cytochemical methods (determination of DNA, RNA and AMPS *in vitro* and the bone salts *in vivo*).

Our results indicate that the Royal Jelly especially significantly accelerated the regenerative process. The favorable influence of Royal Jelly upon regeneration is dependent for its activity on differentiation and synthetic processes of osteoblasts.

Improvements on the Ferritin-antibody Technique

VOGT, ARNOLD, (Hygiene-Institut der Universität Freiburg i.Br., Germany)

A combination of purification procedures was developed yielding a conjugate of satisfactory antibody activity and low toxicity. The nonspecific reaction *in vitro* and the toxic effects *in vivo* could be essentially reduced. The difficulties of the *in vitro* application of ferritin-conjugated antibodies are discussed.

Chemodifferentiation of the Rat Small Intestine: the Development of Enzyme Pattern

VOLLRATH, L. (Anatomisches Institut der Universität Würzburg, 87, Würzberg, Germany)

The present study deals with development of the enzyme pattern in rat small intestine (jejunum) from the 16th day of gestation onwards. The earlier stages were not investigated as the major steps in the morpho logical development of the mucosa (formation of villi, differentiation of enterocytes, enterochromaffin cells, and goblet cells, respectively) take place in the last quarter of gestation. On the 16th day of gestation, several enzymes can be demonstrated histochemically—although in low activity—in the relatively undifferentiated epithelial cells: acid phosphatase, nonspecific esterase, α-glycerophosphate dehydrogenase (DH), glutamate DH, β-hydroxybutyrate DH, glucose-6-phosphate DH, succinate DH, lactate DH, and NADH- and NADPH-diaphorase, respectively. Alkaline phosphatase and leucine aminopeptidase cannot be visualized before the 17th and 20th day of gestation, respectively. In the early postnatal period, the following enzymes decrease in activity: alkaline phosphatase, leucine aminopeptidase, α-glycerophosphate DH, glutamate DH, and β-hydroxybutyrate DH. From the end of the second week of life onwards, activity of these enzymes increases again. The other enzymes investigated show a more or less steady increase postnatally. By the end of the 3rd week of life, the enzyme pattern of the young resembles that of the adult. Factors influencing the development of the enzyme pattern will be discussed.

The Problem of Non-Specific Staining in Immuno-Fluorescent Methods

VON MAYERSBACH, HEINZ (Anatomisches Institut, Medizinische Hochschule, Hannover)

In spite of the frequent and successful use of immunofluorescent methods in bacteriology and virology, the histochemist is faced with several special problems which are involved in the method when he applies it to tissue cells. One of the most important methodical steps is the preparation of the fluorescent antibodies and the prevention of their non-specific staining abilities. The formerly given concept of the electrostatic absorptions of labelled sera as a source of non-specific stainings has sponsored a new investigation on the influence of different labelling-markers, their binding capacities and their action on immuno-globulins for non-specific stainings.

A new method for gaining most specific staining fluorescent marked antibodies will be given. Furthermore, a method for *in vivo* localization will be demonstrated by the example of experimental induced glomerulo-nephritis.

The Significance of Circadian Cycles in Histochemical Work

VON MAYERSBACH, HEINZ (Anatomisches Institut, Medizinische Hochschule, Hannover)

Regardless of the knowledge of physiological circadian differences of animals and men, there is little known about their influence on results of histochemical reactions. Several casual findings have sponsored a se-ries of investigations in which standardized rats have been used. The standardization consisted of maintaining equal experimental conditions (food, external environment of the animals, age, sex, and strain). The experiments were performed in different seasons during several years. In the course of each experiment, animals were sacrificed in an uninter-rupted period of 24 hours; and liver, kidney, and spleen were investi-gated by histochemical and biochemical means for content of glycogen, nucleic acids, esterases, lactodehydrogenases and glucose-6-dehydro-genases. It was revealed that all the substances change drastically in amount as well as in cellular distribution during the 24-hour period. The pattern of the circadian curves are significantly modified by sex and season. The significance of these biological changes for histochem-ical research will be outlined and experimental influences on animals will be discussed.

Cytochemical Studies on Nucleoside Phosphatases Activity in Cell Nuclei

VORBRODT, ANDRZEJ (Department of Tumour Biology, Institute of Oncology, Gliwice, Poland)

The localization of enzyme(s) hydrolyzing ATP, GTP, GDP, GMP and sodium β-glycerophosphate at pH 5.9 was studied cytochemically using light and electron microscopy. The cytochemical method (1) was based essentially on the quantitative, biochemical studies by Siebert (2). The experiments were carried out on isolated rat liver nuclei, as well as on tissue sections fixed in formol-calcium or in glutaraldehyde buffered with sodium cacodylate. The glutaraldehyde-fixed ultrathin frozen sections

286

obtained according to method of Bernhard (3), and Bernhard and Leduc (4) were also used. Some additional control experiments concerning the appearance of diffusion artifacts, adsorption of lead phosphate on nuclear structures and specificity of the reaction were also performed. The results obtained indicate that in the nucleolus and in interchromatinic granules, enzymes hydrolyzing the nucleoside phosphates GTP, ATP and to some extent GDP and GMP, are present. These enzymes seem to be highly insoluble and are bound to ribonucleoprotein components of the cell nucleus. The chromatin shows no enzymatic activity. The highest activity of enzymes hydrolyzing ATP and GTP was observed in the fibrillar zone of the nucleolus. No positive reaction in nuclei and nucleoli was observed when sodium β-glycerophosphate was used as a phosphate ester.

References:

(1) Vorbrodt, A., S. Krzyzowska-Gruca, and Z. Steplewski: Bull. Acad. Pol. Sci. Class II, *12*, 337 (1964).
(2) Siebert, G.: Exptl. Cell Research, suppl. *9*, 389 (1963).
(3) Bernhard, W.: Ann. Biol. *4*, 5 (1965).
(4) Bernhard, W. and E. H. Leduc: J. Cell Biol., *34*, 757 (1967).

The Distribution of Lysosomal Enzymes in the Amphibian Pituitary During Ontogenesis

WÄCHTLER, KLAUS (Institut für Zoologie der Tierärztlichen Hochschule, 3 Hannover-Kirchrode, Germany)

As shown in a previous study, the pars distalis of the pituitary gland of adult caudate amphibians (newts and salamanders) was found to exhibit an exceptionally high activity of several lysosomal enzymes which were particularly abundant in the β-cells (producing gonadotropic hormone) of sexually active females. This paper deals with observations made on the pars distalis and pars intermedia at various stages of development. Acid phosphatase, E-600 resistant esterase, β-glucoronidase, β-glucosaminidase and sulphatase were demonstrated by azocoupling reactions using naphthol-AS compounds as substrates and hexazonium pararosanilin as a coupler. At most stages of development these enzymes were found in a granular localization in the pars distalis. In many cases, acid phosphatase and glucosaminidase activity could be detected in the pars intermedia as well. In stages near or just before metamorphosis, slight to moderate activity was observed. From metaphorphosis to sexual maturity, a gradual increase of lysosomal activity could be noticed particularly in β-cells of the pars distalis. The strong reaction in these cells suggests a close relation between lysosomes and hormone secretion. This relationship will be discussed.

Biosynthesis of Luteinizing Hormone and its Control in Male Rats *

WAKABAYASHI, KATSUMI and BUN-ICHI TAMAOKI (National Institute of Radiological Sciences, Chiba-shi, Japan)

After several treatments which caused various changes in androgenic states of animals, biosynthesis of luteinizing hormone (LH) was estimated from radioactivity incorporated into an immunochemically isolated LH fraction after incubation of the anterior pituitary glands with C^{14}-leucine *in vitro* (1). Orchiectomy or immunization of animals with ovine LH induced a severe androgen deficient state and caused a rapid and specific increase in LH biosynthesis, which was successfully prevented by androgen replacement made throughout the experimental period (2). Moderate and very mild androgen deficient states were induced by chryptorchism and local X-ray irradiation to the testes, respectively, one month after these treatments. In such animals, increase of LH biosynthesis was still obvious, suggesting that LH biosynthesis was very sensitive to androgen deficiency. Daily administration of androgen to normal rats for 2 weeks reduced LH biosynthesis below the normal level. LH biosynthesis seemed to be not so sensitive to an androgen excess state, because daily androgen administration to normal and castrated animals caused a transient non-specific increase of C^{14}-leucine incorporation into pituitary proteins. This included the LH fraction after the first 2-4 injections; the negative feedback effect was observed after 6 injections. This acute promoting action of androgen on radioactive amino acid incorporation seemed to be due, not to protein anabolic action, for a potent anabolic steroid with a weak androgenic action, methylandrostenolone acetate, failed to affect the biosynthesis (3). Chronic administration of L-thyroxine failed to produce any significant effect on LH biosynthesis in normal and various androgen deficient rats.

References:

(1) Wakabayashi, K. and B. Tamaoki: Endocrinology 77, 264 (1965).
(2) Wakabayashi, K. and B. Tamaoki: *ibid. 80,* 409 (1967).
(3) Wakabayashi, K., T. Ogiso, and B. Tamaoki: *ibid.* (1968) *in press.*

* Supported by NIH Grant AM 07715.

An Automatic Microscope for Cytogenetic Analyses

WALD, NIEL and RUSSELL RANSHAW (Radiation Health Division, Graduate School of Public Health, University of Pittsburgh, Pittsburgh, Pa.)

In view of the increasing usefulness of cytogenetic methodology in biomedical research, clinical diagnosis, and biologic monitoring of human populations, we have attempted the development of an automatic system for chromosome analysis. Since the two major time-consuming steps in the present manual methodology are the search for suitable mitotic cells and the numerical and structural analysis of chromosomes of these cells, it was deemed necessary to devise a mechanized microscope to provide rapid input in an automatic cytogenetic analysis system which, in turn, will carry out the chromosome analyses. The system is composed of a PDP-7 digital computer, an ultra-precision flying spot scanner and the mechanized microscope using either an incoherent or a coherent light source.

The microscope was built after a design study using a breadboard model. It will be described in detail, including the mechanical, coherent optical and incoherent optical sub-systems. Initial results of operational tests will be discussed as well.

Metabolism of Intracellular Lipid Droplets in Atherosclerosis

WELLER, ROY O. (Albert Einstein Coll. of Medicine, Bronx, N. Y. and Guy's Hospital Medical School, London S.E. 1, England)

Two main types of intracellular lipid droplet are observed when human and rabbit atherosclerotic lesions are studied with polarized light, histochemical, and electron microscopical techniques (1). One type is an anisotropic, liquid crystalline droplet with a lamellated ultrastructure containing phospholipid (osmium tetroxide-α-naphthylamine (OTAN) method (2) and cholesterol (digitonide precipitation). The other type is an isotropic, truly liquid droplet composed of hydrophobic globular lipid (OTAN), of which cholesterol ester forms the major proportion in atherosclerotic lesions. Electron microscopically the isotropic droplets are osmiophilic but have no detectable ultrastructure.

Most of the droplets are anisotropic in early atherosclerotic lesions whereas, in more advanced plaques, isotropic droplets predominate. Solid crystals of cholesterol also increase in later lesions but these appear to be extracellular. Intermediate stages between lamellated (anisotropic) cholesterol-phospholipid droplets and amorphous (isotropic) cholesterol ester droplets have been observed with the electron microscope. Lakes of amorphous lipid are interposed between the organized lipid lamellae and, in a possible later stage, the droplet is almost totally amorphous with only a few remaining lamellar fragments.

These microscopical observations not only support the biochemical evidence for cholesterol esterification in the aortic wall (3) but also suggest the probable intracellular site where this reaction occurs.

References:

(1) Weller, R. O.: J. Path. Bact. *94*, 171 (1967).
(2) Adams, C. W. M.: J. Path. Bact. *77*, 648 (1959).
(3) Lofland, H. B., D. M. Moury, C. W. Hoffman, and T. B. Clarkson: J. Lipid Res. *6*, 112 (1965).

Ultrastructural Investigations of the Role of Cholinesterases in the Production and Release of Calcitonin by Thyroid C Cells

WELSCH, ULRICH and A. G. E. PEARSE (Anatomisches Institut der Universität Kiel and Royal Postgraduate Medical School, London)

A high content of cholinesterase, most frequently of butyrylcholines-terase (BuChE), is one of the cief characteristics of the thyroid C cells (Carvalheira and Pearse, 1967), which produce the polypeptide hormone, calcitonin (Foster, MacIntyre, and Pearse, 1964).

In order to demonstrate the exact intracellular localization of this enzyme and thus to be able to obtain information about its functional role, a modification of the Koelle technique was transferred to the electron microscopical level. Rat thyroids were fixed with glutaraldehyde or formaldehyde by perfusion through the aorta, which allows very short fixation and rinsing times. The reaction product was observed mainly on membranes of the nuclear envelope and the granulated reticulum. A lesser amount of activity was also regularly found on the plasma membrane, on lateral offbudding end-formations of the Golgi region and inside a small number of the specific vesicles.

As it is known that the C cells respond to high levels of calcium in the blood by discharging their calcitonin (Pearse, 1966a, Matsuzawa, 1967), and that they rapidly decarboxylate certain amino acid derivatives (Ritzen, Hammarström and Ullberg, 1965, Pearse, 1966b), experiments with injected calcium chloride and 5-hydroxyptophan and DL-dihydroxyphenylalanine were carried out before doing the enzyme reaction. It is concluded that BuChE participates in the release mechanism of the dense-cored vesicles and presumably also has a function in protein synthesis.

References:

Carvalheira, A. F. and A. G. E. Pearse: Histochemie *8*, 175 (1967).
Foster, G., I. MacIntyre, and A. G. E. Pearse: Nature *203*, 1029 (1964).
Matsuzawa, T.: Arch. Histol. Jap. 27, 521 (1967).
Pearse, A. G. E.: Proc. Roy. Soc. B. *164*, 478 (1966a).
Pearse, A. G. E.: Nature *211*, 5049 (1966b).
Ritzen, M., L. Hammarström, and S. Ullberg: Biochem. Pharmacol. *14*, 313 (1965).

Eine Autoradiographische Methode zur Darstellung Nichtflüchtiger Stoffe in Biologischem Material

WERNER, GOTTFRIED (Max-Planck-Institut für Hirnforschung, Arbeitsgruppe Neurochemie, 6000 Frankfurt am Main, DBR)

Mit einer von uns (1-6) schon 1963 entwickelten Methode, ist es möglich, Substanzen (ausser allen leicht flüchtigen Verbindungen), die in wasserlöslicher oder lipoidlöslicher Form in Geweben vorliegen, autoradiographisch bestimmten histologischen Strukturen zuzuordnen. Das Verfahren besteht im wesentlichen darin, mit der Gefrierschnitt-Methodik Gewebsschnitte herzustellen und diese nach Gefrier-Trocknung in engen Kontakt mit einer ebenfalls trockenen Photo-Emulsion zu bringen. Nach der Exposition wird der Gefrierschnitt für die färberische Differenzierung von der Photo-Emulsion wieder getrennt und das Autoradiogramm photographisch entwickelt. Dadurch ergibt sich die Notwendigkeit für ein Verfahren der genauen Zuordnung von autoradiographischem Bild zu den entsprechenden histologischen Strukturen. Es wird eine besonders dafür entwickelte optische Anordnung beschrieben, bei der Autoradiogramm und histologisch gefärbter Schnitt optisch zur Deckung gebracht werden. Der Schwärzungsgrad bestimmter Areale des vom Gefrierschnitt getrennten Autoradiogramms wird mittels eines Mikrophotometers durch Messung der Absorption im Durchlicht quantitativ bestimmt.

Es werden an einigen Beispielen die Anwendbarkeit und Leistungsfähigkeit dieser Methode demonstriert.

Kausale Zusammenhänge zwischen autoradiographischer Lokalisation von pharmakologisch oder biologisch wirksamen Substanzen in bestimmten Strukturen des Gehirns oder anderer Organe und der pharmakologischen dzw. biologischen Wirkung dieser Stoffe im Körper werden diskutiert.

References:

(1) Werner, G.: 6th Intern. Congr. Biochem. (New York) Abstr. S. 458 (1964). (1964).
(2) Bosque, P. G. and G. Werner: 2nd Intern. Congr. Histo. and Cytochemistry (Frankfurt/Main) S. 188 (1964).
(3) Werner, G., P. G. Bosque, und J. Carreres Quevedo: C. R. assoc. anatom. [Nancy] (Madrid) S. 1869 und 1871 (1964).
(4) Werner, G., P. G. Bosque, und J. Carreres Quevedo: Abh. dtsch. Akad. Wiss. Berlin, Kl. Chem., Geol. Biol. *3*, 541 und 629 (1966).
(5) Werner, G., H. Werner, P. G. Bosque, und J. Carreres Quevedo: Z. Naturforschg. *21b*, 238 (1966).
(6) Werner, G.: Arzneimittel-Forschg. *18*, (1968) *i. Druck.*

Computer Assisted (TICAS) Tumor Cell Identification

WIED, GEORGE L., PETER H. BARTELS, and GUNTER F. BAHR (University of Chicago, Departments of Obstetrics and Gynecology, and Pathology, Chicago, Ill., University of Arizona, Department of Microbiology, Tucson, Arizona, Armed Forces Institute of Pathology, Washington, D. C.)

A feasibility study of computer assisted cell identification with TICAS (TAXONOMIC INTRA-CELLULAR ANALYTIC SYSTEM) was performed on Papanicolaou-stained material consisting of 30 normal glandular cells (15 with and 15 without cytoplasm), 30 apparent histiocytes, and 30 apparent tumor cells (from adenocarcinoma), all derived from the uterine cavity. Absorption measurements were performed by means of a Zeiss UMSP-I microspectrophotometer connected on-line to a LINC-8 computer. Several discrimination procedures were employed and described.

With the exception of one apparent tumor cell which exhibited degenerative changes (chromatolysis), each tumor cell could be unequivocally distinguished from each normal glandular cell and each histiocyte. From its present developmental direction, it seems that TICAS may evolve into a computerized consultant. It will be of particular assistance in the discrimination of cell types which are so similar that they cannot be unequivocally distinguished by their morphology.

Observations on a New Fluorophore in the Adrenal Medulla

WILLIAMS, VICK and FRAN MORRISS (The University of Texas Southwestern Medical School at Dallas, Dallas, Texas)

Characteristic fluorescence can be obtained from a number of biogenic amines in the presence of dry protein when exposed to ultraviolet light after treatment with formaldehyde vapor (1). The fluorescence of serotonin is yellow, while that of norepinephrine and epinephrine is a yellow-green color. In addition, since epinephrine requires longer exposure to formaldehyde to develop fluorescence, it is at least theoretically distinguishable.

In the course of experiments to standardize procedures for studying various monoamines in the central nervous system, adrenal medullae of rats and mice were freeze-dried, exposed to formaldehyde gas under controlled conditions of temperature and humidity (2) and examined in the fluorescence microscope. In addition to the yellow-green fluorescing cells ordinarily described, these preparations contained scattered islets of cells possessing an orange-brown fluorescence. Since the orange-brown color is not one expected for formaldehyde-treated catecholamines and tryptamines, an attempt was made to characterize the material responsible for the fluorescence. Separate solutions of norepinephrine and serotonin were prepared to simulate tissue concentrations; neither of these, when evaporated on a dry protein substrate and reacted under conditions comparable to those used in preparing the adrenals, produced the orange-brown color. The autofluorescence of lipofuscin can be distinguished by its golden yellow color and smaller size; furthermore, it is predominantly located in the cortex. Ultraviolet examination of freeze-dried, untreated tissue revealed no medullary fluorescence. When exposure of the tissue to formaldehyde was prolonged in an attempt to distinguish the epinephrine-producing cells, no difference in the fluorescence pattern or color was discernible. In Syrian hamsters, in which norepinephrine-producing cells occupy the medullary border, the "orange-brown" cells were also peripheral.

The origin of the orange-brown fluorescence is unknown, but it seems likely that this procedure detects the presence of a compound structurally related to the catecholamines. Of representative derivatives of phenylalanine, the only ones developing intense fluorescence are primary amines with hydroxyl groups at the 3 and 4 positions, the 3-OH being absolutely essential; condensation of formaldehyde and the ethylamine chain converts the molecule into a tetrahydroisoquinoline (3). Pharmacological and histochemical studies are being carried out in an attempt to characterize further the orange-brown fluorescing material in adrenomedullary cells.

References:

(1) Falck, B.: Acta Physiol. Scand. *56*, Suppl. 197 (1962).
(2) Hamberger, B., T. Malmfors, and C. Sachs: J. Histochem. Cytochem. *13*, 147 (1965).
(3) Falck, B., N. A. Hillarp, G. Thieme, and A. Torp: J. Histochem. Cytochem. *10*, 348 (1962).

Enzyme Histochemistry and Prognosis of Human Tumors

WILLIGHAGEN, R. G. J. (Pathological Laboratory, Leiden, The Netherlands)

Grading in several types of human tumors can be based on the morphological differentiation of these tumors. By investigating the histochemically demonstrable enzyme activities in 2500 human tumors, it was found that, in several tumor types, the activities of hydrolytic enzymes show considerable differences. These differences are clearly related in many tumor types with morphological differentiation. It appeared, however, that this phenomenon which we consider to be enzymatic differentiation, does not always go parallel with the morphological grade of differentiation. Therefore, we tried in greater groups of tumors to find a relation between the grade of enzymatic differentiation and the clinical behaviour of these tumors such as survival time, tendency to metastasize, rate of growth, etc. Of the most common tumor types such as mammary gland carcinoma, carcinoma of the stomach, carcinoma of the lung, cervix and colon, a greater number were available for clinical evaluation. But also from more rare human tumors, greater groups could be studied. In this group were included phaeochromocytoma, thyroid gland carcinoma and fibrosarcoma.

In the group of lung tumors, a distinct relation was found between the amount of non-specific esterases. The higher the activity of this enzyme. the better the mean survival time. In the groups of stomach cancers and mammary gland carcinomas, no relation between enzymatic differentiation and prognosis could be found. In colon carcinomas there was a parallelism between morphological and enzymatic differentiation. As in cervix carcinomas, however, there was a great variation in the anatomical extension of these tumors. When we divided these bigger numbers of tumors into groups with the same anatomic extension, they became too small to be evaluated.

The prognosis of thyroid carcinomas seems to be correlated with activity of 5′-nucleotidase, the prognosis of fibrosarcomas with activity of alkaline phosphatase.

Cholinesterases and Inherited Muscular Dystrophy of the Chicken [*]

WILSON, BARRY W. (Department of Poultry Husbandry, University of California, Davis, Calif.)

A progressive form of muscular dystrophy is inherited as an autosomal recessive trait in the chicken. Muscles with a high proportion of white fibers such as those of the breast and wings are the first ones to be affected. Although there is evidence suggesting that development of these muscles differ in normal and dystrophic chicks, the primary events associated with this genetic abnormality have not yet been established. This report concerns results of studies of the cholinesterases of muscles from embryos, chicks and adults of several normal lines and four lines homozygous for the dystrophic gene using acrylamide cell electrophoresis, spectrophotometric enzyme assays and histochemical techniques. The data show that cholinesterase isoenzymes characteristic of the skeletal muscles of normal embryos are maintained in the pectoral and biceps muscles of adult dystrophic birds whereas these enzymes fall to very low levels in the muscles of normal birds after the chicks are a few weeks old. For example, 6 and 12 week old pectoral muscles from dystrophic

294

line 307 birds averaged 39 and 41 times the cholinesterase activities of their normal counterparts. In contrast, line 307 adductor muscles (a "red" muscle of the leg) had activities only 1.8 and 1.5 times that of the normal line muscles. Acrylamide gel electrophoresis studies have, to date, revealed that at least 5 bands of cholinesterase activity are involved. Preliminary histochemical studies indicate that high cholinesterase activity of dystrophic line pectoral muscle is associated with regions of the muscle fibers themselves. The results of studies examining the cellular localization of the cholinesterase activities and the effects of organophosphorus inhibitors and various substrates on the cholinesterase isoenzymes of muscles from normal and dystrophic line embryos and chicks will be presented. The implications of the fact that dystrophic pectoral and biceps muscles seem unable to stop the synthesis of cholinesterases characteristic of normal embryonic skeletal muscle will be discussed.

* Supported by grants from the NIH (NB-07359) and the USPHS (ES-00202).

Zur Spezifitaet des histochemischen Nachweises der L-Gulono-γ-Lactondehydrogenase

WOHLRAB, FRANK (Pathologisches Institut der Karl-Marx-Universität Leipzig, Leipzig, East-Germany)

Die L-Gulono-γ-Lactondehydrogenase katalysiert im Rahmen der Ascorbinsäuresynthese den wichtigen Stoffwechselschritt L-Gulono-γ-Lacton → L-Ascorbinsäure, wobei das mikrosomale Enzym streng spezifisch für lie Lacton-Form ist (1). Die Topik des auch biochemisch charakterisierten Enzyms (2) war an der Rattenleber nach der von Cohen (3) inaugurierten histochemischen Technik nicht reproduzierbar. Die von uns unter aeroben Inkubationsbedingungen an nativen Kryostatschnitten von Ratten-Leber und Frosch-Niere durchgeführten Untersuchungen ergaben keinen Anhalt für einen enzymatischen Abbau des L-Gulono-γ-Lactons. Eine entsprechende Variation bzw. Kombination der einzelnen Bestandteile des Inkubationsmediums (L-Gulono-γ-Lacton, 1 mM—20 mM; o,o6 M Phosphat- bzw. o,2 M Tris-Maleat-Puffer, pH 6,5—7,5; Nitro-BT, 6 mM; Phenazinmethosulfat, o,8 mM; Menadion, o,1—o,5 mM, gelöst in Aceton; Vitamin K_3, o,1—o,5 mM; NADP, 1—5 mM; NaN_3, 10 mM; KCN, 10mM (Jeweils Finalkonzentration)) ergibt keine signifikanten Unterschiede gegenüber Kontrollen ohne Substrat.—Diese im Gegensatz zu ben Befunden von Cohen stehenden Ergebnisse sind nur schwer zu deuten. Es gibt hierfür folgende Erklärungsmöglichkeiten: 1. Wenn die Gulonolactonhydrolase (E. C. 3.1.1.18) aktiv ist, wird L-Gulono-γ-lacton enzymatisch hydrolysiert zur L-Gulonsäure, sodass im

Prinzip Substrat für 4 Enzyme (E. C. 1.1.1.19; E. C. 1.1.1. 20; E. C. 1.1.1. 45; E. C. 1.1.3.8) vorliegt, dessen Konzentration im Inkubationsmedium variiert. 2. Das Substrat wird unter den gewählten Inkubationsbedingungen nichtenzymatisch hydrolysiert. Bemerkenswert erscheint aber, dass Balogh (4) beim histochemischen Nachweis der L-Gulonat: NADP-Oxidoreduktase keine Reaktion mit L-Gulono-γ-Lacton erhielt. 3. Die L-Gulono-γ-Lactondehydrogenase-Aktivität sinkt schon 1 Stunde post mortem um die Hälfte ab (5).—Weitere Untersuchungen zur Histotopochemie der Ascorbinsäuresynthese werden vorgenommen.

References:

(1) Chatterjee, I. B. *et al.*: Biochem. J. *76*, 279 (1960).
(2) Bublitz, C.: Biochem. Biophys. *48*, 63 (1961).
(3) Cohen, R. B.,: Proc. Soc. exp. Biol. Med. *106*, 309 (1961).
(4) Balogh, K.: J. Histochem. Cytochem. *13*, 533 (1965).
(5) Holimann, S. und J. Neubaur: Klin. Wschr. *44*, 722 (1966).

Application of the Indigogenic Principle to the Histochemical Localization of Phosphodiesterases I and II [*]

WOLF, PAUL L., JEROME P. HORWITZ, JOSEF V. FREISLER, ELISABETH VON DER MUEHLL, and JANICE VAZQUEZ (Detroit Institute of Cancer Research Division of the Michigan Cancer Foundation, and Departments of Pathology and Oncology, Wayne State University School of Medicine, Detroit, Mich.)

Previous studies in this laboratory have demonstrated the utility of suitably substituted *o*-indoxyl derivatives for the histochemical localization and demonstration of a variety of hydrolytic enzymes. The application of these substrates derives from the rapid deposition of a highly colored, microcrystalline indigo at sites of reactivity through oxidation of an enzymically released intermediate indoxyl. The efficacy and advantages of indigogenic staining in the cytochemical localization of a variety of hydrolases through appropriately substituted *o*-indoxyl derivative are now well documented in a series of publications by the present authors. We have now successfully extended the indigogenic principle through the histochemical demonstration and localization of the phosphodiesterases (PDases I and II) through syntheses of 5-bromo-4-chloro-3-indolyl thymidine 3'-phosphate (A) and 5-bromo-4-chloro-3-indolyl thymidine 5'-phosphate (B). PDase I activity was observed at pH 9 with B, which affords a granular deposition of the indigo in the cytoplasm of the small intestine, mucosa, liver, and the proximal and distal convoluted tubules of the kidney. There was no evidence of nuclear activity. Substrate A, when incubated at pH 4.8 with fresh frozen sections of spleen, kidney, liver, and intestine of both rat and pig, gave rise to indigo staining of cell nuclei and cytoplasm. At pH 5.9, the staining is almost exclusively confined to cytoplasm and appears to be lysosomal. The observation of both nuclear and cytoplasmic staining with substrate A is in accord with a recent report of Bernardi and co-workers who reported that highly purified hog spleen DNase II shows "phosphodiesterase activity" on a series of *para*-nitrophenylphosphodiesters in addition to its expected action on deoxyribonucleic acid.

The present state of knowledge concerning intracellular localization of enzymes attacking nucleic acids, that is, the polynucleotidases, has in large measure been accumulated from studies involving cellular fractionation techniques. Substrates A and B afford a means of obtaining comparable data on the PDases I and II by standard histochemical techniques.

* This investigation was supported in part by Public Health Service Research Grant No. CA 02624 from the National Cancer Institute and in part by an institutional grant to the Detroit Institute of Cancer Research Division of the Michigan Cancer Foundation from the United Foundation of Greater Detroit.

Odland Body (Membrane Coating Granule): An Epidermal Lysosome?

WOLFF, K., J. TAPPEINER, and K. HOLUBAR (Univ. Vienna, Dept. of Dermatology, School of Medicine, Austria)

Odland bodies of normal and keratin-stripped epidermis were investigated by electronmicroscopic and electron-cytochemical techniques. These membrane limited organelles contain a complex system of closely set lamellae with individual subunits consisting of three parallel membranes separated by 55 Å. Odland bodies are produced by keratinocytes of the upper spinous and the transitional layers of the epidermis, they fuse with the plasma membrane and their contents are extruded into the extracellular space where they give rise to the formation of densely packed lamellar masses. Electron-cytochemically it can be demonstrated that Odland bodies and their extracellular derivatives contain acid phosphatase attached to their lamellar subunits. The enzyme can be removed from these organelles by treatment with Triton X 100. Experimentally induced rapid parakeratotic keratinization results in a retention of Odland bodies in parakeratotic cells; accelerated proliferation of epidermal cells leads to an increased production and rate of extrusion of Odland bodies by keratinocytes. It is concluded that Odland bodies may represent special forms of epidermal lysosomes which are morphologically quite distinct from the lysosomes present in the lower layers of the epidermis.

Histochemistry of Demyelination and Myelination

WOLMAN, MOSHE (Departments of Pathology and of Cell Biology and Histology, Tel-Aviv University Medical School, Government Hospital, Tel-Hashomer, Israel)

Histochemical, histophysical and electron microscopic data indicate that the myelin sheath is a multi-lamellar structure constituted mainly of ordered lipo-proteic units. The presence of mucopolysaccharide has been reported but is debated. The presence of enzyme proteins has also been upheld, but with a few exceptions, has been denied on the basis of further studies.

Differences were observed between central and peripheral myelin in the extractability of various fractions with lipid solvent mixtures and their digestibility with proteolytic enzymes. These differences might explain the different reactions to disease processes of the two types of myelin. There appear to be different types of demyelinative processes. The best studied were those in which demyelination was caused by transection of the axon (Wallerian degeneration), by multiple sclerosis, by experimental allergic encephalomyelitis, certain toxic substances and by ischemia. In these processes, myelin breaks down into globules and ovoids, myelin fragments, which appear to be made of particles of normal myelin, but which differ in their physical characteristics, some being more hydrophilic than others. Many of these fragments are ingested by neighboring cells, others, however, appear to be degraded *in situ*.

The etiology and pathogenesis of myelin breakdown are poorly understood, some progress having been made, with numerous problems remaining unsolved. There is evidence indicating release of lysosomal enzymes from axons during Wallerian degeneration which may explain the myelinolysis. Evidence indicating that Wallerian degeneration is primarily due to hydrolysis of a trypsin-digestible protein has recently been adduced and might be of importance. Knowledge regarding enzymatic activities of myelin breakdown in processes affecting primarily the sheath itself (multiple sclerosis, E.A.E. and diphtheritic neuropathy) is however scanty, although there is some evidence of lysosomal activation in the parent or neighboring cells of the sheaths. The problem how do these enzymes diffuse through the 20-30 Å wide spaces between the layers and in which of the two hydrophilic spaces of myelin do the enzymes act, has not yet been answered.

The studies of our group suggest that trypsin-digestible proteins are present in both layers of myelin, i.e., the layers which continue the surface facing the extracellular space and those which face the continuation of the intracellular medium. The problem whether myelinolysis due to axonal change is due to diffusion of enzymes along the extracellular space, and whether myelinolysis due to changes in the parent cells of myelin is due to diffusion along the intracellular space, has not yet been answered.

Myelination is known to consist of progressive apposition of new lamellae around axons. The problem whether the newly-laid myelin lamellae are mature, the process of myelination thus consisting of a quantitative increase in the number of lamellae, or if the apposition is accompanied and followed by maturation of the lamellae has not yet been satisfactorily answered. Histochemical, morphological and biochemical studies yielded data which in part support the one and, in part, the other hypothesis.

Effects of Phase-transitions in Membranes on Bound Enzymes *

WOLMAN, M., M. KALINA and J. J. BUBIS (Departments of Cell Biology, Histology and Pathology, Tel-Aviv University Medical School, Government Hospital, Tel-Hashomer, Israel)

It has been previously reported (1, 2, 3) that various enzyme activities can be reversibly inhibited by exposing fresh frozen sections to Ca or CNS ions. The effect of Ca was suggested to be due to action of this ion on anionic charges of one side of cellular membranes, whereas the effect of CNS was believed to be caused by its action on cationic charges on the opposite side of membranes. These effects could be reversed by treatment with NaCl.

Further experiments showed that in the case of some enzymes (e.g., acid phosphatase), the results were in part due to a step of drying introduced between the treatment by ions and the histochemical demonstration of enzyme activity. The inhibitory action of the Ca and CNS ions was found to occur at higher concentrations than those previously reported, the higher concentrations being obtained in the sections during the drying process. Treatment by high concentrations of ions caused inhibition which was partly reversible for most enzymes. Drying of sections treated with Ca or CNS ions (but not with NaCl) caused inhibition which was not reversible, probably because of fixation of the "water in oil" phase. With some enzymes, irreversible fixation was not obtained by drying after Ca treatment when the drying was done with nitrogen.

References:

(1) Wolman, M.: Z. Zellforsch. *65*, 1 (1965).
(2) Wolman, M. and J. J. Bubis: Histochemie 7, 105 (1966).
(3) Wolman, M., J. J. Bubis, and H. Wiener: Histochemie *9*, 1 (1967).

---

* Supported by grant No. 4x5109 of the U.S.P.H.S.

A Rapid Recording Microfluorometer and its Application for the Study of Nucleic Acid-dye Binding

YAMADA, MASAOKI (Lab. of Cytochem., Department of Anatomy, School of Medicine, Tokushima University, Tokushima, Japan)

The fluorescence of nucleic acid-acridine orange (AO) compounds was widely studied (1) and applied to measurement, at cellular levels, using a microfluorometer developed by Caspersson's group (2). The author constructed a simplified microfluorospectrophotometer with a high speed recording device. The fluorescence intensity of the compounds varied with the ratio of nucleic acid-P to AO (M/M). The wave length of maximum fluorescence depended upon the mode of binding of nucleic acids with dye. The effect of basic proteins on the fluorescence of nucleic acid-dye compounds was studied and the results were compared with the fluorescence of nuclei and ribosomes. The fluorescence of nucleic acids induced by cooperation with carcinogenic dyes was also studied.

References:

(1) Van Duuren, B. L.: Fluorescence and phosphorescence analysis, pp. 195-205 (D. M. Hercules edit.) New York: J. Wiley (1966).
(2) Riger, R., Jr.: 2nd Intern. Congr. Histo-Cytochem. (Frankfurt/Main) Abstr. p. 87 (1964), Acta Physiol. Scand. *67*, Suppl., 267 (1966).

A Possible Role of Basic Ribosomal Proteins in Liver Nuclei and Ribosomes

YAMADA, MASAOKI and SUNAO IWATA (Lab. of Cytochem., Department of Anatomy, School of Medicine, Tokushima University, Tokushima, Japan)

Nuclei which were active in RNA synthesis were isolated from rat liver (1). The nuclei were used in combination with basic proteins extracted from ribosomes to see how proteins suppress nuclear RNA synthesis. Ribosomes which had been partially deprived of their proteins were used to see whether ribosomal proteins cooperate in ribosomal protein synthesis. Ribosomal proteins were fractionated into three basic protein fractions using a CMC column. The elution pattern of these fractions corresponded well with that of histones extracted from nuclei. ^{3}H-uridine uptake of nuclei was slightly inhibited by addition of these proteins and the uptake was inhibited more by addition of a low, than of a high basic fraction of the proteins. The uptake of ^{14}C-alanine into ribosomes decreased markedly after partial removal of ribosomal proteins, but uptake was restored considerably by treatment causing recombination of ribosomes with their proteins.

These results suggest that these proteins may provide a key to studies on the relationship between nuclear RNA synthesis and ribosomal protein synthesis.

References:

(1) Yamada, M. and S. Iwata: 2nd Intern. Symp. Cell. Chem. (Ohtsu, Japan) Abstr. p. 3 (1966).

Electron Microscopic Localization of Forssman Antigen by the Immunouranium Technique

YAMAGUCHI, HISAO (Keio University School of Medicine, Tokyo, Japan)

The cellular distribution of Forssman antigens in the organs of various species has been studied by Tanaka and Leduc (J. Immunol. 77: 198, 1956) using fluorescent antibodies. We are now reporting the subcellular distribution of these antigens in guinea pig kidney using the immunouranium-T-O technique. One of the advantages of the immunouranium technique is in the application of electron-opaque antibody to the ultrathin section rather than to the tissue prior to embedding. As the antibody molecule does not penetrate every subcellular membrane, only application of antibody after sectioning insures localization of every reacting subcellular antigen site. Hitherto, the immunouranium technique has been used entirely on free cells and fixation in cold osmium tetroxide for 4 minutes was adequate. In our present employment of the immunouranium technique on solid tissue, fixation and embedding were modified to insure penetration of fixative throughout the tissue without destroying antigenic reactivity. Guinea pig kidney was fixed, washed, dehydrated and embedded at 1°C in 6% glutaraldehyde (30 min), 0.1M caccodylate buffer, pH 7.4 (2 hrs), 4 changes of graded ethylene glycols (40 to 90%) for 1 hr each, 100% ethylene glycol overnight, 50% hydroxypropylmethacrylate (H.P.M.A.) in ethylene glycol (30 min), 97% HPMA (30 min), 50% methyl-butylmethacrylate (3:2) in HPMA (30 min) and a solution containing 1% divinylbenzene in a mixture of 3 parts of methyl with 2 parts of butylmethacrylate. The blocks were hardened at 47°C for 36 hrs. After etching of the sections in benzene water for 10 min and rehydration for 1 hr they were treated with rabbit anti-sheep erythrocyte serum, uranium-labeled sheep antibody to rabbit antibody, thiocarbohydrazide and osmium tetroxide. Control sections, always prepared from the same block as the experimental sections, were treated in an identical manner except that normal rabbit serum absorbed with guinea pig kidney was substituted for the rabbit anti-sheep erythrocyte serum.

In glomeruli, Forssman antigen was found along the pedicles and cell membranes of the epithelial cells. With fluorescent antibody, Tanaka and Leduc found the antigen in the cytoplasm of some of the glomerular cells that may have been either endothelial or epithelial cells.

In proximal tubules, uranium antibody failed to reveal antigen in microvilli or in other regions of the apical portion of the proximal tubule cell. The antigen was found however as aggregates of granules in sparse cytoplasmic regions between mitochondria in the basal portions. With light microscopy, Tanaka and Leduc found no Forssman antigen in the cells of the proximal tubules except for small amounts in the brush border of proximal convoluted segments.

In cells of the distal tubules, uranium antibody localized antigen close to the cell membrane which exhibited typical blunted microvilli. Moderate amounts of antigen in the cytoplasm of distal tubules was reported in the light microscopic studies of Tanaka and Leduc.

In cells of the collecting tubules, uranium antibody localized the antigen as granules dispersed throughout the cytoplasm. The fluorescent antibody studies of Tanaka and Leduc revealed abundant antigen in the cytoplasm of these cells.

Microfluorimetric Studies on the Periodic Acid-Schiff Reaction in Blood Cells

YATAGANAS, X., G. GAHRTON and B. THORELL (Institute of Pathology, Department of Internal Medicine, Karolinska Sjukhuset and the Institute for Medical Cell Research and Genetics, Karolinska Institutet, Stockholm 60, Sweden)

A fluorescent periodic acid-Schiff (F-PAS) reaction in neutrophil leukocytes was quantitated by means of microfluorimetry (1) using 2,5 bis (4' aminophenyl (—1')) 1,3,4 oxdiazol, (BAO), CIBA, Basel, Switzerland, as the Schiff type reagent. Parallel absorption measurements of the usual periodic acid-Schiff (PAS) reaction were performed in a scanning microspectrophotometer (2). Emission and absorption spectra (3) of the F-PAS reaction showed peaks at 430 and 340 $m\mu$ respectively both in pure glycogen solution and in cells. There was no influence of absorption on excitation energy and therefore on fluorescence yield in the cells. This was studied by measuring the total absorption at the excitation wave length (365 $m\mu$), in the microspectrophotometer, and fluorescence in neutrophils using different concentrations of the BAO reagent. The F-PAS positive substance in neutrophils was almost completely digestible with α-amylase, which indicated that glycogen was the reactive substance. The rate of fluorescent Schiff reaction in neutrophils was dependant both on the BAO concentration and the glycogen content of the cells. A plateau was reached at different times in high and low glycogen containing cells.

The correlation between fluorescence yield (after complete F-PAS reaction) and the glycogen amount, was investigated using high and low glycogen containing neutrophils (circa 100 per cent difference between lowest and highest mean values as determined by PAS-absorption photometry) from patients with different diseases. There was linearity between the PAS microspectrophotometric and the F-PAS microfluorimetric measurements in the cells (means of 100 cells in each case) for the lowest BAO concentration used (0.00001 per cent BAO). The high sensitivity of the method was shown to be of value in a preliminary investigation of the glycogen content in erythroblasts from patients with thalassemia major.

References:

(1) Caspersson, T., G. Lomakka, and R. Rigler, Jr.: Acta Histochem. Suppl VI. 123 (1965).
(2) Lomakka, G.: Acta Histochem. Suppl. VI, 47, (1965).
(3) Chance, B., R. Perry, L. Akerman, and M. Thorell, Rev. Scient. Instr. *30,* 735 (1959).

Color in Automated Cell Analysis *

YOUNG, IAN T. (Research Laboratory of Electronics, Massachusetts Institute of Technology, Cambridge, Mass.)

Biological stains are principally used to add the "dimension" of color to problems of cell analysis. The ability to describe quantitatively the color contents of white blood cell photomicrographs has been achieved through the use of SCAD, a flying-spot color scanner, and a set of computer algorithms. These algorithms compute chromaticity coordinates, measure spectrally band-passed intensity distributions, and correct for non-uniformities in cell and scanner illumination.

The limitations as well as the capabilities of the system will be discussed in terms of the type of spectral illumination used (narrow-band versus wide-band) and the effect on the results of various color films.

---

* This work was supported principally by the National Institutes of Health (Grants 1 P01, GM-14940-01 and 1 P01, GM-15006-01, and in part by the Joint Services Electronics Programs (Contract DA28-043-AMC-02536(E)).

Immunohistochemical Evolution of the Effects of Surgical and Chemical Bursectomy in the Chicken

ZACCHEO, D., C. E. GROSSI and V. GENTA (Institutes of Human Anatomy, Universities of Genova and Cagliari, Italy)

The role of the bursa of Fabricius in the immunological responsiveness of the chicken has been defined in the last years. This organ, through production of immunocompetent cells and possibly also of a diffusible factor, should control the lymphoid periphery and promote humoral reactions. Immunohistochemical investigations have shown that the bursa of Fabricius has active immunoglobulin (Ig)-synthesizing capacities. After hatching, Ig can be localized within the medulla of bursal follicles. On the other hand, the bursa itself does not show antibody-producing capacities.

Surgical bursectomy has been performed in newly hatched chickens and in animals artificially hatched on the 20th day of incubation. Secondary responses have been produced by repeated antigenic stimulation (BSA). Immunohistochemical localization of Ig producing elements has been investigated by means of fluorescent anti-Ig sera in the thymus, spleen, and caecum.

Chemical bursectomy has been performed by the injection of hormones in the yolk-sac of chicken embryos on the 9th day of incubation. 19-nortestosterone and dihydrotestosterone have been employed in these experiments. Immunohistochemical tests have been applied as previously, after repeated antigenic stimulation.

A comparison of the two experimental procedures shows that chemical bursectomy is more effective in reducing the immunological capacities of the chicken. However, a residual antibody-producing activity is still detectable in secondary responses.

An Electron Histochemical Study of Rat Kidney Lysosomes During Autolysis

ZACCHEO, D. and A. RIVA (Dept. of Normal Anatomy, University of Cagliari, Italy)

By means of techniques for ultrastructural localization of acid phosphatase and arylsulfatase, the behaviour of lysosomes of the epithelial cells of the proximal convolute tubule of rat kidney have been studied during autolysis. Within 15 minutes after death, a diffusion of enzymes into the surrounding cytoplasm is already discernible. Some of the lysosomes are also ruptured showing loss of their content. With progress of autolysis, changes in lysosomes become more evident until the complete or partial fragmentation of many of these organelles.

Tetanus Toxin: Localization in CNS and Skeletal Muscle Using Horseradish Peroxide Labelled Antitoxin

ZACKS, S. I. and M. F. SHEFF (Ayer Clinical Laboratory and Department of Neurology, Pennsylvania Hospital, Philadelphia, Pa.)

Current data (1,2) indicate that the syndrome of tetanus intoxication is the result of toxin binding in both the central nervous system and skeletal muscle. Previous studies from this laboratory have shown marked selectivity for fluorescein-labelled toxin binding by mitochondria isolated from mouse brain and in skeletal muscle (3). To increase the resolution of binding studies, the recently reported technique of labelling proteins with horseradish peroxidase was used (4).

Mice poisoned with a minimum saturation dose (9λ) of purified tetanus neurotoxin (5) were sacrificed when signs of generalized intoxication were well developed. Sections of brain and skeletal muscle were fixed in 10% phosphate buffered formalin for 4 hours, and prepared according to the Pierce and Nakane method (4) for electron microscopy. Controls consisted of normal brain and skeletal muscle incubated with peroxidase-labelled tetanus antitoxin. Immunoelectrophoresis of the labelled antitoxin showed that the antitoxin was labelled and retained its ability to combine with toxin.

The histochemical reaction product indicating sites of binding was found within central nervous system mitochondria and within the sarcotubular system of skeletal muscle. No reaction product was found in CNS synapses or within motor endplates. Control preparations were devoid of reaction product in these sites. The significance of these observations will be discussed.

References:

(1) Brooks, V. B., D. R. Curtis, and J. C. Eccles: J. Physiol. *135*, 655 (1957).
(2) Ranson, S. W.: Arch. Neurol. Psychiat. *20*, 663 (1928).
(3) Zacks, S. I. and M. F. Sheff: Acta Neuropath, *4*, 267 (1965).
(4) Nakane, P. K. and G. B. Pierce, Jr.: J. Histochem. Cytochem. *14*, 931 (1966).
(5) Sheff, M. F., M. B. Perry, and S. I. Zacks: Biochem. et Biophys. Acta *100*, 215 (1965).

Lactic Dehydrogenase Isoenzymes and Their Relationship to Tissue Differentiation

ZAMFIRESCU-GHEORGHIU, M., M. SERBAN, C. VLADESCU, Z. CHIRULESCU and N. MARCUS (Institute of Internal Medicine, Bucharest 10, R. S. Romania)

Starting from the relationship particularly emphasised by Pfleiderer *et al.*, and Cahn, Kaplan *et al.*, between the energetic ambiance of a tissue and its isoenzymes pattern, we followed experimentally (in tissue and serum) and clinically (in serum and leukocytes) the possible changes undergone by these patterns during pathologic subacute and chronic processes.

In toxic hepatitis induced with CCl_4 in the rabbit, a tendency has been noticed in the hepatic parenchyma for an equal percentage distribution of the LDH activity among the fractions, with concentration in the hybrid ones (predominantly in fraction III) and a constant decrease in fraction V. The tissue LDH isoenzymic pattern might constitute an indication of the state and degree of tissue differentiation of the aerobiotic ambiance—the anodic pattern—or of the anaerobiotic one—the cathodic pattern—of the respective system. The serum isoenzymic patterns reflecting that of the impaired tissue, has a character of organospecificity useful for the differential diagnosis of the site of a necrotic process—hepatic diseases and myocardial infarction for instance—with the limitation determined by some morbid associations.

The isoenzymic patterns were obtained after electrophoretic separation in agar-gel (the micro-Scheidegger-variant) of the serum and of the supernatant of tissue homogenates. The visualization of the LDH fractions was achieved with the Van der Helm method (with sodium lactate as substrate, NAD as coenzyme, phenazinmethosulphate as intermediary acceptor and nitro-BT as final acceptor of electrons; the estimation of the fractions being carried out by densitometry and planimetry. Assay of the total LDH in sera and homogenates was carried out by the spectrophotometric method in U.V. at 340 mμ (the CF$_4$DR Optica Milano spectrophotometer) expressed in Wroblewski units on ml for serum and on mg N$_2$ from organ extracts.

Regularities of Acridine Fluorochrome Interaction with Living Cells

ZELENIN, A. V., E. A. KIRIANOVA and N. G. STEPANOVA (Institute of Molecular Biology, Academy of Sciences USSR, Moscow B-312, USSR)

Living cell treatment with fluorescent stains (fluorochromes) is one of the most commonly used approaches in cell physiology and cell biology. The present paper deals with some new data on principal features of the living cell reaction to acridine fluorochromes.

The effect of acridine derivatives on nucleic acid and protein synthesis has been investigated autoradiographically. The investigation has proved 3,6—diaminoacridines (acridine orange, proflavine, acriflavine, coriphosphine) to be strong inhibitors of protein synthesis. This effect has been found (1,2) to be connected with interference in the first steps of protein synthesis (the formation of a complex between amino acids and s-RNA being hindered). The influence of the compounds under investigation on nucleic acid metabolism having been shown to be much less pronounced, the well-known antimitotic effect of 3,6-diaminoacridines (trypaflavine effect of Dusten) has been interpreted as a result of their interference in protein biosynthesis. Some other aminoacridines (atebrin and rivanol) have at the same time been shown to act on nucleic acid synthesis, not affecting directly that of proteins.

The regularities of distribution of acridine derivatives among cell structures was investigated by fluorescence microscopy. It has been previously shown that acridine orange exhibits marked tendency to accumulate within lysosomic structures of living cells (3,4,5,6). It has been found that this process is temperature dependent, energy dependent, passing against concentration gradient and exhibiting competitive inhibition. It has also been found that these features are characteristic for the accumulation in lysosomes of many other substances. These facts allow us to suggest the presence of an active transport mechanism at the lysosome membrane.

Our evidence shows the following conditions to be of essential importance for fluorescence—microscopical investigation of lysosomes in living cells. First, normal functioning of all cell mechanisms required for accumulation of acridine orange in lysosomes. Secondly, selection of a suitable acridine orange concentration which does not result in lysosome changes and ribosome aggregation in the form of so-called krinomes, that also may apparently exhibit with acridine orange red fluorescence hardly distinguishable from that of lysosomes.

The experimental conditions for nucleic DNA and RNA detection in living cells using acridine orange are also discussed.

References:

(1) Liapunova, E. A. and A. V. Zelenin: Voprosy medizinskoi khimii, 224 (1966).
(2) Werenne, J., H. Grosjen, and H. Chantrenne: Biophys. acta *129*, 585 (1966).
(3) Zelenin, A. V. and E. A. Liapunova: *In*: Second International Congress of Histo-and Cytochemistry, p. 217 (1964).
(4) Robbins, E., Ph. I. Marcus, and N. K. Gonatas: J. Cell Biology *21*, 49 (1964).
(5) Zelenin, A. V., V. I. Birjuzova, N. E. Vorotnitskaja, and E. A. Liapunova: Doklady Akademii Nauk SSSR (Russ., Moscow) *162*, 925 (1965).
(6) Zelenin, A. V.: Nature *212*, 425 (1966).

Uses of Recording Gradient-Diver for Single Cells

ZEUTHEN, ERIK (Biological Institute of the Carlsberg Foundation, Copenhagen N, Denmark)

Some years ago I proposed a simple version—the ampulladiver—of the Linderstrøm-Lang and Holter Cartesian diver gasometer.

The ampulla-diver is the broken-off, and suitably shaped, outer end of a braking pipette. It holds a tiny volume of biological medium in which a single cell is allowed clonal growth under sterile conditions.

The ampulla-diver can be manually operated as a Cartesian diver or it can be used as a non-Cartesian diver ("gradient-diver") for automatic recording of gaseous exchanges of single cells and of small clones. Advantages and disadvantages of the two uses of this diver will be discussed with examples of measurements of rates of oxygen uptake and carbon dioxide output during clonal growth of *Tetrahymena* and *Acanthamoeba* and during the division cycles in cleaving frog eggs. The work has been done in collaboration with Drs. Kirsten Hamburger and R. Z. Klekowski.

Histochemische und blutchemische Untersuchungen an der Kaninchenleber nach temporaerer Ischamie *

ZIMMERMANN, HORST (Pathologisches Institut, Stadtkrankenhaus, Frankfurt/M-Hoechst, West-Germany)

Einleitend wird darauf hingewiesen, dass ein unspezifischer Operationsreiz zu einer starken Glykogenverminderung der Leber führt. Eine unspezifische Stressreaktion in Form einer doppelseitigen einstündigen Nierenischämie führt neben einer Glykogenabnahme auch zu einer leichten Abnahme der Succinodehydrogenase-(SD) und der Glycerinaldehyd-3-phosphatdehydrogenase-(GAP) Aktivität der Leber. Im Anschluss an eine Ischämie von 15 bis 40 min waren in der geschädigten Leber Zentralvenen und perizentrale Sinusoide stark erweitert, Nekrosen traten erst nach 25 min Ischämie und 30 Std Manifestationszeit auf. Die SD und GAP nahmen an Aktivität ab. Nach 25-40 min Ischämie ist die Aktivitätsabnahme herdförmig. Der Glykogengehalt der Leber nimmt in jedem Fall nach temporärer Ischämie ab und ist in der Intensität der Abnahme abhängig von der Ischämiedauer. Der Glykogenverlust beginnt in der Läppchenperipherie und schreitet nach zentral fort. Ischämiezeiten jenseits der Wiederbelebungszeit führen zum totalen Glykogenschwund. Eine Neoglykogenie war am Ende der Versuchsdauer nur bei längerer Ischämiedauer (25-40 min) eben sichtbar. Der Blutzuckerspiegel steigt mit Abfall des Leberglykogens an, hat seine Maxima 3-8 Std nach Aufheben der Ischämie und fällt danach kontinuierlich in den Normbereich ab.-Neben der direkten Einwirkung der Ischämie auf die Leber wird auch eine indirekte Stressreaktion als Ursache der Zellschädigung angenommen. Zeitlich gehen Engym- und Glykogenschwund der Nekrophanerose der Leberzellen voraus.

* Supported by a grant of the "Deutsche Forschungsgemeinschaft." The experiments were carried out by Dr. Dieter Ross.

308

The Significance of the Mucopolysaccharides in Atherosclerosis

ZUGIBE, FREDERICK T. (West Virginia University Medical School, Department of Pathology, Morgantown, West Va.)

Histochemical studies of the mucopolysaccharides and lipids of human coronary arteries, cerebral arteries and aortas ranging in age from fetal life to 70 years of age has been made utilizing new histochemical techniques developed in our laboratory. These include new ion association fractionation and enzymatic techniques for identifying and quantitating individual mucopolysaccharides and methods for demonstrating lipids which have obviated the lipid-leaching difficulties observed with other commonly used lipid staining techniques.

No apparent relationship was found between lipid and acid mucopolysaccharides with respect to staining intensity or distribution in any of the groups studied. An ostensible relationship however, was observed between chondroitin sulfate B and collagen. The ratio of chondroitin sulfate B to condroitin sulfate A or C increased with severity of changes of elastic fibers in the area of the internal elastic membrane and media with a concomitant increase in the ratio of coarse to fine collagen. The earliest morphological alteration observed was subendothelial aggregates within muscle fibers and macrophages, or as extracellular deposits in the coronary arteries and aortas but rarely in the cerebral arteries. In the latter, the most frequent site was within the internal elastic membrane and reduplicated fibers. Intraelastic lipid was rarely observed in the internal elastic membrane of the coronary arteries, but was occasionally seen in the aortas. The significance of these findings will be discussed.

ABSTRACTS RECEIVED LATE

Microspectrophotometric Quantitation of Metallic-Sulfide Granules in Tissue Sections

ABRAHAMSON, DEAN E. and JOSEPH L. RIGATUSO (University of Minnesota, Minneapolis, Minn.)

A number of histochemical enzyme reactions have as their end product the deposition of metallic-sulfide granules in the tissue. These granules are thought to be deposited at the site of the enzyme activity. Since the granules are of a brown-black color, and can be considered as essentially opaque, the usual photometric absorption techniques cannot be used to estimate their quantity. The two-wavelength method of microphotometry is also not applicable as the granular material has essentially infinite absorbance at all wavelengths in the visible spectrum. If, however, the granules are uniform in size within a given tissue section, and their number density is sufficiently low such that there is no appreciable shadowing of one granule by another, it might be possible to use photometric techniques to estimate the total quantity of granular material in tissue sections. The light loss between the incident and emergent beams would be due to some combination of absorption and scattering. The results of a series of measurements of lead sulfide granules in liver resulting from a histochemical reaction for glucose-6-phosphatase are presented. The correlation between the photometric estimates of granule quantity and the chemical determinations of glucose-6-phosphatase activity indicates that the photometric method is valid in the system used. Other possible applications of the photometric technique are presented.

The Technique of Zonal Centrifugation and its Application to Cytochemistry

ANDERSON, NORMAN G. (Oak Ridge National Laboratory, Oak Ridge, Tenn.)

(no abstract submitted)

Newer Instrumentation for Nuclear Cytochemical Studies

CASPERSSON, TÖRBJORN (Institute for Cell Research, Karolinska Institute, Stockholm 60, Sweden)

(no abstract submitted)

Phago-lysosomes in Developing Suckling Rat Ileum

CORNELL, RICHARD (National Institutes of Health, National Cancer Institute, Laboratory of Biology, Tissue Culture Section, Bethesda, Md.)

In suckling rats, passive immunity is transferred from mother to offspring by selective passage of colostral proteins across the distal small intestine. Cytochemical, biochemical, and ultrastructural studies conducted in our laboratory have demonstrated that the mechanism of protein absorption by postnatal intestinal epithelium is related to the presence of a highly developed phago-lysosomal system in these neonatal cells. Absorbed protein first appears in superficial small vacuoles free of acid phosphatase activity. It then is found in deeper droplets and a large supranuclear body which are basophilic, PAS positive, and rich in acid phosphatase activity. Other cytochemically demonstrable enzymes associated with this cytoplasmic absorptive system include esterase, ATP'ase, alkaline phosphatase, and thiamine pyrophosphatase.

Ultrastructural Localization of Acid Mucosubstances

DOUGHERTY, WILLIAM (Medical College of South Carolina, Anatomy Department, Charleston, S. C.)

Studies involving the ultrastructural localization of acid mucosubstances by selective staining have been performed by several workers who have generally utilized cationic or micellar heavy metals. Adaptation of iron containing histochemical stains such as dialyzed iron (DI) or colloidal iron (CI) to electron microscopy has revealed the presence of acid mucosubstances intracellularly in cytoplasmic vesicles (6, 7), in Golgi lamellae (6), in mouse Paneth cell granules (5), in heterophil and basophil granules (4), in mucous globules occurring within intestinal goblet cells and colonic "deep crypt" mucous cells (6), and in the sarcoplasmic reticulum of vertebrate skeletal muscles (2). Acid mucosubstances appear to be present also on the surfaces of various cell types (3, 5, 6, 7) and blood platelets (1), to be associated with extracellular collagen fibrils (6, 7), and to occur as interstitial connective tissue material (6). Deposition of micellar iron has been reported to occur on nonperiodic filamentous material located both intra- and extracellularly in fibroblast tissue cultures (7) and in association with I band thin filaments in vertebrate skeletal muscles (2). Fine structure studies indicate that hyaluronate, sialomucins (3), and probably sulfomucins (5) and sulfated mucosubstances (4, 5, 7) are revealed by Hale stain modifications adapted to electron microscopy. Various factors influencing the selectivity of DI staining as well as other methods for revealing mucosubstances by electron microscopy will be discussed.

References:

(1) Benke, O.: Anat. Rec. *158*, 121 (1967).
(2) Dougherty, W. J. and S. S. Spicer: 25th Ann. Electron Microscopy Soc. Am., p. 52 (1967).
(3) Gasic, G. and L. Berwick: J. Cell Biol. *19*, 223 (1963).
(4) Hardin, J. H., S. S. Spicer, W. B. Greene, and R. G. Horn: 19th Histochem. Soc. Meet. (New Orleans), Abstr. p. 17 (1968).
(5) Spicer, S. S., M. W. Staley, M. G. Wetzel, and B. K. Wetzel: J. Histochem. Cytochem. *15*, 225 (1967).
(6) Wetzel, M. G., B. K. Wetzel, and S. S. Spicer: J. Cell Biol. *30*, 299 (1966).
(7) Yardley, J. H. and G. D. Brown: Lab. Invest. *14*, 501 (1965).

Development of the Steroidogenous Organs of the Human Fetus

JIRASEK, J. E. (Inst. for the Care of Mother and Child, Prague)

Histochemical investigations of the human fetal endocrine glands permit the following correlations: 3β-ol steroid dehydrogenase (substrate DHA and pregnenolon) was demonstrated in the syncytiotrophoblasts of all specimens of normal chorionic tissue. The youngest *chorion* studied was from a presomite embryo of ca. 20 days of age. In the fetal *adrenal cortex*, activity was shown in epithelial cells of the central zone in embryos 20-22 mm long (45-50 days of age). Its activity raises in fetuses 60-70 mm long (10-11 weeks old). At this stage, activity is present in cells of the fetal adrenal cortex with the exception of cells of the subcapsular zone. In the *fetal testicles* it was found only in the epitheloid Leydig cells occurring in fetuses 30 mm long (ca 60 days of true age) and older.

In *fetal ovaries*, the cells containing the 3β-ol steroid dehydrogenase were classified as epitheloid thecal cells of the growing follicles. They are present in ovaries of fetuses from the VIIth month and older.

HCG—glycoprotein with a specifical immunopheorescency—was shown in syncytiotrophoblasts and in trophoblastic giant cells of an implanted blastocyst ca 14 days old and in the trophoblastic syncytium of all older specimens. Its presence in the maternal blood was confirmed immunologically and by the hyperemic test. In the fetal tissues and fetal tissue extract, a positive hyperemic test was found only in embryos longer than 30 mm CR length.

Gonadotrophic hormones of the fetal hypophysis were visualized immunologically. However, this method did not permit any conclusions concerning secretion.

Microfluorimetric Studies on the Formaldehyde-Induced Fluorescence of Noradrenaline in Adrenergic Nerves of Rat Iris

JONSSON, GÖSTA (Department of Histology, Karolinska Institutet, Stockholm, Sweden)

The formaldehyde vapour technique of Falck and Hillarp for the histochemical demonstration of biogenic monoamines at the cellular level is based on the principle that the amines can be condensed with formaldehyde to yield strongly fluorescent 3,4-dihydroisoquinolines. The fluorescent compound thus formed from noradrenaline is 4,6,7-trihydroxy-3,4-dihydroisoquinoline which in a dried protein layer—as in freeze-dried and air-dried tissues—is in a pH-dependent equilibrium with its tautomeric quinoidal form, responsible for the strong fluorescence at 48 mμ (see Corrodi and Jonsson, 1967).

The present study was undertaken to investigate the possibilities of quantifying microfluorimetrically the formaldehyde-induced fluorescence of noradrenaline in adrenergic nerves of rat iris, and with a view to obtaining information on the fluorescence characteristics when the amine is granule-bound or extra-granularly distributed in the adrenergic neuron.

The formaldehyde vapour technique is able to demonstrate both granule —and extra-granular stored noradrenaline, and the fluorescence spectra of the fluorophor are identical during the both types of storage. The fluorescence-concentration relationship is linear up to a value corresponding to about 25 per cent of normal noradrenaline levels. Above this value a concentration-dependent quenching of the fluorescence occurs. This value is somewhat lower when the amine is granule-bound compared with extra-granular storage.

Circadian Influences on Histochemical Demonstration of Glycogen

LESKE, REGINA (Anatomisches Institut, Medizinische Hochschule, Hannover, Germany)

Many studies devoted to histochemical glycogen demonstration emphasise the necessity of choice of fixatives while others state that the influence of fixation is overestimated. The same is true for methods of mounting sections. Finally, the chemical configuration of glycogen and its linkage to proteins was held responsible for histochemical demonstrability.

Our investigations in which biochemical determinations of glycogen have been performed and compared with histochemical findings have revealed the following:

1) It should be considered that in histochemistry there exists a definite circadian rhythm of glycogen assimilation and dissimilation in the liver, in which the glycogen content varies considerably. The charts of these daily rhythmic changes in amounts which can be demonstrated easily by chemical glycogen determinations, are influenced by sex and seasons.

2) Fixatives for glycogen have very different effects, depending on the biological state in which the fixative meets the tissues. Therefore, we observed great variability in glycogen preservation using recommended fixation techniques, while when repeating these fixation experiments at other times of day and season, the same fixatives show very little or no difference in their action. Similarly, sections of livers harvested at different hours of the day respond differently when treated; especially the liver sections of animals sacrificed in summer can be mounted without any precaution. On the contrary, glycogen was easily extracted during mounting of sections when the livers were obtained in the winter.

3) It can be concluded that an ideal glycogen fixative does not exist. The reason is that glycogen by itself is not fixed but only trapped by cell proteins. Since the cell proteins undergo daily rhythmic changes, the trapping reaction through protein precipitation will be different.

Lactate Dehydrogenase Isozymes in the Study of Development

MOYER, FRANK H. (University of Missouri, St. Louis, Mo.)

Of many enzymes known to exist in isozymic forms, lactate dehydrogenase (LDH) has been studied most extensively, and has proved most useful in studies of development. In most animals studied, it is a tetramer existing in five forms: A_4, A_3B, A_2B_2, AB_3, and B_4. The subunits A and B are under separate genetic control and probably associate at random. The isozymes are commonly detected by gel electrophoresis, a technique of simplicity and sensitivity. Studies of LDH isozyme distribution have provided: 1. evidence of differential gene action during development, 2. sensitive means for early detection of histodifferentiation, and 3. evidence of activation of the embryonic genome.

In animals studied thus far, distribution of LDH activity among isozymes undergoes marked changes during development. These observations have been reviewed frequently and will not be discussed in detail. They provide evidence for both time and tissue specificity of gene action during development, and indicate that the loci controlling the A and B subunits act independently.

Since LDH subunits probably associate at random, a bimodal distribution of activity among the isozymes of a single tissue probably indicates cellular heterogeneity. The bimodal pattern of kidney LDH, for example, is the sum of the patterns of the distal and proximal convoluted tubules. The appearance of a bimodal LDH pattern, therefore, is a sensitive indicator of histodifferentiation. In mammals, for instance, the appearance of such a pattern in the diaphragm is correlated with the differentiation of "fast" and "slow" muscle fibers.

Isozymes of LDH may also be used to determine the time at which the embryonic genome becomes active. Backcrossing and outcrossing of frogs heterozygous at the locus controlling the B subunit provides embryos in which "hybrid" LDH appears shortly after muscular movement begins (Shumway stage 18). At this time most of the LDH activity of the embryo is in the "maternal" isozymes which are not entirely replaced until 11 days after the tadpole begins to feed. Appropriate crosses show that the maternal enzyme is made during oogenesis and not by a "masked" mRNA after fertilization.

At present much effort is directed toward elucidation of the metabolic role of LDH isozymes. As clearer understanding of their role emerges, we may expect them to be used as sensitive indicators of metabolic changes during development. Further research along these lines may be expected to extend existing knowledge of genetic and epigenetic interactions during development.

The Centrifugal Isolation and Some Properties of Subcellular Components

SCHNEIDER, WALTER C. (Laboratory of Biochemistry, National Cancer Institute, National Institutes of Health, Bethesda, Md.)

The centrifugal isolation of nuclei, mitochondria, microsomes, and other subcellular components will be described. The basic principles of the procedures and the attendant difficulties in their execution will also be discussed. The chemical composition and the biochemical functions of isolated subcellular components will also be summarized.

Histoenzymology of Normal and Diseased Gastric Surface Epithelium in Man

SIEGEL, HOWARD I., ROBERT LEV and GEORGE B. JERZY GLASS (Section of Gastroenterology and Department of Pathology, New York Medical College, New York, N. Y.)

Human gastric mucosa from over 100 patients with various gastric disease and controls was studied for 10 lysosomal, Golgi apparatus, membrane and mitochondrial enzymes, with special reference to surface epithelium. The material examined was obtained by suction biopsy or from fresh gastrectomy specimens. It was quick frozen ($-70°C$) and either briefly post fixed or directly incubated in substrate. The distribution and localization of the enzymes were compared with the histology of the gastric mucosa.

In normal surface epithelium and in 7 cases of superficial gastritis, no staining for lysosomal enzymes and minimal staining for LDH, and DPN and TPN diaphorases was noted. No change in the qualitative pattern of surface epithelium was observed in 7 patients with duodenal or gastric ulcer without atrophy, nor in 6 patients with acute hemorrhagic gastritis, nor in 5 cases with superficial gastric erosions following augmented histamine test.

In 26 cases of early atrophic gastritis, normal surface epithelium was replaced partially or diffusely by cuboidal epithelium of a regenerative type characterized by appearance of a moderately intensive but diffuse cytoplasmic staining for AcP, not seen in the normal surface epithelium as well as LDH, and DPN and TPN diaphorases.

In 28 cases of advanced chronic atrophic gastritis, gastric epithelium in turn was superseded by an intestinalized epithelium with goblet cells. The cytoplasm of this metaplastic epithelium showed discrete supra-nuclear lysosomal granules best visualized with AcP reaction, and displaying a more diffuse staining for dehydrogenases and diaphorases. For the first time, AlP and ATPase activity was found in the surface epithelium and this was localized mainly on the cell surface. This is in line with the absorptive properties which this new epithelium has acquired.

Administration of 6-12 weeks course of intensive cortisone therapy did not change the enzymatic and histologic patterns of the atrophic mucosa in 5 patients treated in this way.

The histoenzymatic activity of 15 gastric carcinomas was found to be generally qualitatively inferior to that exhibited by intestinal metaplasia, but greater than that found in normal gastric surface epithelium. In adenocarcinoma, variability of staining was observed. ATPase and AlP were found only in one case of well differentiated carcinoma. AcP and LAP were often present in most cases, even in undifferentiated malignant cells. Diffuse cytoplasmic staining for LDH and both diaphorases was found in many cases of the carcinoma. These enzymatic patterns strengthen the link between certain gastric adenocarcinomas and intestinalized gastric mucosa.

Quantitative Electron Immunocytochemistry of Lysozyme in Monocytes

STERNBERGER, LUDWIG, ELLIOTT F. OSSERMAN and ARNOLD M. SELIGMAN (Physiology Department, Medical Research Laboratory, Edgewood Arsenal, Md., Francis Delafield Hospital, New York, N.Y., Dept. of Surgery, Sinai Hospital of Baltimore, Depts. of Microbiology and Surgery, The Johns Hopkins University School of Medicine, Baltimore, Md. and Dept. of Pathology, Columbia University College of Physicians and Surgeons, New York, N.Y.)

In the immunouranium-TO technique, the application of electron-opaque antibody to ultra-thin sections rather than to tissue prior to embedding permits evaluation of immunohistochemical reactions in terms of optical density (OD) units because: 1) the thinness of the sections assures accessibility of all subcellular sites to antibodies, 2) the thinness of the sections assures that any non-specifically adherent materials are removed by washing, and 3) experimental and control sections are prepared from the same tissue block and are treated with the same electron-opaque

316

reagents. In the indirect technique, experimental sections are exposed
to specific, unlabeled antiserum and control sections to non-specific,
unlabeled antiserum.

In qualitative histochemistry, results are presented by illustrative micrographs chosen from a larger sample of micrographs. In quantitative
histochemistry, it is possible to report the aggregate of all observations
made and evaluate their significance. Expression of contrast in terms of
ratios of OD's of specific subcellular areas to OD's of adjacent regions
devoid of cellular material eliminates errors from variations in thickness of sections and photographic plates.

We have recently applied electron immunocytochemical technique to
the analysis of lysosomes in monocytes in monocytic leukemia. When
these lysosomes were stained with an antiserum specific for human
lysozyme the mean ratios of OD's were unequivocally higher than upon
staining with either antiserum for human fibrinogen or normal serum.
In contrast, the α-granulomeres of normal platelets when stained with
these same antisera showed higher mean OD ratios with antifibrinogen
than with antilysozyme or normal serum. The probabilities of significance of these differences exceeded 99.9%.

The antiserum to fibrinogen was a gift of Dr. N. F. Rodman.

Sites of Synthesis and Migration of Glycoproteins and Mucopoly-
saccharides. A Review

WEINSTOCK, A. (Department of Anatomy, McGill University, Montreal, Canada)

Glycoprotein and mucopolysaccharide synthesis has been traced by
radioautography after administration of labeled precursors. Early light
microscopic investigations in this department showed labeling of the
Golgi region of a variety of rat cells at early intervals after injection
of ^{3}H-glucose or ^{3}H-galactose. Five minutes following ^{3}H-glucose injection, silver grains were located with the electron microscope over Golgi
saccules of colonic goblet cells. Within one hour, label appeared in
mucus itself. Five minutes after injecting ^{3}H-galactose, silver grains were
seen overlying Golgi elements of ameloblasts, and by one hour were
located over enamel matrix. A comparable secretory pathway was traced
in thyroid follicular cells and hepatocytes. Similar results have been
observed *in vitro* with ^{3}H-glucosamine in synovial cells.

When ^{35}S-sulfate was administered, label was first seen localized over
the Golgi region of chondrocytes, indicating that sulfation of mucopolysaccharides also occurs in this organelle. Label was later observed
over cartilage matrix.

These findings were interpreted as showing synthesis of glycoproteins and/or mucopolysaccharides in the Golgi complex, and their subsequent migration and release from the cells as secretions.

While it is also possible that some carbohydrate incorporation takes place in the rough endoplasmic reticulum during glycoprotein synthesis, as suggested by biochemical evidence, the Golgi complex appears to play the major role in the attachment of carbohydrate to protein.

References:

(1) Barland, P., C. Smith, and D. Hamerman: J. Cell Biol. *37*, 13 (1968).
(2) Fewer, P., J. Threadgold, and H. Sheldon: J. Ultrastruct. Res. *11*, 166 (1964).
(3) Godman, G. C. and N. Lane: J. Cell Biol. *21*, 353 (1964).
(4) Herscovics, A., *In press.*
(5) Neutra, M. and C. P. Leblond: J. Cell Biol. *30*, 119 (1966).
(6) Neutra, M. and C. P. Leblond: J. Cell Biol. *30*, 137 (1966).
(7) Weinstock, A.: Anat. Rec. (Proc.) *157*, 341 (1967).
(8) Whur, P. and A. Herscovics: Anat. Rec. (Proc.) *160*, 450 (1968).

Histochemistry
Histochemie
Histochimie

HISTOCHEMIE is an international journal, covering the entire field of cell and tissue chemistry. The journal concentrates mainly on basic research in cell and tissue chemistry and tissue physiology with particular reference to methodology (including fractionation, homogenization, autoradiography, polarization optics and fluorescence microscopy). It also includes the areas marginal to biochemistry and ultrastructure research. Papers on applied histochemistry are accepted if they are fundamental in nature and if they tend to contribute to the advance of research in cell and tissue chemistry.

Great stress is placed on rapid publication and high-quality reproduction of illustrations.

Histochemie will be published irregularly to ensure rapid publication of all new material received
Each volume consists of approx four single issues.
1968, Volumes 12-14, DM 420,—;
US $ 106.20 (Incl. postage)
(Subscriptions are entered with prepayment only)

SPRINGER-VERLAG
BERLIN·HEIDELBERG·NEW YORK

Springer-Verlag New York Inc.
175 Fifth Avenue · New York/N.Y. 10010

An Atlas of
Mammalian Chromosomes

By T. C. Hsu,
Houston, Texas
and
Kurt Benirschke,
Hanover,
New Hampshire

In recent years, because of advances in karyological techniques, there has been a remarkable renewal of interest in studies of mammalian chromosomes. These techniques, generally involving the use of tissue culture, colchicine and hypotonic solution pretreatments, allow for a much clearer display of metaphase chromosomes of mammalian cells than the classic direct squash or tissue section methods. Consequently, what was known about the chromosome complement of most mammals must be revised. Many animals are being examined cytologically for the first time.

The findings are now extensive and scattered; they appear in numerous periodicals and newsletters, or they are kept in cytologists' file drawers without being published. It is difficult to have access to pertinent data for comparison among related species or for evaluation of various karyological characteristics within a karyotype. Such evaluations can be done only when reasonably uniform material is collected and placed side by side for comparison, accompanied by relative references. *An Atlas of Mammalian Chromosomes* is planned to fulfill such a need.

Two volumes containing 100 plates have been published to date. Standing orders are invited to assure prompt receipt of each volume as released. One volume per year is scheduled.

Volume 1:   50 plates. X, 200 pages. 1967
                Loose-leaf boxed DM 37,60; US $ 9.40

Volume 2:   50 plates. XX, 208 pages. 1968
                Loose-leaf boxed DM 37,60; US $ 9.40

■ Prospectus
on request!

Loose-leaf binders may be purchased separately for the preservation and rearrangement of the Folios:
DM 8,—; US $ 2.00

SPRINGER-VERLAG
BERLIN·HEIDELBERG·NEW YORK

The Cellular Aspects of Biorhythms

Symposium on
Rhythmic Research
Sponsored by the
VIIIth International
Congress of Anatomy,
Wiesbaden
8.—14. August 1965

Edited by Professor
Dr. H. v. Mayersbach
Universiteit van
Nijmegen,
Faculteit der
Geneeskunde,
Laboratorium voor
cytologie en histologie

With 101 figures
VIII, 198 pages 8vo.
1967
Cloth DM 52,–
US $ 13.00

■ Prospectus
on request!

Contents: E. Bünning: Circadiane Rhythmik. F. Halberg: Circadian System Phase. T. Petrèn, A. Sollberger: Developmental Rhythms. R. B. McHugh: Validity and Efficiency in the Design of Transverse Physiologic Periodicity Experiments. H. v. Mayersbach: Seasonal Influences on Biological Rhythms of Standardized Laboratory Animals. Chr. Pilgrim: Autoradiographic Investigations with 3-H-thymidine on the Influence of the Diurnal Rhythm on Cell Proliferation Kinetics. W. Eling: The Circadian Rhythm of Nucleic Acids. Chr. Jerusalem: Circadian Changes of the DNA-content in Rat Liver Cells as Revealed by Histophotometric Methods. O. Bucher, P. Suppan: Amitose et fusion nucléaire au cours du rhythme circadien. R. Leske: Technical Factors Involved in the Histochemical Determination of the Circadian Changes of Liver Glycogen. P. Yap: Circadian Phase Differences of Lyo- and Desmoenzymes. Chr. Jerusalem, O. Müller, H. v. Mayersbach: Circadian Ultrastructural Changes in Liver Cells. J. J. Chiakulas, L. E. Scheving: The Effects of the Presence or Absence of the Pituitary Gland on the Daily Rhythmicity of Mitotic Rates in Urodele Larval Tissues. L. E. Scheving, J. E. Pauly: Effect of Adrenalectomy, Adrenal Medullectomy and Hypophysectomy on the Daily Mitotic Rhythm in the Corneal Epithelium of the Rat. J. M. Echave Llanos, E. G. Bade, A. F. Badran: Circadian Rhythms in Growth Processes. F. Gerritzen: Constancy and Rhythm. A. Sollberger, H. P. Apple, R. M. Greenway, P. H. King, O. Lindan, J. B. Reswick: Automation in Biological Rhythm Research.